The De Gruyter Handbook of Artificial Intelligence, Identity and Technology Studies

De Gruyter Handbooks of Digital Transformation

Series Editor
Anthony Elliott

Volume 1

The De Gruyter Handbook of Artificial Intelligence, Identity and Technology Studies

Edited by
Anthony Elliott

DE GRUYTER

ISBN 978-3-11-072125-6
e-ISBN (PDF) 978-3-11-072175-1
e-ISBN (EPUB) 978-3-11-072184-3
ISSN 2940-7249

Library of Congress Control Number: 2024934689

Bibliographic information published by the Deutsche Nationalbibliothek
The Deutsche Nationalbibliothek lists this publication in the Deutsche Nationalbibliografie; detailed bibliographic data are available on the internet at http://dnb.dnb.de.

Cover image: Paul Campbell/iStock/Getty Images Plus
Typesetting: Integra Software Services Pvt. Ltd.
Printing and binding: CPI books GmbH, Leck

www.degruyter.com

Contents

List of Contributors

Jan K. Arganski is a Senior Postdoctoral Researcher at the Sano Centre for Computational Medicine and an Assistant Professor at the Institute of Applied Computer Science, Jagiellonian University. His interests include artificial intelligence, affective computing, computational neuroscience, and VR/AR.

Katarzyna Baliga-Nicholson is Senior Business Developer at the Sano Centre for Computational Medicine. Her research is dedicated to advancing our understanding of how to effectively harness the potential of advanced technologies and bring them from the laboratory to society. She is particularly interested in designing frameworks that enable AI technologies to create sustainable human-in-the-loop relationships that move beyond simple implementation. Her research challenges commonly held assumptions in areas such as functionality, design, aesthetics, and technological intimacy.

Thomas Birtchnell is an Associate Professor in Geography and Sustainability at the University of Wollongong, Australia. Before this, he was a Research Associate at Lancaster University in the UK in a project funded by the Economic and Social Research Council (ESRC) and led by Distinguished Professor John Urry. The research project (ES/J007455/1) examined the past and future impacts of 3D printing on transport and society. His latest book is *A New Industrial Future? 3D Printing and the Reconfiguring of Production, Distribution, and Consumption* (Routledge).

Elise Bohan is a Senior Research Scholar at the University of Oxford's Future of Humanity Institute (FHI). She holds a PhD in evolutionary macrohistory (Big History) and wrote the world's first book-length history of transhumanism as a doctoral student. At FHI, she is part of a cohort of scholars who are dedicated to understanding, and tackling, humanity's most pressing problems.

Ross Boyd is a Research Fellow based in the JMCOE, Justice and Society Academic Unit, University of South Australia. His current research draws attention studies together with his background in social theory and economic sociology to explore, among other things, the possibilities for critique immanent to complex socio-technical systems, the intersections of social theory and algorithm design, and the problem of indifference.

Anthony Elliott is Bradley Distinguished Professor of Sociology and Executive Director of the Jean Monnet Centre of Excellence at the University of South Australia, where he is Dean of External Engagement. He is Super-Global Professor of Sociology (Visiting) at Keio University, Japan and Visiting Professor of Sociology at University College Dublin, Ireland. He is a Fellow of the Academy of Social Sciences in the UK, Fellow of the Academy of the Social Sciences in Australia, and Senior Member of King's College, Cambridge. In the 2023 King's Birthdays Honours, he was made a Member of the Order of Australia in recognition of his significant service to education, social science policy and research. He is the Series Editor of the De Gruyter Handbook of Digital Transformations and the author and editor of over 50 books, including most recently *The Culture of AI: Everyday Life and the Digital Revolution* (Routledge, 2019), *Making Sense of AI: Our Algorithmic World* (Polity, 2021), *Algorithmic Intimacy: The Digital Revolution in Personal Relationships* (Polity, 2023) and *Algorithms of Anxiety: Fear in the Digital Age* (Polity, 2024).

Louis Everuss is the Coordinator of the Jean Monnet Centre of Excellence at the University of South Australia, where he also lectures in sociology. Louis' research has covered an array of subjects, including how systems of mobility are incorporated into representations of sovereign outsiders, the way borders are racially constructed through everyday performances of the law, and how public opinions of climate change are impacted by national context. Across much of this work, Louis has sought to develop new

https://doi.org/10.1515/9783110721751-203

conceptions of state sovereignty that take account of the way it is produced both through and in response to mobilities. Louis' research has been published in Political Geography, Distinktion: Journal of Social Theory, the Journal of Sociology and Applied Mobilities. Louis' inaugural book, *Digital Mobilities and Smart Borders: How Digital Technologies Transform Migration and Sovereign Borders*, is being published by De Gruyter in 2024.

Tim Gorichanaz is an Assistant Teaching Professor in the College of Computing & Informatics at Drexel University in Philadelphia, where he teaches and conducts research in human–computer interaction, philosophy of information and information behavior. His current research is exploring the connections between virtue ethics and digital design, including questions at the intersection of selfhood, the good life and artificial intelligence. He received a Ph.D. from Drexel University, an M.A. from New York University, and a B.A. from Marquette University

Elisabeth Hildt is a Professor of Philosophy and the Director of the Center for the Study of Ethics in the Professions at the Illinois Institute of Technology in Chicago. She is interested in philosophical and ethical aspects of science and technology. Her research focuses on bioethics, ethics & technology, neurophilosophy, and philosophical and ethical aspects of Artificial Intelligence.

Eric L. Hsu is a Lecturer in Sociology at the University of South Australia, where he also presently serves as a co-leader of a work package on 'risk, emergency management and global pandemics' at the UniSA Jean Monnet Centre of Excellence. He is most recently co-author of *Technosleep: Frontiers, Fictions, Futures* (2023) published by Palgrave Macmillan. More information about his work, including the *Sociology of Everything* podcast he produces, can be found on his website, www.ericlhsu.com.

J. Michael Innes is an Adjunct Research Professor at the EU Jean Monnet Centre of Excellence, University of South Australia and currently a Fellow, Churchill College, Cambridge. He has taught at the Universities of Birmingham, Edinburgh, Michigan, and Adelaide. He is a social and political psychologist, with interests in the psychology of social influence, psychological methodology and the social consequences of technological development, especially artificial intelligence.

Masataka Katagiri is now emeritus professor at Chiba National University in Japan and adjunct professor at University of South Australia. His recent books include *Humans, AI and Animals: Sociology of the Post Humans* (in Japanese), Maruzen Publishing Company in Tokyo, 2022 and *Sociology of the Precarious Selves: On the Origin and Future of Individualization* (in Japanese), Minelva Publishing Company in Kyoto, 2017. His main sociological works have been introduced in English in the paper:" Modernity, Postmodernity, and Late Modernity: A Review of Sociological Theories in Contemporary Japan" by Shigeki SATO, *International Sociology Reviews*, 36–2, 2021.

Ben Lyall (Monash University) is a digital sociologist and Lecturer in Digital Media at the School of Media Film and Journalism, Faculty of Arts. His research explores how social lives are impacted by digitalisation in areas of digital health, social policy, and employment futures.

Ben Morrison is a Senior Lecturer and Organisational Psychologist at Macquarie University, Australia. Ben's research investigates how new and emerging technologies can aid human-centered processes like decision making, and how human capabilities, limitations, and preferences can inform the design of effective technological systems. Ben is particularly interested in the nexus between human expertise and advanced technologies such as artificial intelligence.

Sven Nyholm is Professor of the Ethics of Artificial Intelligence at LMU Munich in Germany. Nyholm is also a Principal Investigator in the Munich Center for Machine Learning and an Associate Editor of the journal Science and Engineering Ethics. He has a PhD in philosophy from the University of Michigan and has previously held academic positions in Cologne, Eindhoven, and Utrecht. His books include Humans and Robots: Ethics, Agency, and Anthropomorphism (2020) and This is Technology Ethics: An Introduction (2023).

Ludovica Paseri is a postdoctoral researcher in Philosophy of Law and Legal Informatics at the Department of Law, University of Turin. She holds a PhD in Law at the Alma Mater Studiorum University of Bologna, in co-tutelle with the Faculty of Science, Technology and Medicine (FSTM) of the University of Luxembourg. Fellow of the Nexa Center for Internet and Society at Politecnico and the University of Turin, her current research mainly concerns the right to science, open science and research data, privacy and data protection, law and technology, AI & law.

Luca M. Possati serves as an Assistant Professor at the University of Twente in The Netherlands. Trained as a philosopher, he has held positions as a researcher and lecturer at Delft University of Technology in The Netherlands, the University of Porto in Portugal, and the Institut Catholique in France. His research primarily explores the philosophy of technology and AI, technology assessment, and AI ethics, and he also engages in the field of software studies. He is the author of *The Algorithmic Unconscious* (Routledge 2021).

Rina Yamamoto is Associate Professor at Seijo University in Tokyo. She obtained her PhD in advanced social and international studies from the University of Tokyo. Her expertise lies in the theory of contemporary society and the sociology of housing. She received a distinguished book award from the Association of Urban Housing Sciences for her book titled *Mai Hōmu Shinwa no Seisei to Rinkai* ([Generation and Criticality of My Home Myth], Iwanami Shoten, 2014; https://www.u-tokyo.ac.jp/biblioplaza/en/B_00185.html). Additionally, she has served as a committee member for central and local governments as well as private research organisations, advising on housing policies and issues in Japan.

Anthony Elliott

1 Mapping AI, Identity and Technology Studies

The De Gruyter Handbook of Artificial Intelligence, Identity and Technology Studies provides a sweeping, systematic guide to the leading developments and trends in artificial intelligence (AI) and its consequences for both personal and cultural identities in the digital age. Part of my motivation for producing this volume has been to bring together, in the one place, a comprehensive and accessible guide to the key issues, concepts, theories and studies on how AI is transforming identity, selfhood and the wider parameters of our technological culture. Breaking new ground through integrating the social science of technology with the latest developments in AI, computer science and machine learning, this *Handbook* focuses, as it were, on the 'emotional side' of people's relationships with technology—from computers and social media to chatbots and smart algorithms.

The *Handbook* covers a wide array of phenomena associated with AI, and provides the most up-to-date coverage of developments in supercomputing, deep learning, neural networks and much else. Topics addressed where AI currently transforms, or in the future promises to transform, social, economic, cultural and psychological processes and mechanisms of identity include:

- AI as a new context for self-exploration, social encounters and social relations.
- AI mobile technology including apps and bots, embedded within operating systems, impacting social networking.
- Sociable robotics throughout the home and impacting lifestyle change.
- AI technologies, governance and ethics.
- AI conversations, talk and creativity.
- Advances in machine learning and productivity of the self.
- Advances in AI, big data and digital research methods.
- Advances in machine cognition systems and identity theft.
- Race, ethnicity and gender politics in human–machine interfaces.
- Automated futures, advanced robotics and self-driving vehicles in the context of smart cities.
- Advances in AI, such as the current phase of deep learning, and public policy, transparency and governance.

Anthony Elliott, University of South Australia, Australia

https://doi.org/10.1515/9783110721751-001

The *Handbook* provides representative coverage of the full range of social science engagements with the AI revolution, from employment and jobs to education and new digital skills, to automated technologies of military warfare and the future of ethics.

In my role as the editor of *The De Gruyter Handbook of Artificial Intelligence, Identity and Technology Studies*, I have been acutely mindful of the intricate tapestry of identity in the context of the ongoing digital revolution, significantly influenced by artificial intelligence. One of the central aims of this *Handbook* is to navigate the intricate terrain of identity, where complexities and paradoxes abound, and to understand how they intersect with, and are impacted by, digital technologies, automated decision making and machine-learning predictive algorithms. Among the key objectives of the volume is to elucidate how the puzzles and contradictions of identity have far-reaching implications for the very organising categories of artificial intelligence, as well as how divisions between artificial and human intelligence impact upon the ordering, reordering and transformation of identity. This journey through the multifaceted landscape of identity is intended to offer a comprehensive introduction to the prevailing themes, traditions and domains of contemporary approaches to the analysis and critique of identity in the digital age. The breadth of approaches covered in this *Handbook* mirrors the surge of interest in digital identities in recent times, catalysed in no small part by the pervasive presence of artificial intelligence. Authors in the *Handbook* explore a spectrum of theoretical frameworks in the framing of human–machine interaction. In Chapter 2, Ross Boyd examines the rise of complex hybrid systems, focusing on a range of conceptual developments including 'agential realism' (Barad, 2007) and 'flat-' or 'object-oriented ontologies' (Ash, 2020), through to 'new materialisms' (Gamble, Hanan and Nail, 2019), 'assemblage theory' (De Landa, 2016), 'actor-network theory' (Latour, 2007), 'more-than-human theory' (Lupton and Watson, 2021), the 'theory of evocative objects' (Turkle, 2007) and theories of 'non-conscious cognition' (Hayles, 2017). Boyd, as well as other contributors to this volume, rehearses some of the conceptual gains and losses of these approaches, along with consideration of the stakes of reframing the politics of identity in the wider frame of human–machine interactions. The *Handbook* also introduces theories of data selves (Cheney-Lippold, 2017; Lupton, 2020; McQuillan, 2022), conceptual studies of algorithmic disindividuation (Stiegler, 2017), theories of complex digital systems (Elliott, 2019, 2021), governance (Rouvroy and Berns, 2013; Rouvroy, 2020), and personalisation (Lury and Day, 2019). These multiplex perspectives align with the equally extensive diversification in how we perceive and navigate the lived realities of personal identity in the age of AI. Within these pages, you'll find explorations of racial, ethnic, gendered, consumerist, cosmopolitan and global identities, all deeply intertwined with the digital revolution and the advancements in AI that shape our contemporary world.

AI and identity, then, mix and tangle in unexpected, complicated ways. To be inside and outside at the same time of the discourse of AI—to launch a critique of advanced technology from the standpoint of identity studies, while simultaneously positioning tech as essential to the regrooving of identity—is often from where the most innovative ideas emerge, both inside and beyond the academy. An exemplary instance of this is provided by the doyen of Japanese social theory Masataka Katagiri, who in Chapter 3 manages to scoop up interpretive traditions in the social sciences (especially cognitive sociology, phenomenology and symbolic interactionism) to reconsider the problem of how people interact with robots and AI. According to Katagiri, "humans read intentions and emotions into AI/robots just as they do with other humans". What matters in human–machine interaction is to grasp that this 'reading' by people into the frame of non-human others of intentions and emotions is always "situationally dependent and diverse"—a standpoint which, as Katagiri highlights, underscores the essential creativity of how humans interact with other persons as well as understand subjectivity itself.

Automated human–machine interfaces significantly impact decision making in various aspects of people's lives, especially in the domains of gender, race, ethnicity, age, disability and social inequalities. Access to digital technology, particularly automated smart machines, confers considerable advantages on many citizens, often with extensive benefits to society. The growing prevalence of decision making facilitated by human–machine interfaces is contributing to the democratisation of everyday life. However, there is a darker side to this technological advancement. AI has the potential to exacerbate inequalities rapidly and on a large scale, particularly evident in automated decision-making systems that reinforce disparities based on class, race and gender. This trend gives rise to new forms of algorithmic oppression, which are integral to advanced AI societies, effectively excluding individuals and communities. These exclusions result from the pervasive use of powerful digital technology by corporations, governments, interest groups and non-state actors. In Chapter 4, Ludovica Paseri positions identity in the context of discrimination and inequality stemming from automated computational systems. Linking Foucault's account of technologies of the self, Floridi's reflections on digital technologies and Elliott's work on the culture of AI, Paseri turns to developments in Science and Technology Studies (STS) to reconsider the role that scientific research plays in the development and deepening of the digital revolution. The chapter also explores the emergence of the Open Science paradigm, which signifies a novel approach to conducting science directly influenced by the digital revolution. Paseri delves into how Open Science can serve as a potential avenue to address and alleviate biases, inequalities and discrimination stemming from the utilisation of emergent technologies.

Identity in the Aftermath of AI

As the great wave of digital technology breaks today across the world, AI creeps increasingly into our lives. From personal virtual assistants and chatbots to self-driving cars and tele-robotics, AI threads its way into the very fabric of our existence. Its transformative impact extends across society, the economy and, notably, the individual self. Klaus Schwab, the founder of the World Economic Forum, characterises this AI revolution as unprecedented in human history. AI isn't just another technological leap; it's the complete transformation of technology itself. The most fervent public debate has revolved around the spectre of 'robots taking our jobs'. Recent technological leaps have empowered advanced AI to reshape manufacturing, the service sector and business platforms, profoundly impacting nearly every job, including professions once considered immune to digital disruption. Across the USA, UK, Japan and Australia, academic reviews and government inquiries estimate that a staggering 40–50% of existing jobs could succumb to AI technology and automation within the next 15 to 20 years, with a simultaneous rise in job polarisation trends. Yet in this whirlwind of change, AI's economic potential is breathtaking, with predictions suggesting it could infuse a colossal $16 trillion into the global economy by 2030.

AI ushers in a world of both opportunities and challenges for personal identity, existing side by side. From hailing an Uber to tracking exercise on a Fitbit to using virtual assistants like Google Home and Amazon's Alexa, the self is intricately woven into digital technologies. AI fundamentally alters how we perceive and construct our identities. In the realm of advanced AI, digitally mediated elements exert an increasingly fundamental influence on social relationships and individual lives. In this connection, social media, cloud computing and the broader spectrum of digital communication serve as linchpins. Equally instrumental are smart products, 3-D printing, intelligent ecosystems, virtual and augmented reality and machine-learning algorithms. Notably, these changes steer us away from traditional face-to-face interactions, altering the temporal and spatial coordinates that have defined social organisation. Intelligent machines, especially chatbots and softbots, bring forth new communication methods, spreading through workplaces, organisations, homes and private spaces. This ongoing transformation profoundly reshapes the concept of the self.

In the twenty-first century, personal identity and the self have come to mean, among other things, social networking, status updates, posts, blogs, virtual reality apps, cloud computing, big data and digital media. This transformation has been propelled by the pervasive influence of AI in reshaping social interactions and the very essence of identity. As a result, novel conceptions of selfhood have arisen within the context of a global economy increasingly reliant on digital technologies. In this evolving landscape, the role of psychology becomes paramount in understanding these new cultural ideas of selfhood. It prompts us to question whether psychology, in the wake of AI developments, has adapted and developed a new perspective for comprehending our frustrations. This entails delving into the processes of identifying and resolving these issues, as well as fostering introspection and effective communication in a world that has undergone a remarkable technological metamorphosis. In Chapter 5, J. Michael Innes and Ben Morrison examine AI from the field of psychology. They probe the complex, contradictory ways in which complex technological systems are reorganising the very tissue of human psychology as well as the central organising categories of psychological thought. Developments in the social psychology of identity, according to Innes and Morrison, are deeply interwoven with the development of AI and social media. The chapter illustrates how these transformations culminate in the creation of a diverse and polarised society. Within this framework, multiple social groups emerge, each catering to the social needs and objectives of individuals. However, it's essential to recognise that these very social forces have the potential to fragment the social fabric, with potential repercussions for political groups and their positions of influence within societies.

The development of digital technologies, and especially the rise of robotics and AI, is not therefore simply an 'out there' phenomenon, as if technologies operate with straightforward given properties and causal effects upon institutions, organisations and networks. Technologies, on the contrary, are always mediated through personal experience and the self, entangled in our everyday ways of doing things and the living of lives. Examining the introduction of AI and robots into work and employment, Sven Nyholm in Chapter 6 probes the likelihood of human–machine relationships acquiring qualities of collegiality which are characteristic of many human–human relationships that form in work environments. Specifically, he seeks a path beyond binary approaches to the question of whether robots can be good colleagues or co-workers. This results in a multiplex understanding of collegiality, one that gathers together factors ranging from the operational capabilities and characteristics of given technologies and the ways they are positioned in work processes, through to the dynamics of work groups or teams and the meaning of the work for the people involved. The possibility that people can and do come to view robots and AI collegially leads Nyholm to then ask if it is ethically problematic to do so.

If digital technologies transfigure institutions, they also reach profoundly into individual identities. If AI impacts work and employment, so, too, does it transfigure intimate life and social relations in the broadest sense. In Chapter 7, Elisabeth Hildt explores the impacts of human–robot or human–AI interactions upon people's self-experience and self-understanding, as well as how users are perceived by others and the ways people interact with other people. Drawing upon specific research on the deployment of companion and therapeutic robots in elderly care and the education of autistic children, along with several large-scale human–media interaction studies, Hildt identifies a number of key factors that can prompt emotional entanglement in persons with these technologies. Given that most of these involve features of the technologies designed to facilitate interaction through eliciting emotional and other responses, and that the technologies themselves are reactive and at best only simulate basic human-like responses themselves, Hildt concludes by reflecting on the prospects for a future world of artificial consciousness, and the possible consequences for human–machine interaction and understandings of selfhood.

The digital is both around us and inside us. From this angle, AI and digital technologies become raw materials for the production and performance of the self. On this view, if AI fundamentally transforms the economy, it also rewrites the emotions. We need to see that AI penetrates all the way down, reconstituting personal life, passions and the affects. In Chapter 8, Luca Possati takes up this challenge by situating AI in relation to psychoanalysis—especially Lacan's Freud. According to Possati, AI is both the cause and effect of a crisis of identity in global capitalism. Predictive algorithms, from this angle, recondition the very identifications which constitute identity. Pleasure, desire, identification, repression, paranoid defences and delusional foreclosures: a single term to sum up all of these would be *the unconscious*, which Possati argues has become claustrophobically coded, administered and overdetermined by automated machine learning in our own time. He seeks in this eye-opening discussion to also open psychoanalytic criticism of AI to various currents in contemporary social theory, including theories of social acceleration, new individualism and social reflexivity.

AI, as both discourse and actuality, has come to mean the rise of tech platforms in which individuals share their everyday social practices and engage in forms of self-automation, where biometric sensors connect bodies to the Internet of Things, and where machine-learning algorithms process data to offer personalised recommendations, eligibility assessments and risk profiling. Notably, these developments in AI are fundamentally linked to the algorithmic construction of individual identities from big data. In Chapter 9, Ben Lyall turns to consider the rise of 'lifelogging', 'self-tracking' and the 'quantified self'—that is to say, the spread of intimate, 'always on' tracking technologies which collect personal information and recalibrate identity in the process. This certainly addresses the boom in 'fitness trackers' and 'smartwatches' throughout the technology marketplace, as multinationals including Apple, Huawei, Samsung and Xiaomi offer consumers an ever-expanding menu of gadgets to record and analyse daily life (covering everything from daily steps to nightly sleep to calories consumed). Lyall makes the central point that such data is *tradable*: technology companies buy, sell and transform data, while corporate takeovers radically reshape the responsibilities of those who own data. But more than that, the chapter also highlights that self-tracking technologies generate new ways of seeing and understanding bodies and behaviours, documenting and sharing experience, and thus *marking identity*. This relationship between data and the self is taken up further by Tim Gorichanaz in Chapter 10, who addresses the fate of 'data doubles' in the frame of identity theft. Data double profiles assembled by tech companies to represent and predict the behaviour of specific users —from shopping and browsing data to demographic characteristics and beyond—fundamentally recontour the relationship between identity and technology. This creates new opportunities for the self, but also new burdens. Identity theft is such a key risk. As Gorichanaz demonstrates, identity theft is when one party obtains somebody's personal data unlawfully, usually selling that data or using it to commit fraud. This is a phenomenon, he suggests, which is skyrocketing (and most likely will continue to skyrocket) in the age of AI.

The Future(s) of AI

In *What is the Future?*, the late sociologist John Urry (2016) declared that the social sciences needed to reclaim ownership of the field of future studies as a crucial move for restoring robust public debate—both for thinking about the likely trajectories of society and also for grasping the importance of competing visions of the future and the role that such imaginings have in bringing real futures into being. A notable aspect of much writing about the future of AI is the close attention paid to the characteristic features and personal experiences of change as it is occurring in specific domains.

In our own time, there has been a rapid explosion in enforced migration, mostly in terms of asylum seekers attempting to find safe passage from various political hotspots around the globe. Alongside such developments, freely chosen mobilities also continue to develop apace. In this context, critical social science has staged an important encounter with rethinking the role of digital technologies in terms of the rethinking of human movement—both freely chosen mobilities and blocked paths of movement, or immobilities. In Chapter 11, Thomas Birtchnell considers how golden proclamations regarding driverless vehicles by industry, policy, media and academic enthusiasts stack up against the uncertainties and complexities that constitute the real-world mobility environments within which individuals, companies and organisations will be required to operate. Drawing upon theoretical and empirical work on social assemblages, social practices, socio-technical transitions and the mobilities paradigm, Birtchnell argues that before AI can be safely and effectively integrated into mobility systems, it must overcome challenges associated with, among other things, the heterogeneity of human routine and habitual interaction, unforeseen consequences of its own operations and the requirements of mobility justice. In Chapter 12, Louis Everuss and Eric Hsu examine digital border identities in the wider frame of the political forces of globalisation and deglobalisation, emphasising that automated smart machines often code prejudice and bias into sovereign decisions in complex and compressed ways. Unpacking what the algorithmic processing of big data actually entails in terms of contemporary bordering systems, Everuss and Hsu contend that the nation-state is chronically reconstituting itself as 'datafied' through the registration, tracking and tracing of 'composite' digital identities.

Elderly care has emerged as a key site of transformation in the current age, driven by the adoption of various AI technologies and sensor-equipped environments in the broader context of 'smart homes'. These technologies automatically collect data on aged users without requiring their direct involvement, contributing to the emergence of new identities shaped by human–machine interaction. Experts in the fields of preventive medicine and health promotion emphasise the benefits of smart digital technologies and social robots. They also highlight the importance of automated health data as indicators in helping elderly patients prevent disease and illness while navigating these evolving identities. In Chapter 13, Rina Yamamoto deftly traces out the massive changes which older people face in urban areas, with specific focus on AI and social robotics for aged care. Yamamoto carefully assesses developments surrounding smart homes from the perspective of aged care, with particular stress on transformations of identity through drawing upon recent social theory.

In Chapter 14, Ross Boyd asks what an identity lens might add to contemporary debates over the implications of artificial intelligence for the future of work and employment. Creatively appropriating Paul Ricœur's notion of narrative identity, Boyd argues the case for research that takes as its object the lived experiences of digital disruption in the workplace—specifically, the "processes through which working people are seeking to find a place for themselves within, comprehend and apprehend these more or less fluid situations in order to render them intelligible and manageable". Noting affinities between this standpoint and E. P. Thompson's classic history of the Industrial Revolution, he indicates some of the directions this research might take through reference to a number of contemporary studies of experiences of disruption on the part of those who have made significant identity investments in their technological expertise.

Conversational user interfaces—chatbots—constitute one of the most ubiquitous contemporary forms of AI. Looking beyond the now well-documented commercial and public service delivery applications of conversational AI, Katarzyna Baliga-Nicholson and Jan Argasinski reflect in Chapter 15 on their participation in the Panefka project. The project aimed to develop a chatbot that would represent LGBTQI+ experiences and perspectives within a broader university student community, and in so doing function as an agent of progressive cultural change. Assessing the project Baliga-Nicholson and Argasinski identify a range of practical and ethical concerns that need to be considered by designers of both ethically aligned and commercial intelligent agents.

Perhaps the most widely discussed, and possibly anticipated, social futures opened by the age of AI has been that of the technological singularity. This is a world consisting of computational superintelligence, where AI has effectively rendered biological humans obsolete. A new high-tech global society predicated on supercomputers surpasses humans, with automated machines so intelligent that they recursively design and improve themselves. This is a hyper-technological world, where systems of superintelligence cover everything from unsolved problems in neuroscience to eradicating the destructive effects of climate change. Ray Kurzweil's *The Singularity is Near* is arguably the best-known account of the contours of such a world. In Chapter 16, Elise Bohan traces the social, cultural, economic and political implications of such a world where machine intelligence dramatically transcends human intelligence. From the contributions of mathematician John von Neumann to futurist Kurtzweil, Bohan reviews the plausibility of the technological singularity and focuses on key themes such as social acceleration and technological innovation in the coming decades.

References

Ash, J. (2020). Flat ontology and geography. *Dialogues in Human Geography, 10*(3), 345–361.

Barad, K. (2007). *Meeting the universe halfway: Quantum physics and the entanglement of matter and meaning.* Duke University Press.

Cheney-Lippold, J. (2017). *We are data.* New York University Press.

DeLanda, M. (2016). *Assemblage theory.* Edinburgh University Press.

Elliott, A. (2019). *The culture of AI: Everyday life and the digital revolution.* Routledge.

Elliott, A. (2021). *Making sense of AI: Our algorithmic world.* John Wiley & Sons.

Gamble, C. N., Hanan, J. S., and Nail, T. (2019). What is new materialism? *Angelaki, 24*(6), 111–134.

Hayles, N. K. (2017). *Unthought.* University of Chicago Press.

Latour, B. (2007). *Reassembling the social: An introduction to actor-network-theory.* Oxford University Press.

Lupton, D. (2020). *Data selves: More-than-human perspectives.* Polity Press.

Lupton, D., and Watson, A. A. (2021). Towards more-than-human digital data studies: Developing research-creation methods. *Qualitative Research, 21*(4), 463–480.

Lury, C., and Day, S. (2019). Algorithmic personalization as a mode of individuation. *Theory, Culture & Society, 36*(2), 17–37.

McQuillan, D. (2022). *Resisting AI: An anti-fascist approach to artificial intelligence.* Policy Press.

Rouvroy, A. (2020). Algorithmic governmentality and the death of politics. *Green European Journal*, March, 1–5.

Rouvroy, A., and Berns, T. (2013). Algorithmic governmentality and prospects of emancipation. *Réseaux, 177*(1), 163–196.

Schwab, K. (2016). The Fourth Industrial Revolution. World Economic Forum.

Stiegler, B. (2017). *Automatic society: Volume 1.* Polity Press.

Turkle, S. (EDd) (2007). Evocative Objects. MIT Press.

Urry, J. (2016). What is the Future. John Wiley and Sons.

Ross Boyd

2 AIs, Hybrid Systems, Agency and Identity

Abstract: This chapter commences with an exploration of the historical career of the concept of agency, from an initial formulation as the capacity for self-action on the part of possessive individuals (and human collectives) through to more recent understandings that centre on the effects of social engagements among humans and non-humans. I then critically examine the ways this history is reflected in representations of the agency of artificial intelligence (AI), be it as a property of a Cartesian thinking thing, or an effect of a network of collaborative relationships constituting hybrid socio-technical systems. The chapter closes with some reflections on the implications of new understandings of agency for the question of identity in an age of AI.[1]

Keywords: AIs, agency, replicants, hybrid systems, centredness/distributedness, automated self, eXperimental self

2.1 Introduction

Identity, agency, AIs and hybrid systems—quite a confection! Permit me to temporarily bracket out 'artificial intelligence', as both a term of technical art and synecdoche for complex hybrid systems (Holton and Boyd, 2021), in order to rough out some initial conceptual and historical reference points regarding well-established connections between identity and agency. These can then serve as both primer and platform for the discussion that follows. Three points should be sufficient here.

(1) Considered initially (and inadequately) as capacities for self-action, agency has long been regarded a defining quality of 'modern' conceptions of selfhood. That is, understandings of self that emerge, haltingly, tentatively in particular places (mostly across Europe and subsequently in the settler colonies of the Americas and beyond) and at particular times (from the 1500s on). Places and times wherein, as Reinhart Koselleck in *Futures Past* (1985) observed, people started to reorient themselves away from the well-worn ways of tradition and towards 'open futures' in which received understandings, present experiences and expectations of what might and should come would be subject to negotiation with the results not given in advance. And wherein identity, in keeping with the emerging mood of the times, would come to be seen as something

1 Preparation of this chapter was assisted by support from Australian Research Council Discovery Grant DP220102630, *Investigating Conversational AI: Development, Usage and Governance*.

Ross Boyd, University of South Australia, Australia

https://doi.org/10.1515/9783110721751-002

achieved, realised through one's own rational design and effort—the exercise of one's agency—rather than ascribed—assigned (fixed, immutable, beyond will) by nature and the 'natural' social order at birth. Hence the figures that still wield considerable normative power as models of the human-as-agent-achieving-full-(adult)-personhood, within the enabling mythologies of Euro-American+ societies: (a) the unique, bounded individual, integral, purposeful, capable of planning and choosing between better or worse plans, hence accountable (Taylor, 1985), and above all possessive, 'the proprietor of their own person and capabilities, owing nothing to society for them' (Macpherson, 1964, p. 3); and (b) the autonomous, transcendental, rationally cognising Cartesian subject, standing apart from and over the world of matter, its own body included, that is its object (Lury, 2013).

(2) Agency, and the ways it gets coupled with identity, needs also to be considered part of the experience of modernity in collective terms—the experience of an 'us' rather than an 'I'. The agents referred to above as engaged in an ongoing finessing of the differences between received understandings and future expectations have, as Bayly (2021, p. 3) notes, also been known by a multiplicity of collective terms: 'the nation, the empire, The West, the Aryan race, the commonwealth, the demos, the masses, the proletariat, the people, the public, the multitude, the crowd, the party, the part who have no part, the subaltern, the 99 per cent, humanity . . . '. The capability for pursuing individual self-assertion (and accordingly 'full personhood') tends, as a rule, to closely follow the situation ante distribution of economic, political, cultural and other social resources. Yet the same hierarchically organised categories invoked to legitimate these distributions—historically class, gender and race have figured prominently—have, precisely because of this, served to galvanise collective identities and mobilise forms of collective self-assertion. More, the self-image and self-experience of privileged classed, gendered and raced sections of society as autonomous (self-directed and self-sufficient) individual agents has long been revealed as a myth—if a very powerful one—by generations of critical scholarship that have sought to drag into view the enormous amount of unacknowledged physical, emotional and symbolic labour put in by other differently gendered, raced and classed people that was actually constitutive of this agency (Mbembe, 2019; Collins, 1990; Minh-Ha, 1989).

(3) For some time now scholars have been (re)discovering the agential powers of things (Erofeeva, 2019). In the opening chapter of his *Pandora's Hope*, Bruno Latour (1999, pp. 1–23), for example, critiques the founding and compounding folly of the dominant tradition of Western epistemology for the past three centuries or more. Here the original folly is the notion of the transcendental Cartesian self: that is, Descartes's unplugging of the human mind from its corporeal ties to its terrestrial surround, and the subsequent demand that this disconnected organ furnish certain knowledge of an 'outside' world. This figure of a detached mind gazing out at an external world that slips further beyond its ken forms something of a constant across the otherwise diverse and divergent epistemologies that followed in the Cartesian train: the empiricist divining

the outlines of the world by recognising patterns amid the ruck of dumb stimuli; the Kantian reaching into the structures and processes of mind itself for those rational (*a priori*) categories that would hold its world pictures steady; the social constructivist pointing up the workings of culture—the biases, traditions, standpoints—within the grain of the *a priori*; the infamous prison house of language . . . Latour's is but one of a number of substantial contributions seeking to conceptually reconnect mind to body and, through body, mind to its normal ecology, thereby enabling a knowledge of the world grounded in the number of relations with the non-human stuff of the world engaged with in daily practice (e.g., Harman, 2018; Barad, 2007; Garcia, 2014). This work has done much to precipitate a reconsideration of notions of agency outside of those anchored solely in human capacities for action, and a recognition of the co-constitution of the agency of things and the agency of humans with significant ramifications for understandings of identity.

The next section of this chapter picks up on my initial (parenthesised) comment about the inadequacy of conceptualising agency as simply capacities for self-action. Commencing with Emirbayer and Mische's (1998) observation that 'agency', while a central idea in social thought for many years, has "maintained an elusive, albeit resonant, vagueness" (p. 962), I briefly follow these authors' attempts to develop an analysis of human agency as a category in its own right, a property of social engagement, irreducible to the purportedly unique attributes of some or other self-capable individual as well as not unduly bound to social structure. This discussion is, in turn, placed alongside significant moves being made around the same time to extend the concept to where the agency of non-humans is recognised, leading to an elaboration of agency as a relational phenomenon. Taken ensemble, these moves create new challenges for thinking agency as it bears upon questions of identity. In section 2.3 I introduce AIs into the mix, further compounding these challenges. I focus on two developments here. The first is the way AI research has been, and in many significant ways still is (e.g., Chollet, 2019), explicitly or tacitly referenced against models of cognition that bear a striking resemblance to the Euro-American+ Cartesian/possessive individualistic self, leading to discussions about the necessity of 'decolonising' AIs. The second entails the elaboration of an increasing number of models of AIs in terms of hybrid human–machine systems, and consequent reformulations of agency and identity as relational and distributed. In section 2.4, I return identity to the discussion and explore some new research directions on the coupling of agency and identity in the age of AIs.

2.2 Agency: Some Relatively Recent and Important Developments in the Career of the Concept

Ideas of agency have long held a central position in social theory and sociological analysis, often serving as the principal prop supporting the 'individual' or 'action' side of dynamic individual–society or action–structure couplings, as well as a placeholder for many of the qualities (will, intentionality, choice, purpose, volition, freedom) taken to characterise modern ideals of agentic individualism (Hitlin and Elder, 2007; Emirbayer and Mische, 1998). Positioned thus, the notion of agency has been called upon to do much of the heavy lifting, conceptually speaking, in theoretical accounts of social change and the historicity of structure, the possibility of politics, creativity and invention, or the ways structural constraints are constitutive (but never fully determinative) of the very action that is itself constitutive of structure, among other things (e.g., Beck, Giddens and Lash, 1994; Joas, 1996; McNay, 1999). More, the concept carries out this work across a wide variety of social theoretical traditions, including early action or conflict theories and their subsequent derivatives, through symbolic interactionism, practice and social movement theories, to the more recent development of theories of performative (Butler, 1997) and postcolonial (Jabri, 2014) agency, among others.

Given its apparent longstanding significance it is therefore noteworthy that in one of the most frequently cited papers in the *American Journal of Sociology*, Mustafa Emirbayer and Ann Mische (1998) would claim that failures on the part of successive generations of social theorists to distinguish and develop agency as an analytical category in its own right had resulted in "a flat and impoverished conception that, when it escapes the abstract voluntarism of rational choice theory, tends to remain so tightly bound to structure that one loses sight of the different ways in which agency actually shapes social action"(p. 963).

They respond to this deficit by detailing an account of agency centred on the capacities drawn upon by, accorded to and generated by people through the emergent processes by which they practically and normatively orient themselves within a range of everyday social situations that are both mutable and coterminous. And the ways agentic capacities interplay with, reproduce or transform the structural contexts in which these processes play out as people creatively engage problems thrown up by changing historical situations. Specifically, Emirbayer and Mische (1998) identify three core elements that constitute the internal composition of agency or, more accurately, agentic orientations:

(1) *The iterational*: the selective recall and reactivation of past patterns or schemes of thought and action which are routinely embedded in practical activity and provide stability to social worlds through sustaining identities and institutions.

(2) *The projective*: the imaginative invention or improvisation of possible courses of action, including the creative appropriation of extant schemas, that can reconfig-

ure received structures of thought and action in line with actors' aspirations, desires, fears and hopes for the future.

(3) *The practical-evaluative*: the communicative exercise of both pragmatic and normative judgements as to better or worse, preferred or unwanted, trajectories of action in response to the demands, dilemmas and ambiguities of evolving situations as these arise in the present moment.

Agency, on this account, is neither analytically nor empirically reducible to social structure conceived as temporal-relational (i.e., relatively durable and extensive) contexts of action—although these contexts certainly shape and influence agentic orientations. Nor, however, are there moments where agency can be considered wholly free from structure, as if it were an intrinsic quality of acting individuals: "agency itself remains a dimension that is present in (but conceptually distinct from) all empirical instances of human action, hence there are no concrete agents but only actors who engage agentically with their structuring environments" (Emirbayer and Mische, 1998, p. 1004).

More bluntly, given temporal-relational contexts of action support or conduce to particular agentic orientations, that is, specific ways of configuring possibilities for deliberative re-evaluation of, imaginative distancing from or iterative recall of received structures. And through giving direction and form to the orientation of actors within and towards their environments, they afford actors differing degrees and types of transformative leverage over these contexts. Agency, therefore, appears as a property of the interactive relationship between actors and between actors and structures, that is, of *shared situations*, in contradistinction to being regarded, as it had most often appeared to be prior to Emirbayer and Mische's intervention, a property of actors (or, indeed, structures). This shift in the locus of agency is a significant, thoroughly sociological, advance in the theorisation of the concept.

Or at least a significant advance for its time. While insisting on a distinction between agency and actors, it is clear that Emirbayer and Mische do intend their model of agency to apply solely to human action. This is most evident in the centrality they accord conscious reflection in their discussion of the internal composition of agency, be this a consciousness that is dialled up in the case of critical deliberation or dialled down in the more habitual elements of iterative recall. That this might even be considered an issue is very much due to developments in thinking about agency that resolve upon a more capacious view of the relationality of agency through a focus on the active role of non-human things within social interaction.

In the latter decades of the twentieth century scholars across the social sciences and humanities, including some with a strong grounding in the physical sciences (e.g., Karen Barad, N. Katherine Hayles), started paying attention to the idea of agency as it applies in the case of non-human 'things' (Erofeeva, 2019). Some of the earliest significant statements in this area include Gibson's (1979) work on the ecology of perception which yielded the highly influential concept of 'affordances'; the collection of essays on the 'social life of things' convened by Appadurai (1988); Latour's (1992) pathbreaking essay on the 'missing masses of society'; probably Deleuze's re-reading of Spinoza's notion of 'conatus' and Liebniz's 'vital power immanent to things', and most certainly Haraway's (1991) problematisation, in the figure of the cyborg, of the hitherto jealously guarded boundaries separating the human from other organisms and between the organic and the artificial/machinic. Moving forward into the twenty-first century, we have witnessed the emergence of a wide range of schools of thought that share (at times little more than) an interest in rethinking the concept of agency to incorporate things—from 'agential realism' (Barad, 2007), 'flat-' or 'object-oriented ontologies' (Ash, 2020) and 'new materialisms' (Gamble, Hanan and Nail, 2019) through to 'assemblage theory' (DeLanda, 2016), 'actor-network theory' (Latour, 2007), 'more-than-human theory' (Lupton and Watson, 2021) and theories of 'non-conscious cognition' (Hayles, 2017).

It cannot be my purpose to provide an exhaustive mapping of these various positions relative to one another here, but rather just piece together a rough working demonstration of different moments in recent thinking of the concept of agency in conjunction with objects or things, one that will be helpful moving forward to where I reintroduce AIs and resume discussion on the agency–identity coupling. For purely expository reasons I arrange these various positions on a line running from those making the most modest claims to those proposing something far more ambitious.

In his posthumously published *Art and Agency* Alfred Gell (1998) put forward an account of agential qualities attributed to non-human things that, although gaining considerable recognition for this move as breaking new anthropological ground, is in actuality relatively limited in terms of its ambition. Limited, that is, to a particular class of things, artworks, and by an understanding of agency that remains largely tied to a free-standing agent's 'will' or 'intent' (pp. 16–21). Artworks here possess, at best, secondary agency as media through which the primary agency of an artist's intentions is manifested and realised. Yet this is still agency of a sort, inasmuch as artworks, for Gell, 'move' others to make abductive inferences, interpretations and so forth about the ideas, intentions and so on of the producers (or exhibitors) of the work.

Superficial similarities notwithstanding, the collection of essays assembled in *Evocative Objects* (Turkle, 2007), along with the conceptual framework roughed out by Sherry Turkle in the introductory and concluding chapters, are considerably more nuanced and ambitious than Gell. Agency is approached here by way of evocative qualities attendant upon objects: the ways, that is, particular objects prompt emotional and intellectual resonances, and lead or arouse thoughts, feelings, memories, theories and so forth. In the essays that comprise the volume, in which authors recount their often-intimate but not always congenial relationships with a wide array of objects—a cello, knots, Le Corbusier's archive, an elite glucometer, a yellow raincoat, an ancient axe head, a 1964 Ford Falcon, a stuffed bunny called Murray, a rolling pin and apples, among many other things—it becomes clear that the agentic quality of evocation emerges as a property of the complex interactions between objects, people and situations. Elaborating further on Turkle's framework, Graham Harman (2008) pushes back against any glib suggestions that these 'moving powers' attributed to objects involve little more than individual projections, informing us solely about the inner lives of the particular persons doing the projecting while adding nothing to any understanding of the properties objects may contribute to shaping and conditioning interactions. Instead, Harman argues, evocative objects (and not all objects are or become evocative) possess certain features that are unlocked or aroused through the activity of evocation. These features can be summed thus:

- *Depth*: evocative objects are not blank slates to which traits may be arbitrarily assigned; "We humans never know exactly what an object is capable of, and must enter into a kind of dialogue with it" (Harman, 2008, p. 456).
- *Richness*: once in view evocative objects differ, more or less, from their 'class' or 'type', and it is these often-subtle variations that provide points of contact for the kinds of human commerce with them that fall under the rubric of evocation.
- *Autonomy*: it is in the ways that objects may partially evade our attempts to press them into human service or remain recalcitrant to our plans that they (*qua* Heidegger's tool-analysis) can emerge from the invisibility of smoothly functioning tools and acquire a noticeable presence.

With his meditation on the Mexican artist Damián Ortega's celebrated installation *Cosmic Thing*, John Tresch (2007) allows for this notion of 'evocative objects' to be extended to where it touches upon some of the most creative and adventurous contemporary thinking about the agency of things. The installation itself consists of a 1989 Volkswagen Beetle, disassembled, suspended but with its parts arranged in a way that the vehicle retains a clear resemblance of its original familiar form. Yet for all this it still appears strange, uncanny and as such more available for thought than might otherwise be the case (for images of Ortega's installation see https://arthur.io/art/damian-ortega/cosmic-thing/4). Tresch 'follows' this disconnected machine across space—from the near side of harassed drivers and the errands of the daily metropolitan round along unpaved country roads, past the VW factory at Puebla Mexica that birthed this *Vocho* and thousands like it for American markets, and on to the Institute of Contemporary Art in Philadelphia. And over time—from the industrial fascism of Nazi Germany to Kerouac, Cassady and 50s/60s counterculture. In the process he traces out, explicitly and in allusion, an extensive webbing that enjoins, in more or less fragile planet-wide affiliations, production line workers, expert engineers, mechanics, drivers, profiteers and other powerful individuals, artists, gallery curators, corporations, governments and governing institutions, fossil fuels, crankcases, torsion beam suspensions, rear-mounted air-cooled engines, rubber, bitumen and concrete . . . all of which exercise (and acquire) different kinds of agency through ongoing processes of assemblage and which, collectively, bring about the unique event of the *Cosmic Thing*.

Ortega's/Tresch's *Vocho* thus transports us a considerable distance away from a notion of agency understood in terms of intrinsic capacities for self-action on the part of idealised free-standing (very human) subjects. Agency (and indeed identity) presents here, rather, as decentred, distributed and emergent, a property not just of localised interactions (human–human, human–non-human or non-human–non-human) and their situations but oftentimes vast, dynamic heterogeneous networks through which all manner of entities acquire agentic capabilities that, when acted out, also enact (configure and reconfigure) the given network itself. Which means this battered jalopy deposits us in close proximity to what is known as actor-network theory and other contemporary schools of thought, prominent among them feminist science and technology studies, in which agency inheres within specific socio-material arrangements.

For the pioneers of actor-network theory, notable among them Bruno Latour, Madeleine Akrich, Michel Callon, Annemarie Mol and John Law, a central objective was to shift the focus of social analysis away from society, considered in terms of interactions among humans, serving as a source of explanation towards an approach to society as something in need of explanation itself. And in need of more-than-human factors as a crucial component of this explaining. As Latour (2000) puts it,

> The great import of technology studies to the social sciences is to have shown, for instance, how many features of the former society, durability, expansion, scale, mobility, were actually due to the capacity of artefacts to construct, literally and not metaphorically, social order . . . including the infamous agent/structure dilemma. . . . They are not reflecting it as if the 'reflected' society existed somewhere else and was made of some other stuff. They are in large part the stuff out of which socialness is made. (pp. 113–114)

If human–human interactions are all there is to social life then, on this account, social life would be fleeting and fugitive in the extreme (a perfect spot market, as it were), of no greater extent or duration than the interactions that formed it. Non-humans work, whatever else they do, to stabilise the social.

The overall actor-network theory project chimes, in important respects, with the concern of feminist science and technology studies to worry the problem of whose—or what's—agency is recognised (and whose/what's is not) when accounting for the success of complex socio-technical arrangements (Suchman, 2019a; Rhee, 2018). Together these frameworks have been applied to an extraordinary range of empirical instances to yield a wealth of knowledge about agency as relational, collective and distributed across human and non-human actants. Lucy Suchman, for example, illustrates this well through reference to Dawn Goodwin's study *Acting in Anaesthesia: Agency, Participation and Legitimation.* Here, patients in surgery are seen to be transitioned through a range of anaesthetic states in which complex socio-technical arrangements comprising machines, medical experts, routinised practices, monitoring devices and so on, together produce an ongoing progressive reconfiguration of their capacity for action.

> [S]he demonstrates how the technologies of anaesthesia are joined to the patient's body, in ways that render the latter highly dependent and vulnerable, but nonetheless intensely (albeit sometimes ambiguously) communicative. This joining is analyzed as a delicate choreography involv-

> ing patients, medical practitioners and machines. Over the course of an anaesthesia, agencies involved in the sustenance of vital bodily functions are progressively delegated from 'the patient' as an autonomously embodied entity, to an intricately interconnected sociomaterial assemblage, and then back again. (Suchman, 2019a, p. 75)

Of course, this analytical movement away from the normative ideal of agency as the capacity for self-action of a wholly rational, self-sufficient, self-determining, individual to something that is an emergent property of interactive networks or assemblages that are themselves emergent is by no means shared by all, be this analytically—as will be addressed in the next section that returns AIs to the discussion—or in the popular imagination.

2.3 Bringing AIs Back In

Artificial intelligence? There are far too many definitions and descriptions of AI to canvass here, and indeed, as will be indicated in the discussion that follows, there are good technological as well as social scientific reasons to avoid approaching AI as a unitary phenomenon. The term itself was initially coined primarily as an organising device designed to gather together a diverse body of existing studies into a coherent research programme based on "the conjecture that every aspect of learning or any other feature of intelligence can in principle be so precisely defined that a machine can be made to simulate it" (McCarthy et al., 1955, p. 2). Subsequent definitions range from the technical to the sociological. Walsh et al. (2019), for example, refer to AI as "a collection of interrelated technologies used to solve problems that would otherwise require human cognition. Artificial Intelligence encompasses a number of methods, including machine learning (ML), natural language processing (NLP), speech recognition, computer vision and automated reasoning" (p. 14).

Elsewhere, a wider view of the components that comprise AI systems—for instance machine learning and other algorithms, big data, engineers, code libraries, ghost workers, budgets, sensor systems, data scientists, organisational configurations, lawyers, software architectures, actuators, angel investors and so on—are described in terms of social, political, economic and cultural projects in addition to purely technological ones (Holton and Boyd, 2021; Crawford, 2021; Boyd, 2021).

Much of this variation can be put down to either standpoint—for instance whether one is viewing AIs as would a machine-learning practitioner as opposed to a social scientist (Roberge and Castelle, 2021)—or to processes of mutual learning that inhere within technology design practice conceived of as an ongoing "set of interactions between the designer(s) and the materials being worked with . . . a two-way relationship" in which the accommodations and resistances that Pickering referred to as the "dance of agency" come into play (Ihde, 2008, p. 57). Here the history of technology design is seen to be as much one of iterative learning (individually and collectively) about ourselves and our understandings of the projects in which we are involved, as it is about the materials, their properties and possibilities for innovative combinations with which we are working (Marres, 2019).

This notion of technology as a mirror of the human is not a particularly new one, in its 'modern' iteration reaching back at least as far as la Mettrie's *L'homme machine* or Vaucanson's automata in the eighteenth century. It has, nonetheless, gained considerably in salience with the advent of AIs. Here I will organise my observations around two very different ways thoughts about the agency of AIs have served as reflecting surfaces for a consideration of human agency and, through this, identity; as well as the ways very different ideas about human agency (and identity) have informed understandings of the entailments, development and consequences of AIs.

2.3.1 The replicant ambition

The figure of the replicant, the genetically engineered human simulation introduced in the film *Bladerunner*, along with a range of other film androids and bots (e.g., Ava in *Ex Machina*; see also Wilson, 2006), quite aptly captures much of the ambition and controversy attending some of the most significant AI research programmes from the earliest days through to the present. As Natale and Ballatore (2020) have demonstrated, the original and subsequent AI researchers understood and discursively represented their scientific and engineering projects in terms of broad claims about the human mind, "relying on daring analogies among humans, animals and machines" (p. 8) that drew upon, variously, cybernetics, philosophies of mind, control theory, neuropsychology and operations research.

Specifically, most of the pioneers of AIs understood their projects as entailing the development of machines that replicated processes of human thought. This was certainly clear from the original statement of intention for an AI research programme drafted at the Dartmouth College Summer Workshop (see McCarthy et al. 1955) which asserted (following directly on from the conjecture reproduced above), "An attempt will be made to find how to make machines use language, form abstractions and concepts, solve kinds of problems now reserved for humans, and improve themselves" (p. 2). In the paper "Computing Machinery and Intelligence", in which he proposed the imitation game that would subsequently become the famous 'Turing Test', Alan Turing (2012 [1950]) looked to machines that could imitate human thought—including learning, searching and making decisions—to solve problems in mathematics and the philosophy of mathematics. For Turing, it was not the biological composition of the brain that was important to understand but rather the logical structure of its operations, for once grasped this structure could be reproduced in non-biological, machine media (Hodges, 1983). In the 1960s leading AI researchers Herbert Simon and Marvin Minsky both asserted that in the proximate future, once the technical shortcomings in the AIs of their time had been overcome, machines would have matched most human intellectual capabilities.

This emphasis on replicating human cognitive functioning, or significant aspects of it, would guide the lion's share of AI research for the next 70 years. In part this was due to participants at the Dartmouth workshops, Minsky and Simon included, establishing many of the world's most prominent academic and industry AI research programmes and laboratories. But it was also due to the enduring power of 'replication' as an organising narrative enabling so many researchers working in the field of AIs and related technologies development to locate and make sense of their specific projects in terms of a broader, deeper programme. For example, in a large-scale oral history project involving interviews with over 100 researchers who were leading pioneering work on AIs and robotics in the academy and industry, participants regularly made sense of and represented their endeavours in terms of fundamental research into the constitution of human intelligence, language, emotion and physical capability (Cheon and Su, 2016, pp. 378–379; see also Sabanovic et al., 2015). In a more recent, heavily cited, paper, François Chollet (2019), a Google-based AI researcher and software engineer, proposed a psychometric test—explicitly grounded in the psychological traditions of intelligence testing that go back to Binet. This is designed to measure an AI system's general-purpose abstraction and reasoning capabilities (in contrast to task-specific skills) in order to measure progress in AI research towards the overarching goal of human-like intelligence.

The 'replicant ambition' has, in turn, been subject to criticism targeting, in different ways, the narrowness of the representation of human cognition entailed. This has led to both a questioning of the technical plausibility of the ambition of achieving human-like intelligence, and hence agency, itself, as well as claims that the project unreflexively replicates hierarchical notions of intelligence that have historically done much to undergird raced, gendered and classed forms of power and privilege—subjugation and disadvantage (Adams, 2021; Rhee, 2018).

The first line of questioning concerning the plausibility of the project as a technical endeavour is nearly as old as the project itself, going back at least as far as Mortimer Taube's (1961) questioning of the validity of machine translation on account of the erroneous assumptions about human language upon which it was based. Arguably the most influential of the early criticisms were, however, those levelled by Hubert Dreyfus (1979). Dreyfus argued that attempts to encapsulate human thought in terms of explicit rules seriously neglected the way intelligence was irreducibly contextual—was not merely a set of formal mechanical operations occurring within people's heads but, rather, emerged within the complex interactions that constituted real-life situations. And while there now exist advanced machine-learning systems deployed in finance trading, public and private service delivery, security and marketing, among other things (Pink et al., 2022), that demonstrate many of the qualifying conditions commonly taken to represent agency—they make decisions, they initiate real-world actions, they modify their operating rules (learn) with little in the way of apparent human direction—the prospect of human-level/-like machine intelligence still remains, in the main, futuristic. In a recent critique, Holton (2023) challenges claims that artificial intelligences can, by themselves, address the economic, political and cultural dimensions of the problem of radical uncertainties about present trends and future possibilities better than the institutions human societies have developed and evolved over time to meet these challenges. He argues that there are significant epistemological and methodological difficulties (e.g., 'the problem of underspecification', 'the problem of data saturation') that continue to seriously limit what artificial intelligences might achieve in comparison with human practical and creative capacities for abductive reasoning or the negotiation of shared meaning as responses to complex, dynamic and uncertain surrounds.

Put another, blunter way, the ambition to replicate human-like intelligence would appear to rely on a narrow, exclusively mentalistic and rationalistic understanding of intelligence of a Cartesian 'ghost in the machine' type–that characterised (above) as unplugged from corporeal ties to its terrestrial surround. This is, I would note, an understanding entirely consistent with Turing (2012 [1950]), who considered the task of engineering artificial equivalents for human mental operations far easier to accomplish than the mechanical reproduction of the human sensorium and bodily capabilities. It also squares with modern notions of identity as an accomplishment of the exercise of agency conceived of as an intrinsic quality of individual actors, notions that certainly retain considerable purchase on popular understandings in historical contexts marked by strong cultural, economic and political investments in individualistic worldviews. Yet it would seem here that we need to seriously consider granting the term 'artificial' an equal weighting to that usually accorded 'intelligence' in debates over AIs. I will return to this.

Here I want to open discussion out to a second line of critique in which questions about whether AIs have or might technically achieve human-like intelligence, and the attributions of agency and identity attendant upon this, are replaced by an interrogation of the figure of 'the human' that informs understandings of just what constitutes 'human-like'. Here, the replicant ambition of AI researchers is seen not just as technically problematic but, in addition and more importantly, as normatively troublesome. In this regard Gunkel's (2012) observations regarding the act of characterising moral agency (in the context of an extended discussion of AIs and ethics) apply equally well to attempts to lay out precise specifications for human-like intelligence. Gunkel notes: "[T]he very act of characterizing ['human-like'] already and unavoidably makes a decision between who or what is to be included and who or what is to be excluded—who is a member of the club and who is marginalized as its constitutive other" (p. 23).

Gunkel then proceeds to note that these decisions have, historically, had significant material as well as legal impacts for those excluded by them from full participation in the community. Or, to reformulate this, for those who are excluded from the benefits of membership of a society but who must still be considered as integral to it, of necessity, inasmuch as the social order owes its ongoing existence in its given form to their labours, symbolic status and so forth (Patterson, 2018). The line of questioning thus changes from whether or when AIs will achieve human-like capabilities to how 'human-like' itself is defined, who decides, according to what grounds and with what consequences? In effect AIs can be considered here as furnishing reflective surfaces upon which a vast and violent history of struggles for recognition of a humanity withheld continues to play itself out.

Decolonial and feminist critiques of the replicant ambition most often commence with observations of a striking homology between the mentalistic/rationalistic aspirations held for AIs by Turing, the Dartmouth alumni and Chollet, among many others, and the Cartesian self/bounded individual encountered in my introductory comments. And they document how this idealised self-understanding of Western subjecthood was positioned as the normative Archimedean point for a hierarchical ordering of intelligence in which proximity to a model 'White adult male' served as the measure of humanity—this, in turn, functioning to underwrite ongoing practices of gendered devaluation and racialised dehumanization (Broussard, 2023; Rhee, 2018; Castañeda and Suchman, 2014).

Decolonial and feminist critiques of AIs have yielded a wealth of significant empirical studies, from Marie Hicks's (2018) history of the sidelining of women in AI research and Wendy Hui Kyong Chun's (2021) detailing of a strong eugenicist influence on machine-learning theory and practice, to Ruha Benjamin's (2019) account of algorithmic rehearsals of racial bias and Couldry and Mejias's (2019) study of data colonialism. Yet it is arguably in tethering the AI project and its replicant ambitions to the epistemological assumptions, knowledge practices and teleology (AIs as progressing humanity's inevitable mastery of world) of Western imperial imaginaries that these critiques cut most deeply (Adams, 2021). In part this is because these critiques work to invite questioning as to whether AIs could, "as it is currently imagined, ever be 'good', 'benevolent' or 'decolonial'?" (Adams, 2021, p. 191). But it is also because through destabilising the AI project in this fashion these critiques open out to alternative ways of thinking with AIs—ways that hold a very different mirror up to human agency and identity.

2.3.2 The hybrid system

In a relatively recent reflection on the victory of IBM's Deep Blue over world chess champion Garry Kasparov in 1997 (prompting hyperbolic media claims about an ascendent machine intelligence) Lucy Suchman (2019b) directs our attention to an Associated Press photo of the final, decisive game of the six-game match. The image features Kasparov hunched studiously over the chessboard, across which sits another man at a computer terminal—not Kasparov's opponent, Suchman is quick to add, although conceivably a component, of sorts, of Deep Blue—a mediator between it and the board. Deep Blue itself extends, spatially and temporally, well beyond the man and the terminal that represent it in the image, the terminal itself being networked to an array of 512 PowerPC processors, positioned somewhere off-stage and out of sight, combined with a raft of specialised chips designed to evaluate positions and coordinate data in real time, which in turn are maintained by a staff of technicians. The code being processed, moreover, is the product of extensive labours of software and electrical engineers, informed by chess experts consulting with IBM as part of the team. Far from a singular entity, in other words, Deep Blue is the nominalisation of a network of labouring bodies, including humans as well as multiple machines (Suchman, 2019b, pp. 41–42).

The agency (and identity as a discrete 'autonomous' entity) of Suchman's Deep Blue is thus presented as a particular effect of a network of collaborative relationships with affiliated technologies and humans. We can view it, in other words, as both a socially extended hybrid and a locally condensed network (see Strathern, 1996).

This returns us to a place near those discussions on the agency of things (above), in so doing prompting the question, what difference might AI make to these more established, although still hardly uncontroversial, understandings of 'thing-agency'? One way of approaching this question is by fastening attention on to characteristics or operational capabilities of the particular technology in question—its affordances or, as it were, the particular ways it might act agentically in the world. Here analyses of the connective capabilities of algorithms by Francis Lee and colleagues are instructive (Lee et al., 2019).

Specifically, this work highlights the varied processes by which algorithms, across a legion of specific and multiple applications, relate and order all manner of things—data, social relations, materials, analytical methods, values, computations. This includes the ways that algorithms connect things across space and time that would otherwise not be connected: for instance, "a mathematical formula for aftershock prediction might be folded into a system for predictive policing . . . or health statistics from the USA in the 1960s might be folded into German health recommendations in the 2010s' (Lee et al., 2019, p. 3). And the ways that they set apart things that might ordinarily seem close, rendering them effectively irrelevant to one another.

It is through their inclusion as increasingly pivotal parts of the ordering processes of the social, the natural and, indeed, the normative that, it is argued, algorithms acquire their powers to configure and reconfigure parts of the world. So, for example, national and international strategies for HIV prevention planning and patient infection management continue to be shaped and conditioned by combinations of (a) the algorithmic rearrangement of AIDS visualisations away from geographical population proximities and towards non-geographical transmission and risk topographies; and (b) the Recent Infection Testing Algorithm calculating major time points in an AIDS infection process, and the subsequent (problematic) folding of the aggregated population dynamics of an epidemic onto individual patient assessments and treatment plans. Of course, algorithms are not the only actors connecting and ordering things.

Humans, too, order, forge bridges between algorithms and things, or fold things into algorithms, yielding at times routine and other times novel effects.

Critiques of the attribution of agency to non-human things will often direct themselves at points like these. The staunchest critics contest this attribution of agency itself. Prominent claims here include: that these attributions demonstrate that social scientists are far less competent when speaking on the capabilities of things than are physical scientists and engineers, and that non-human things have no 'social life', as it were, worth attending to (Collins and Yearley); that treating things as actors is a form of hylozoism, the imputation of wilfulness or intent to the inanimate stuff of the world (Schaffer); or that it ontologically undoes the subject–object dualism that epistemologically grounds all knowledge, inasmuch as knowledge requires that there be a superintending subject and an object that is known (Bloor) (for a tidy summation of and response to these forms of critique, see Sayes, 2014). To put it bluntly, on this view even the most sophisticated technologies at best function, and only humans, precisely because they are possessed of conscious will and purpose, can be considered agents. These contests turn on fundamental differences in the ways that society/science and society/nature interfaces are construed, and on the significance of intentionality to our understanding of agency. As I cannot do justice to these debates here beyond that which has been touched on already, I will eschew any direct engagement with them.

A second critique is, however, worth taking the time to consider more fully. Here the notion of agency as relational, distributed and inclusive of non-humans is not at issue. Rather, it is the need to draw distinctions between types and degrees of agency, including between humans and non-humans, as well as among non-humans, without lapsing into an essentialist dismissal of any collective, relational basis to agentic capabilities. Werner Rammert (2012), for example, takes aim at the illusion of autonomy, be it applied to the agency of humans or technology, while at the same time refusing to "follow actor-network ontology into the flatlands where all things are endowed with the equal power of agency" (p. 100). By "flatlands" Rammert is referring here to what Michel Callon called 'the generalised symmetry principle'—the principle that demands we treat agency as of the same order whether we are referring to humans, machines, materials, nature and so forth in order to recall and re-member the often overlooked contributions of non-humans to human accomplishments (I would note that flat ontologies are by no means exclusive to actor-network theory: see Ash, 2020).

The call for greater attention to the distinctions between degrees of agency and agencies of different kind, Rammert insists, is entirely consistent with a relational or distributed view of agency. Drawing upon Pragmatist traditions of thought, he argues that all distinctions are the product of interrelationships, conceived of in terms of resistances, distancings and so forth. Indeed, following Pickering (1995), he maintains that it is in the recalcitrance of technologies to human expectations, the ways they thwart our designs and 'suggest' unexpected pathways, that the interdependent na-

ture of agency is best observed. Rammert then proceeds to outline a gradual and multi-level model of agencies of different intensities and kind. These range across:
- *passive* or *semi-active agency* (the basic tool moved from the outside, the habitual or reflex action);
- *contingent* or *directive communicative agency* (reacting to changing environments, selecting among options—including self-opening doors and technological systems with self-monitoring/activating programmes);
- *interactive communicative agency* (including where the architecture of multi-agent machine systems is coordinated by mechanisms of information seeking, problem solving, cooperation and communication—for example, the Internet of Things—and where human users must adopt a role-taking perspective if they are to engage successfully with others, including technology).

Returning to our question 'what difference might AI make?', we can talk of AI in terms of increased levels of agentic capabilities for those digital technologies that, as Lee et al. (2019) demonstrated, are threaded through the processes that order our social worlds. As Holton and Boyd (2021) observe, we can clearly draw distinctions between

> technologies that are more passive and do not routinely cause us to consider their agency—e.g., a hammer—once it is balanced in our hand we do not think of it as providing force or direction to our fist, and those that are more interactive and oblige us to acknowledge their agency as collaborators—our computers, electronic personal assistants, games, and social robots. (p. 186)

These distinctions can be extended to cover the ongoing evolution of specific technological systems. For example, where the internet was once a 'static labyrinth' to be navigated through in the manner of the adventurer, the contemporary use of machine-learning algorithms now allows the internet to configure and reconfigure itself around the individual user (Garvey, 2022).

These increased capabilities have often led to the impression that these technologies function as if they were independent actors (Natale and Guzman, 2022). That they are not independent is evidenced in Suchman's Deep Blue example—as soon as one looks beyond the immediate setting of the chess match. And where once considerable effort was required to summon the tracery of all the hidden non-human (and disenfranchised human) contributors to human agency, we are now seeing similar efforts being made to render visible the hidden human labour required for AI systems to function as promised (Tubaro, Casilli and Coville, 2020; Crawford, 2021). For my purposes here, however, what matters most is the implicature of these complex hybrid human–machine (including AI) systems in the processes through which we—that is, us humans—make sense of our worlds and our experiences in and of it, including the ways we conceive of and give expression to ourselves.

2.4 Bringing Identity Back In: Current Directions and Their Possible Futures

Garry Kasparov, to return briefly to Suchman's account, while celebrated as an individual, presents as something other than a singular agent, a bounded individual possessed of a self-contained genius: "Kasparov's genius is an effect of his relations with, among others, the long lineage of chess players and games through which his expertise is constituted . . . a very specific cultural practice, for which he has a demonstrable 'gift' and at which he has trained with great commitment and discipline" (Suchman, 2019b, p. 42). In making this point Suchman leans heavily upon Hélène Mialet's (2019a, 2019b) concept of the 'distributed-centred subject', detailed in case studies of the scientist/engineer François Montel and physicist Stephen Hawking. Mialet's concept is worth attending to further here, although not just for those aspects Suchman looks to.

In its initial stages the Montel analysis (Mialet, 2009, 2019b), based on ethnographic studies at the applied research laboratory Montel ran at the French oil company Elf Aquitaine, reads like a well-executed yet standard exercise in the tradition of science, technology and society studies—one bent on demonstrating how scientific research is social through and through, with little to be gained by investigating the specific qualities of the individual scientist. This certainly comprises an important part of the analysis and is deployed to good effect in challenging the classic tropes of invention found in the laboratory—those geared to the innate creative powers of Montel, divining original computer models for simulating petroleum fluids using nothing but his own raw, but brilliant, mental capabilities. Here Mialet (2009, 2019a) undercuts this figure, dispersing the boundaries of Montel's agency throughout a collective of humans using and improving his models, instruments, devices and machines that both extend and constitute Montel (and the production of knowledge), objects of research, webs of practices, institutional procedures, and so on and so forth.

Yet if we follow Mialet's analysis further something interesting happens. Just as Montel's agentic capabilities and identity threaten to sink into something resembling Rammert's 'flatlands' they recuperate and we see "at the centre of a web of practices, collectivities and technologies, an individual who acts, that is, who 'creates'" (Mialet, 2009, p. 257). To be sure, not the heroic figure of the bounded, purposeful and possessive individual that informs the classic tropes of invention (along with the *Replicant Ambition*, populist understandings of individualism and the adolescent libertarianism of Silicon Valley). But one who invents and singularises himself through identifying with the objects of his research, his institution and the needs and aspirations of his co-workers, among other things. Montel, in effect, distributes himself. And the more distributed he becomes the more he emerges as a specific presence in the midst of it all:

> [T]he more he was enmeshed with his object of research, his institution, and his colleagues, the more he was able to produce new knowledge; the more people from different fields were coming

> to ask him questions because he was 'an expert', the more he was investigating fields he was not familiar with, and the more he was able to make unknown connections and stabilize his status as the main expert in the company. (Mialet, 2019b, p. 377)

While Mialet's primary interest is in understanding innovation, this notion of the 'distributed-centred subject' might be useful for thinking identity, and its connection with agency, as this is attendant upon attachments to machines considered in some way intelligent. In the context of this task what it offers is not entirely new, as befits a conceptualisation of singularisation that works by way of individual variations on collective procedures—of merging with a system in order to use it and in so doing distinguish oneself within it. It shares more than a little in common with groundbreaking innovations in thinking about distributed cognition (as in the neuropsychology of Merlin Donald, or the neuroanthropology of Ed Hutchins or Andy Clark) and extended selves (Russell Belk) established quite some time ago. And it allows for the sense that identity is something that one 'does' (is enacted, performed), rather than something one 'is', that characterises most of the major contemporary statements that constitute the now well-established field of identity studies, their often-substantial disagreements notwithstanding (for an incisive, creative and critical elaboration of these major statements see Elliott, 2020).

There are, in addition, strong affinities between Mialet's concept and recent explorations of the agency–identity couplet as this occurs in contexts mediated by new smart technologies. Gillian Rose's (2017) Stiegler-inspired view of individuation as the enactment-of-difference-through-repetition, for example, shares much with Mialet's theorisation of singularisation, including the capacity to account for the emergence of inventive human activity within urban environments in which digital technologies are heavily implicated in the standardising production of spatial and temporal relations. But might we also creatively adapt the idea of distributed-centred subjectivity as a heuristic device that enables us to line up a quite diverse array of positions on the identity–agency couplet now emerging on the advancing front where identity studies meet AIs? Might we, in other words, play around with different types and gradients of 'distributedness' and 'centredness' (and in ways that do not necessarily square with Mialet's intentions)? Here I will just note, by way of a concluding comment, two emergent tendencies in the research. These by no means exhaust the range of possible conceptualisations of identity that may occupy identity researchers engaging with the challenges of AIs in the future, but they can, when taken together, provide a rough frame for identifying some key concerns.

The *automated self* builds initially upon accounts of data selves (Lupton, 2020; McQuillan, 2022; Cheney-Lippold, 2017) as well as theorisations of algorithmic disindividuation (Stiegler, 2017), governance (Rouvroy, 2020; Rouvroy and Berns, 2013) or personalisation (Lury and Day, 2019). The term 'Ais', in this context, refers to the bundle of technologies that constitute those platforms upon which people declare and share their personal behaviours, characteristics and preferences, the worn biometric sensors that position bodies as nodes on the Internet of Things, and the machine-learning algorithms that process the mass of data so generated in order to provide feedback in the form of 'personalised' recommendations, eligibility determinations, risk profiling and the like. Here the distributed-centred dynamic works predominantly through the processes by which individual identities are algorithmically assembled as collections or sequences of datapoints.

As Rouvroy and Berns (2013), following Alain Desrosières, describe it, earlier societal uses of mass data entailed statistical analyses modelling the 'average' citizen, consumer and so on, in order to (pre)establish categories against which individual behaviours could be subsequently referenced and regulated, but which, as public categories, were themselves subject to critique, contest, reinterpretation and revision. There is, they argue, little need for such averages in the new algorithmic dispensation. Here the key reference points are patterns of user behaviour algorithmically quarried in the data traces that document a user's past activity, these subsequently being correlated with parcels of data traces from other users (those that qualify on the grounds of some or other always provisional criterion of likeness). This then serves as the basis for system anticipations of future behaviours that occur at processing speeds far in excess of human cognition. This 'real time' feedback allows the prompts and notifications of algorithmic anticipatory pre-emption (e.g., recommendations) to be increasingly woven into the rhythms, routines and habits of everyday life, integrating people ever more deeply into automated solicitation-and-response loops (see also Day, Lury and Ward, 2023).

Automation can be conceived of here not as any eclipse of centring by distributedness, but rather the delegation of more and more of the work of centring to machine systems. An always contingent 'you' emerges as a sort of personalised open-ended pathway generated by recursive feedback loops, a "you that is 'a you' [that] is similar to other 'yous', that is, nearly but not quite the same as other 'yous', and never quite able to be consolidated as an 'us'" (Lury and Day, 2019, p. 28). For some this represents a disturbing prospect, a calculative foreclosure of much of the individual lived experience and affective range that characterises embodied human being (McQuillan, 2022), or the recursive reduction of the future as an, individually and collectively, open horizon of possibilities (Stiegler, 2017, 2019). For others some experience of an agential self that aspires, desires and feels, and is always situated in a historical and material surround, remains integral to any practice such as that we are referring to as centring. Bucher (2017), for example, documents recursive processes of mutual influence between users and the Facebook algorithm, specifically in the ways users adapt their behaviour in line with how they imagine the algorithm works, or endeavour to 'train' it, this in turn affecting the way the algorithm functions. Elsewhere Dourish (2021) emphasises how contemporary computational infrastructures are experienced compositionally, that is, ensemble, as a multiplicity of functional affordances that people orchestrate as they craft their own experience with and by way of the technology. This is very much akin to Lupton's (2020) 'agential cuts', the meanings people draw from or attribute to their personal data as part of their relating to and through this.

The figure of the *eXperimental self* bears considerable resemblance to the experimental self in recent social thought (Baudrillard, 1994; Elliott, 2020), with the 'eX' indicating, as is now commonplace in much technical, humanities and social science literatures, the technological extension of the manifold of human experiencing. AI here refers to new augmented and virtual reality technologies, immersive surrounds such as the Metaverse (and specific Metaverse sites) and digital avatars, among other things (Floridi, 2022; Gorichanaz, 2023; Papacharissi, 2019). These technologies promise (portend?) a radical progression of the tendency towards experimentation that Lury (2013) detailed in her *Prosthetic Culture*. While prosthesticity, as an aspect of the human condition, has a long and rich provenance (Draycott, 2018; Foster, 2006), Lury argues that a prosthetic culture only fully takes shape with the shift from a plural to a postplural society. In plural societies social descriptions are generated through an analogous play of difference and similarity. For instance, individual personhood emerges—is 'identified'—by way of partial analogies to society and nature: that is, is construed as a unique synthetic whole composed more or less from, and as such is both similar to and different from, various parts of society and nature. Society and nature and their constituent domains are here also wholes of which individuals too form parts.

The current transition to a postplural society, so it is said, entails a shift from the synthetic to the prosthetic, this transition in part occasioned by cultural developments, especially digital visual technologies destabilising the faculty of memory that previously worked to narratively ground the integrity and relational continuity of individual, social and natural wholes. In prosthetic cultures the individual is not made of and hence analogically integrated with parts of other natural/social contexts, but is rather more concatenated, consisting "of parts accreting with other parts, strategic assemblages, adjacent, coeval parts within the frame of a single form: memories substituting for one another, organs transplanted from one body to work alongside those of another, hues of cloth juxtaposed with colours of skin, hair and eyes" (Lury, 2013, p. 16).

In the stead of synthesis, the meaningful integration of biographical, historical and social experience, we find prosthetic extension wherein the social and natural attributes that were once considered inherent constituents of the self (genetics, consciousness, embodiment, culture, place) are made available as largely context-free resources for technologically assisted choice and/or experimentation. Self-knowledge ('I think, therefore I am') is progressively displaced by self-extension ('I can, therefore I am') (Lury, 2013, p. 3). Citing Barthes's *Camera Lucida*, Lury sees in the experimental self the moment when a subject experiences itself becoming an object: a property subject to ceaseless processes of hypothesis, evaluation, revision and rebooting.

With the advent of extended experience technologies, this prospect—at least if the rhetoric is to be believed—will be taken to a new level. Gorichanaz (2022), for example, has usefully identified the key themes rehearsed in public statements about the Metaverse by Mark Zuckerberg, most notably that it will be a natural and inevitable evolution of the internet, one that will render people more fully present alongside others inside of digital experiences playing—working, making, consuming, learning . . . — and that the cardinal unit for the Metaverse is the person, each participating as their digital avatar. Beyond this, Gorichanaz notes, the development of the Metaverse will, for Zuckerberg, result from corporate cooperation, will give rise to a whole new economy centred on digital goods and will give expression (if *sotto voce*) to an orientation towards frontierism and anti-statism characteristic of much Silicon Valley culture.

Any understandings of the full implications of these developments for identity, as well as agency, responsibility, recognition, belonging and so forth, are, as things currently stand, effectively speculative. As Floridi (2022) points out, the rhetoric runs way ahead of a reality confronted by multiple technical (how to engage more than just the visual and acoustic aspects of the human sensorium) and techno-commercial (interoperability) challenges. More, unless these technologies are to be the first in human history to emerge and establish themselves in social worlds exactly as their designers intended, then social and cultural factors will surely influence their futures—with often surprising effects (including, one must assume, the type of mediations Chun discusses under the head of "red pill toxicity, or liberation envy"; Chun, 2021, pp. 29–34).

Here Mialet's concept of the 'distributed-centred subject' can again be of use. To say that the experimental self—or even the experimental self—is a distributed self is relatively unproblematic. Centredness is a different proposition, however, despite any rhetorical nods and winks in the direction of person centredness. In Lury's (2013, p. 22) analysis, we do get occasional glimpses of individuals who emerge from, and take as their own, characteristics of their environment as 'strategic couplings of choice'. But they remain occasional . . . and only glimpses. In this regard Mialet's Montel is instructive for the way the acts of self-distribution through which he singularised himself entailed empathetic and committed identifications. As Gorichanaz (2022) argues, the Metaverse may, conceived as an alternative reality, prompt a turning away from the world and those who dwell within it; however, if conceived more as a supplement it might serve as an invitation to eXperiments with forms of care and concern. Likewise, Floridi (2022) notes the limitations posed by a context in which socialisation is optional, while acknowledging the possibilities the Metaverse offers for gaining an understanding of what it is like to walk in another's shoes, to be synesthetic, to see the world as a cat . . .

References

Adams, R. (2021). Can artificial intelligence be decolonized? *Interdisciplinary Science Reviews*, *46*(1–2), 176–197.
Appadurai, A. (ed.) (1988). *The social life of things: Commodities in cultural perspective*. Cambridge University Press.
Ash, J. (2020). Flat ontology and geography. *Dialogues in Human Geography*, *10*(3), 345–361.
Barad, K. (2007). *Meeting the universe halfway: Quantum physics and the entanglement of matter and meaning*. Duke University Press.
Baudrillard, J. (1994). *Simulacra and simulation*. University of Michigan Press.
Bayly, S. (2021). There's something about things: Feeling around for object-disoriented politics. *Angelaki*, *26*(5), 3–19.
Beck, U., Giddens, A., and Lash, S. (1994). *Reflexive modernization: Politics, tradition and aesthetics in the modern social order*. Stanford University Press.
Benjamin, R. (2019). *Race after technology: Abolitionist tools for the new Jim code*. John Wiley & Sons.
Boyd, R. (2021). Work, employment and unemployment after AI. In A. Elliott (ed.), *The Routledge social science handbook of AI* (pp. 74–90). Routledge.
Broussard, M. (2023). *More than a glitch: Confronting race, gender, and ability bias in tech*. MIT Press.
Bucher, T. (2017). The algorithmic imaginary: Exploring the ordinary affects of Facebook algorithms. *Information, Communication & Society*, *20*(1), 30–44.
Butler, J. (1997). *The psychic life of power: Theories in subjection*. Stanford University Press.
Castañeda, C., and Suchman, L. (2014). Robot visions. *Social Studies of Science*, *44*(3), 315–341.
Cheney-Lippold, J. (2017). *We are data*. New York University Press.
Cheon, E., and Su, N. M. (2016). Integrating roboticist values into a Value Sensitive Design framework for humanoid robots. In *2016 11th ACM/IEEE International Conference on Human-Robot Interaction (HRI)* (pp. 375–382). IEEE.
Chollet, F. (2019). On the measure of intelligence. *arXiv preprint arXiv:1911.01547*.
Chun, W. H. K. (2021). *Discriminating data: Correlation, neighborhoods, and the new politics of recognition*. MIT Press.
Collins, P. H. (1990). Black feminist thought in the matrix of domination. In P. H. Collins (ed.), *Black feminist thought: Knowledge, consciousness, and the politics of empowerment* (pp. 221–238). Routledge.
Couldry, N., and Mejias, U. A. (2019). The costs of connection. In *The costs of connection*. Stanford University Press.
Crawford, K. (2021). *The atlas of AI: Power, politics, and the planetary costs of artificial intelligence*. Yale University Press.
Day, S., Lury, C., and Ward, H. (2023). Personalization: A new political arithmetic? *Distinktion: Journal of Social Theory*. doi: 10.1080/1600910X.2022.2098352.
DeLanda, M. (2016). *Assemblage theory*. Edinburgh University Press.
Dourish, P. (2021). The allure and the paucity of design: Cultures of design and design in culture. *Human-Computer Interaction*, *36*(1), 52–72.
Draycott, J. (Ed) (2018). *Prostheses in antiquity*. Routledge.
Draycott, J. L. (ed.) (2019). *Prostheses in antiquity*. Routledge.
Dreyfus, H. L. (1979). *What computers can't do: A critique of artificial reason*. Harper & Row.
Elliott, A. (2020). *Concepts of the self*. Polity.
Emirbayer, M., and Mische, A. (1998). What is agency? *American Journal of Sociology*, *103*(4), 962–1023.
Erofeeva, M. (2019). On multiple agencies: When do things matter? *Information, Communication & Society*, *22*(5), 590–604.
Floridi, L. (2022). Metaverse: A matter of experience. *Philosophy & Technology*, *35*(3), 73. doi: 10.1007/s13347-022-00568-6.

Foster, H. (2004). *Prosthetic gods*. MIT Press.

Foster, H. (2006). *Prosthetic Gods*. MIT Press

Gamble, C. N., Hanan, J. S., and Nail, T. (2019). What is new materialism? *Angelaki*, *24*(6), 111–134.

Garcia, T. (2014). *Form and object*. Edinburgh University Press.

Garvey, S. (2022). Predators in the attention economy: Identification and defense. In A. Hanemaayer (ed.), *Artificial Intelligence and Its Discontents: Critiques from the Social Sciences and Humanities* (pp. 251–268). Springer International Publishing.

Gell, A. (1998). *Art and agency: An anthropological theory*. Clarendon Press.

Gibson, J.J. (1979). The ecological approach to visual perception. Psychology Press.

Gorichanaz, T. (2023). Being at home in the metaverse? Prospectus for a social imaginary. AI and Ethics, *3*(2); 647–658

Gunkel, D. J. (2012). *The machine question: Critical perspectives on AI, robots, and ethics*. MIT Press.

Haraway, D. (1991). *Simians, cyborgs, and women: The reinvention of nature*. Routledge.

Harman, G. (2008). Zeroing in on evocative objects. *Human Studies*, *31*, 443–457.

Harman, G. (2018). *Object-oriented ontology: A new theory of everything*. Penguin.

Hayles, N. K. (2017). *Unthought*. University of Chicago Press.

Hicks, M. (2018). *Programmed inequality: How Britain discarded women technologists and lost its edge in computing*. MIT Press.

Hitlin, S., and Elder Jr, G. H. (2007). Time, self, and the curiously abstract concept of agency. *Sociological Theory*, *25*(2), 170–191.

Hodges, A. (1983). *Alan Turing: The enigma*. Simon and Schuster.

Holton, R. J. (2023). Artificial intelligence and the problem of radical uncertainty. In S. Lindgren (ed.), *Handbook of critical studies of artificial intelligence* (pp. 56–66). Edward Elgar.

Holton, R., and Boyd, R. (2021). "Where are the people? What are they doing? Why are they doing it?" (Mindell) Situating artificial intelligence within a socio-technical framework. *Journal of Sociology*, *57*(2). doi: 10.1177/1440783319873046.

Ihde, D. (2008). *Ironic technics*. Automatic Press/VIP.

Jabri, V. (2014). Disarming norms: Postcolonial agency and the constitution of the international. *International Theory*, *6*(2), 372–390.

Joas, H. (1996). *The creativity of action*. University of Chicago Press.

Koselleck, R. (1985). *Futures past: On the semantics of historical time*. MIT Press.

Latour, B. (1992). Where are the missing masses? The sociology of a few mundane artifacts. In W. E. Bijker and J. Law (eds), *Shaping technology/building society: Studies in sociotechnical change* (pp. 225–258). MIT Press.

Latour, B. (1999). *Pandora's hope: Essays on the reality of science studies*. Harvard University Press.

Latour, B. (2000). When things strike back: A possible contribution of 'science studies' to the social sciences. *British Journal of Sociology*, *51*(1), 107–123.

Latour, B. (2007). *Reassembling the social: An introduction to actor-network-theory*. Oxford University Press.

Lee, F., Bier, J., Christensen, J., Engelmann, L., Helgesson, C. F., and Williams, R. (2019). Algorithms as folding: Reframing the analytical focus. *Big Data & Society*, *6*(2), 205395171986264. doi: 10.1177/2053951719862647.

Lupton, D. (2020). *Data selves: More-than-human perspectives*. Polity.

Lupton, D., and Watson, A. A. (2021). Towards more-than-human digital data studies: Developing research-creation methods. *Qualitative Research*, *21*(4), 463–480.

Lury, C. (2013). Prosthetic Culture. Routledge.

Lury, C., and Day, S. (2019). Algorithmic personalization as a mode of individuation. *Theory, Culture & Society*, *36*(2), 17–37.

Macpherson, C.B. (1964), The Political Theory of Possessive Individualism. Oxford University Press.

Marres, N. (2019). As ANT is getting undone, can pragmatism help us re-do it? In A. Blok, I. Farias and C. Roberts (eds), *The Routledge companion to actor-network theory* (pp. 112–120). Routledge.

Martin, E. (1995). *Flexible bodies*. Beacon Press.

Mbembe, A. (2019). *Necropolitics*. Duke University Press.

McCarthy, J., Minsky, M. L., Rochester, N., and Shannon, C. E. (1955). A proposal for the Dartmouth summer research project on artificial intelligence. http://www.formal.stanford.edu/jmc/history/dartmouth/dartmouth.html.

McNay, L. (1999). Subject, psyche, and agency: The work of Judith Butler. *Theory, Culture & Society*, *16*(2), 175–193.

McQuillan, D. (2022). *Resisting AI: An anti-fascist approach to artificial intelligence*. Policy Press.

Mialet, H. (2009). Making a difference by becoming the same. *International Journal of Entrepreneurship and Innovation*, *10*(4), 257–265.

Mialet, H. (2019a). The distributed-centered subject. In J. Prinz and N. Weidtmann (eds), *Thinking in the world* (pp. xx–xx). Bloomsbury.

Mialet, H. (2019b). Becoming the other: The body in translation. In J. E. Montoya and L. P. Kajikawa (eds), *The Routledge handbook of language and science* (pp. 375–384). Routledge.

Minh-Ha, T. (1989). *Woman, native, other: Writing, postcoloniality and feminism*. Indiana University Press.

Natale, S., and Ballatore, A. (2020). Imagining the thinking machine: Technological myths and the rise of artificial intelligence. *Convergence*, *26*(1), 3–18.

Natale, S., and Guzman, A. L. (2022). Reclaiming the human in machine cultures: Introduction. *Media, Culture & Society*, *44*(4), 627–637.

Papacharissi, Z. (ed.) (2019). *A networked self and birth, life, death*. Routledge.

Patterson, O. (2018). *Slavery and social death: A comparative study* (with a new preface). Harvard University Press.

Pickering, A. (1995). *The mangle of practice: Time, agency and science*. University of Chicago Press.

Pink, S., Berg, M., Lupton, D., and Ruckenstein, M. (eds) (2022). *Everyday automation: Experiencing and anticipating emerging technologies*. Routledge.

Rammert, W. (2012). Distributed agency and advanced technology. In W. Rammert (ed.), *Agency without actors? New approaches to collective action* (pp. 89–112). Routledge.

Rhee, J. (2018). *The robotic imaginary: The human and the price of dehumanized labor*. University of Minnesota Press.

Roberge, J., and Castelle, M. (2021). Toward an end-to-end sociology of 21st-century machine learning. In *The cultural life of machine learning: An incursion into critical AI studies* (pp. 1–29). Springer.

Rose, G. (2017). Posthuman agency in the digitally mediated city: Exteriorization, individuation, reinvention. *Annals of the American Association of Geographers*, *107*(4), 779–793.

Rouvroy, A. (2020). Algorithmic governmentality and the death of politics. *Green European Journal*, 27 March. algorithmic-governmentality-and-the-death-of-politics.pdf (greeneuropeanjournal.eu).

Rouvroy, A., and Berns, T. (2013). Algorithmic governmentality and prospects of emancipation. *Réseaux*, *177*(1), 163–196.

Sabanovic, S., Milojevic, S., Asaro, P., and Francisco, M. (2015). Robotics narratives and networks [history]. *IEEE Robotics & Automation Magazine*, *22*(1), 137–146.

Sayes, E. (2014). Actor–Network Theory and methodology: Just what does it mean to say that nonhumans have agency? *Social Studies of Science*, *44*(1), 134–149.

Stiegler, B. (2017). *Automatic society: Volume 1*. Polity Press.

Stiegler, B. (2019). *The age of disruption: Technology and madness in computational capitalism*. John Wiley & Sons.

Strathern, M. (1996). Cutting the network. *Journal of the Royal Anthropological Institute*, *2*(3), 517–535.

Suchman, L. (2019a). Agencies in technology design: Feminist reconfigurations. In J. Scholz, S. Völker and E. Tuider (eds), *Aktuelle Herausforderungen der Geschlechterforschung: Beiträge zur ersten gemeinsamen internationalen Konferenz der Fachgesellschaften für Geschlechterforschung/-studien aus Deutschland, Österreich und der Schweiz* (pp. 72–88). doi: 10.25595/1356.

Suchman, L. (2019b). Demystifying the intelligent machine. In T. Heffernan (ed.) (2019). *Cyborg futures: Cross-disciplinary perspectives on artificial intelligence and robotics* (pp. 35–61). Springer Nature.

Taube, M. (1961). Computers and common sense. The myth of thinking machines. Columbia University Press.

Taylor, C. (1985) Philosophical Papers: Volume 1, Human agency and language. Cambridge University Press.

Tresch, J. (2007). Technological world-pictures: Cosmic things and cosmograms. *Isis*, *98*(1), 84–99.

Tubaro, P., Casilli, A. A., and Coville, M. (2020). The trainer, the verifier, the imitator: Three ways in which human platform workers support artificial intelligence. *Big Data & Society*, *7*(1), 2053951720935146. doi: 10.1177/2053951720935146.

Turing, A. M. (2012 [1950]). Computing machinery and intelligence. In B. J. Copeland (ed.), *The essential Turing: The ideas that gave birth to the computer age* (pp. 433–464). Clarendon Press.

Turkle, S. E. (2007). *Evocative objects: Things we think with*. MIT Press.

Walsh, T., Levy, N., Bell, G., Elliott, A., Maclaurin, J., Mareels, I., and Wood, F. (2019). The effective and ethical development of artificial intelligence: An opportunity to improve our wellbeing. *Australian Council of Learned Academies*. https://acola.org/hs4-artificial-intelligence-australia/.

Wilson, E. G. (2006). The melancholy android: On the psychology of sacred machines. SUNY Press.

Masataka Katagiri

3 On the Possibility of Interaction Between Humans and AI/Robots: From a Relational Approach

Abstract: In this chapter, we will introduce post-human problems from a broad perspective and then discuss conventional research trends related to these issues. Afterwards, narrowing our focus, we advocate for a 'relational approach' instead of a 'properties approach' as a foundational method for researching the possibility of interaction between social robots and humans. Based on this basic standpoint, we examine whether we can understand social robots in particular from the interpretive sociology, including cognitive sociology, which shares the same perspective as Coeckelbergh's relational approach. And then, in the later part of the chapter, we explore the Turing test using Goffman's theory, which we define as one of interpretive sociologies. In particular, we employ Goffman's concepts of 'passing' and 'stigma' as corresponding to the relational approach. The key point is to view the Turing test as one of the passing problems, as Goffman points out.

Keywords: cognitive sociology, relational approach, Turing test, Coeckelbergh, Goffman

3.1 Introduction: The Theoretical Significance of the Post-human

The question of whether artificial intelligence (AI) surpasses that of humans is burning amid contemporary discourse. R. Kurzweil's singularity theory has contributed greatly to this debate. His book, *The Singularity is Near* (2005), not only raises questions such as 'Does AI surpass human intelligence?' or 'Will the singularity come?', but also positions the development of advanced science, including AI, within the larger framework of the history of humankind in the universe. Kurzweil (2005) points out six stages of evolution. The first (epoch 1) is the stage of physics and chemistry. This is the stage in which physical things and energy were solidified on the Earth. The second stage (epoch 2) is the stage of biology and DNA when, hundreds of thousands of years ago, life on Earth first emerged. The third stage (epoch 3 or the brain stage) is characterised by the birth of animals with brains. Epoch 4 (technology stage) is defined by the human invention of technology, which is closely followed by epoch

Masataka Katagiri, Chiba University, Japan

https://doi.org/10.1515/9783110721751-003

5, in which human technology and intelligence fuse. The final stage is epoch 6, which Kurzweil refers to as the universe waking up, when the fusion of human intelligence and technology ventures into space. Singularity, much discussed, corresponds to epoch 5 in Kurzweil's definition. In other words, the singularity is considered to be just one stage in the magnificent history of humankind. The point that should be emphasised here is not the question of whether AI exceeds human intelligence, but rather that reaching the singularity is considered by Kurzweil as one stage of a long history of humankind that changes humans themselves as a species (pp. 14–21).

In contrast, Y. N. Harari, an Israeli historian and author of the best-selling book *Sapiens: A Brief History of Humankind* (2011), describes the entire history of humankind through the lens of humans as *Homo sapiens*. According to Harari, humans differ from non-human animals in that they have 'cognitive abilities'. The key cognitive ability that differentiates humans from other animals is the ability to invent fiction. The process of invention leads to the creation of complex societies and structures such as trading networks and political institutions. Harari calls the acquiring of this cognitive ability the 'cognitive revolution'. Beginning with this cognitive revolution, the agricultural revolution, the unification of humankind through the empire, and the scientific revolution have happened successively through to the modern age. Moving forward, Harari predicts the end of *Homo sapiens* or the arrival of 'Homo deus' (human god) in which humans become cyborgs and surpass the existing species of humans (Harari, 2011, 2017).

Kurzweil (2005) and Harari (2011, 2017) have both foretold the era of post-humans but from different points of view. While much attention has been paid to these images of the future, what is important for us is not the feasibility of these concepts but that these concepts of human history show the importance and appeal of thinking humans beyond the previous framework of humanities and social sciences which have not paid attention to humans as a biological species.

Additionally, Braidotti's theory deserves attention, as it indicates the importance of re-questioning humans and their societies beyond traditional frameworks. Braidotti (2013) regards the post-human problem not only as one of humans and social changes due to advanced technology but also as a criticism of the traditional view of humans in the West. The former argument focuses on anthropocentrism, while the latter deals with the problem of humanism. Post-anthropocentrism addresses how humans and society change through the advancement of science, which corresponds to the popular concept of the post-human. Based on this argument, Braidotti discusses the post-human problem as a combination of post-anthropocentrism and post-humanism.

Bringing these two theoretical frameworks together, Braidotti (2013) asserts that the conventional view of humans or subjects created by modern Western society should be criticised. Modern Western societies have surely emancipated humans from traditional social bonds by dismantling particular communities and developing scientific and rational thinking. This process is considered 'the discovery of humans' and also the starting point of humanism (cf. Katagiri, 2022, pp. 6–7). However, according to Braidotti, this process produced the problematic system of categorisation of humans. More specifically, Braidotti clearly pointed out that, from a Western viewpoint, humans have been defined as 'Man': men, adults, white people, people without disabilities, Christians, and so on. Therefore, women, children, people of colour, people with disabilities, people of non-Christian faith and so on have been excluded and othered from 'human society'. However, the criticism of humanism overlooks humans as a category that does not contain non-human organisms or inorganic things. The post-anthropocentric theory, on the other hand, is distinguished from the post-humanism theory in that it questions the position of humans as species among other organic creatures like animals and inorganic things. Such discussions have emerged as the development of various advanced sciences has shaken the boundaries between humans and other organic creatures and inorganic things. In these discussions, it has been criticised not only that 'Man' has been considered dominant over other humans but also that humans as a species have been considered central and superior to other organic creatures and inorganic things (Braidotti, 2013, ch. 1, 2).

These criticisms of both humanism and anthropocentrism distinguish Braidotti's theory of the post-human from other post-human theories or transhumanism like that of Kurzweil. Regardless of whether humanism and anthropocentrism are criticised or not, AI is deeply related to the issue of the post-human. Accordingly, when considering the problems of humans and their societies, it is important to focus not only on humans but also on the problems of AI, machines and other living animals. Although in this chapter, we will focus only on AI/robots (=AI or robots) but not on other living animals, and discuss the possibility of interaction between humans and AI/robots, it is desirable to be positioned in the broader framework of post-anthropocentricity (cf. Katagiri, 2022).

3.2 Viewpoints on the Possibility of Interaction Between AI/Robots and Humans

3.2.1 The frame problem in AI in relation to Schutz's theory of relevance

When considering the relationship between humans and AI, it has been asked whether AI has the same interaction or communication ability as humans. This question is often prompted by 'the frame problem', often explained by an example provided by the philosopher D. Dennett. In his paper, Dennett (1984, p. 129) raises a situation where a robot with AI is instructed to bring a battery out of a room before a bomb in the same room explodes. The battery is on a wagon in the room, and the bomb is also on top of the wagon. Although the robot succeeds in bringing the battery out of the room as instructed, it also brings the bomb out along with the wagon, because it does not know that if it brings the bomb out, it will explode. In other words, the robot cannot understand the relationship between bringing out a battery and bringing out a bomb. As a result, the robot is unable to safely bring the battery out of the room. It is important to note that most humans can carry out such tasks without difficulty, while some humans may refuse to accept such dangerous tasks.

Referring to the frame problem as 'the relevance problem', H. L. Dreyfus considered the frame problem a question of whether AI can act adequately according to the context of interaction (Dreyfus, 2007, p. 248). Dreyfus regarded the frame problem as the problem of knowing which definitions or actions are dependent on each situation. To define a situation, humans use a frame to identify what is relevant and act accordingly. In Dennett's example above, bringing the battery out of the room is the relevant action, and bringing a bomb out is not, which the robot was unable to determine.

Others explain the frame problem from A. Schutz's theory of relevance. Schutz (1964), a phenomenological sociologist, explains which objects in a situation are important based on four layers of relevance. The first layer of relevance includes the object most important to the immediate purpose. The second layer encompasses objects defined as conditions and means for the immediate purpose. The third layer involves objects that do not directly relate to the immediate purpose unless a circumstantial change occurs. The fourth and final layer comprises objects that have nothing to do with the immediate purpose. These four layers of relevance are arranged in order of importance to the subject's purpose (Schutz, 1964, pp. 123–124; Katagiri, 1993, ch. 2). Based on this argument, we will now re-examine Dennett's example.

First, the robot's purpose in Dennett's example is to bring the battery out of the room; therefore, the battery is the most important object (first layer of relevance). Next, as the danger of an explosion exists if the robot brings the bomb out with the battery, the bomb is an important condition for the purpose of bringing the battery out safely. The wagon carrying the battery is also an important object because it is the means for bringing out the battery. It may also be necessary to check whether it works properly (second layer of relevance). Any lighting equipment in the room may not be relevant for the immediate purpose; however, if something were to trigger the lighting equipment to fall on the battery, the robot would have to remove these objects (third layer of relevance). Lastly, the ceiling, windows, the world outside the room, and so on, are of least importance to the robot's immediate purpose.

In this way, when the robot is given the purpose of bringing a battery out of a room, various objects inside and outside of the room are ranked according to degree of relevance. According to Schutz's concepts, objects thus 'thematised by motivation' are in turn interpreted by a 'stock of knowledge' that has piled in someone's own life. Contextual actions are built on this chain consisting of three other different relevant processes: 'motivations', 'thematisation' and 'interpretations' by the stock of knowledge (Schutz and Luckmann, 1984, ch. 3). Considered from another viewpoint, we can say that a robot may not have its own purpose or motivation. Even when a purpose or motivation is given, the robot cannot understand which object to focus on or thematise, how to interpret it by stock of knowledge, or how to act according to the purpose. This is the frame problem newly explained by Schutz's theory of relevance.

3.2.2 The potential interaction between humans and AI/robots via relationship between them

When considering the potential interaction between humans and AI/robots, it is common to assess whether AI has the same interaction abilities as humans. As discussed earlier, the frame problem serves as a typical example of how AI/robots lack the logical and practical abilities of humans.

Various researchers of AI study human intelligence to develop AI systems equivalent to human intelligence. In the fields of engineering and philosophy, several theorists, including Kurzweil (2005), have attempted to understand the structure of the human brain to create AI with equal or greater abilities than humans. Furthermore, various studies in AI/robots, rooted in cognitive science and psychology, explore whether AI/robots can interact with humans. This potential interaction is referred to as human–robot interaction (cf. Bartneck et al., 2020).

These studies share a common basis in that they explore the possibility of interaction between humans and AI/robots based on the abilities or properties of AI/robots. This viewpoint can be called the properties approach. However, there is another viewpoint for considering the possibility of interaction between humans and AI/robots—one based on the relationship between humans and AI/robots.

When considering the potential interaction between humans and AI/robots, the first thing that comes to mind is the relationship between humans and social robots. In contemporary Japan, many people talk to and sometimes seek healing through interaction with social robots, such as AIBO (AI Robot) and LOVOT (Love Robot). It is also well known that the industrial robots introduced in Japanese factories decades ago were named after popular singers of the day. In this way, robots have existed in close relationship with humans for years.

As seen in their involvement with social and industrial robots, humans sometimes read emotions into robots. When you see a robot being hit or kicked by other humans, you might think that the robot is being abused. For example, a Canadian robot called Hitch-bot travels by itself via hitchhiking. While the Hitch-bot was hitchhiking, it was destroyed and abandoned on the side of the road. Can we call this incident an abuse of a robot by humans? Judging by the present state of AI technology, it is unlikely that robots themselves can experience emotion. Nevertheless, humans read emotions into robots.

This emotional attachment also applies to relationships between humans themselves. The sociology of emotion proposes the idea of 'vocabularies of emotions'. According to this idea, emotions are not inherent in the mind but are instead normative social acts defined as situationally appropriate expressions. For example, when someone close to us dies, we are expected to express sadness. At a wedding ceremony, we are expected to express joy. This does not change in the case of robots. Humans read emotions into robots the same way they read emotions into other humans. In other words, when considering the potential interaction between human and AI/robots, the relational perspective is more important than the properties approach. We refer to the former perspective as the 'relational approach', whose concept and terminology was inspired by Coeckelbergh's relational-phenomenological approach; however, the broadly defined relational approach is not limited to Coeckelbergh's applications. The characteristics of Coeckelbergh's relational-phenomenological approach will be discussed in detail later in this chapter.

3.2.3 Research and development by Ishiguro and Okada from a relational approach

Before discussing Coeckelbergh's relational-phenomenological approach, we will look at the research and development of H. Ishiguro and M. Okada, notable robot researchers and developers in Japan who base their work on a relational approach. While Ishiguro does not develop autonomous robots, he is famous for creating an android that resembles a human being. Ishiguro says that his purpose in developing such an android is to investigate how humans interact with robots. Ishiguro also designs robotic dramas where robots are programmed to perform plays. They enact drama together with humans or without humans. From observing these dramas, Ishiguro (2009) concludes that robots do not need to have the same interaction abilities (or properties) as humans when interacting with humans; rather, these interactions are guided by 'humans' reading the minds of the robots, or interpreting and responding to the robots as if they had human minds. This viewpoint is similar to the dramaturgy of O. Hirata, playwright and co-producer of Ishiguro's robot drama, who says that actors do not need to have genuine empathy for the characters they portray and interact with, but that the external appearance of emotion in the drama is more important. According to him, when performing a sad scene the actor does not need to experience the same sadness as the character, but they need to imitate the external expression of sadness, such as crying. In other words, it is not important for the actor to feel emotions but it is crucial for the audience to perceive or feel the actor's emotions regardless of their authenticity. Similarly, robots do not need to experience emotions; rather, they just need to be able to adequately convey the external appearance of emotion in the scene of a drama (Ishiguro, 2009, ch. 6).

Japanese robot developer Michio Okada (2017) shares a similar opinion. Rather than requiring AI/robots to have the same interaction abilities as humans, Okada researches and develops robots based on a relational approach. Okada's robots are called 'weak robots' or 'human-dependent robots'. His ideas can be understood by considering the example of a cleaning robot. Cleaning robots have been developed for one specific purpose (cleaning) and not to have a 'mind'. Therefore, the cleaning robots are just physical objects. However, humans may interact with cleaning robots as if they are more than that. For example, we may try to remove obstacles to make it easier for a cleaning robot to clean, either in advance of its performance or its collision with obstacles. In other words, we read intentions and emotions into the cleaning robot. The same can be said for various weak robots/human-dependent robots developed by Okada, such as his 'trash can robot' that cannot pick up trash on its own (Okada, 2017, ch. 3) but instead persuades or even seduces people to help it pick up trash by exhibiting certain gestures. These examples show that the boundary between mere physical objects and objects that humans read intention and emotion into is fluid.

3.3 Theoretical Standpoints on Interaction Between Humans and AI/Robots

3.3.1 Coeckelbergh's relational-phenomenological approach

As mentioned earlier, Coeckelbergh has emphasised researching robots using the relational approach rather than the properties approach. This perspective is known as the relational-phenomenological approach. Although his standpoint is considered one of the (broadly defined) relational approaches, the two approaches are not completely identical. The rationale for taking a relational approach rather than a properties approach is that humans already have some forms of relationship with robots, regardless of their abilities to interact.

One of the key concepts in Coeckelbergh's discussion is 'appearance'. This does not necessarily refer to external or objective appearance. In his discussion, appearance indicates the subjective experience of humans when robots are perceived as beings that intentionally or emotionally appeal to humans. Since the human relationship with robots is not simply determined by robots' properties, the relationship depends on the subjective experience of humans who come into contact with robots. Furthermore, interaction with robots cannot be explained by human subjective experience alone. It is also defined by both the situational context and the system of language, on which we now elaborate.

Coeckelbergh (2011) first considers a situation in which the interaction between robot and human occurs. He then focuses on the linguistic function within the scenario. When we consider whether animals are viewed as members of a human society, the answer depends on whether they are placed at home as pets or domesticated on a farm. The same is true for robots depending on whether they are placed at a home, a facility for the elderly, a factory or a battlefield. The meaning of each situation also depends on cultural background. For instance, the meanings of robots are perceived in relation to humans differently in Japanese and Western societies. Generally, in the West, robots are considered only physical objects or tools but not members of human society. On the other hand, in Japanese society, robots are regarded as a close relation to humans, as can be seen in the active development of entertainment robots. Moreover, when considering the different meanings attached to robots, a question taken into consideration is whether robots exist outside of human society or not (Coeckelbergh, 2011, p. 201; Coeckelbergh and Gunkel, 2014, p. 727).

The second factor that determines humans' perception of robots is the function of language. Language is an important element in understanding Coeckelbergh's relational-phenomenological approach. Consider the example of domesticated pets and farm animals. The distinction between them is expressed through language. The act of designating the same kind of animal as either a pet or a farm animal defines the appearance of the animal to humans. The same is true for the relationship between robots and humans. Language defines the social and moral grammar of the relationship. For example, the appearance of a robot differs greatly depending on whether the robot is referred to or named as 'it' or 'you'. The naming and linguistic expression of robots have great impact on the moral position of robots in human society. In other words, robots' positions as members of human society depend on how they are categorised, such as companions, tools, slaves and son on (Coeckelbergh, 2018, p. 151; Coeckelbergh and Gunkel, 2014, pp. 725–726).

3.3.2 Cognitive and interpretive sociology

Coeckelbergh's relational-phenomenological approach is compatible with the theoretical standpoint of my own cognitive sociology. (Katagiri, 2006; cf. Sato, 2021) Cognitive sociology is rooted in Schutz's phenomenological sociology and symbolic interactionism, which was derived from G. H. Mead's social psychology and has been developed further from it. Therefore, let us first explore the basic idea of interpretive sociology before explaining cognitive sociology.

3.3.2.1 Interpretive sociology

Interpretive sociology focuses on how people living in the 'everyday world'—which is opposed to the 'world of science'—interpret and interact with various 'objects', including people, physical things, events, situations and so on. Typical theories that stem from an interpretive sociological viewpoint include symbolic interactionism derived from Mead's social psychology and Schutz's phenomenological sociology. People living in the everyday world interpret and interact with objects based on their practical or pragmatic interests and motivations. This process is explained typically by the concept of 'perspective' by Mead (1932) and other symbolic interactionists and 'type and relevance' by Schutz (1964).

The concept of 'perspective' originally refers to the perspective of drawing method. When drawing a picture, the perspective would dictate the drawn object since it is seen from the drawer's point of view. Therefore, the same thing is drawn larger if the object is located close to the drawer and smaller if it is far away from the drawer. Depending on the standing position, the drawer can only see the front side of the object and may not see the back side. In other words, the physical things can only be seen from the viewer's point of view. This concept of drawing perspective is extended to the everyday world. Based on their own interests and motivations, people living in the everyday world focus on specific aspects of objects and interact with each other from their own individual perspectives. In considering humans and their society, the perspective humans have in their actions is the starting point for Mead and symbolic interactionists (Mead, 1932, part IV; Shibutani, 1955).

This discussion is similar to Schutz's concept of 'type and relevance'. According to Schutz, those who live in the everyday world do not focus on all aspects of an object. Based on their own interests and motivations, they focus on specific aspects in a particular situation. When interpreting an object in a particular situation, relevance determines the aspect of the object on which we choose to focus. In other words, the aspect of the object that is focused upon depends on what is important or relevant to interests and motivation of people living in the everyday world. Based on the relevance, different types for interpreting each object are formed (Schutz and Luckmann, 1984, ch. 3; Katagiri, 1993, ch. 2). The explanation of the frame problem by Schutz's concept of relevance, as discussed earlier, shows that each type is classified according to the three processes of motivation, thematisation and interpretation by stock of knowledge. Since perspective and type are similar, as we have discussed, we can say that the frame problem is not only a relevance problem but also a perspective problem. In other words, we interact with each other based on the definition of situations and the subsequent adequate action according to the context.

3.3.2.2 Cognitive sociology

Interpretive sociology, as highlighted by Mead's social psychology and symbolic interactionism, and Schutz's phenomenological sociology, focuses on the process of interpretation by people living in the everyday world. Cognitive sociology focuses on the fact that the process of interpretation is caused by categorisation of various objects. Surely, symbolic interactionism also considers that perspective and interpretation depend on the naming of the object, and Schutz's phenomenological sociology considers that interpretation depends on typification of objects according to degree of relevance. However, cognitive sociology and interpretive sociology differ in that the former generally emphasises that categorisation of various objects constructs self and mutual interactions that make up society. In addition, the role of categorisation is focused upon in the overall process of construction of self and society, including narrative construction of the identity, consequences of self and interaction, or society from the past to present. In other words, the function of categorisation has been conceptualised for considering self and society (Katagiri, 2006).

Selves are categorised in numerous ways. First, 'human' and '*Homo sapiens*' are two different categories, which many post-human theorists attach importance to, as seen above. There are also various subcategories of selves based on roles, such as gender, sexuality, ethnicity, race, and various members of families, organisations, nations and cosmopolitan world society as the most comprehensive system. While some of these subcategories have been excluded from the definition of 'humans', as discussed by Braidotti, others (such as white people and men) have been positioned as synonymous with the category of humans. Further, the categorisation of selves also involves concepts such as (inner) mind, subject, brain and self *per se*. Psychologisation has decomposed and divided the category of humans into (inner) self and mind. In AI theory, whether humans can be reduced to their brains is one of the main questions. In summary, the subcategories of humans are not obvious. This is a source of conflict and negotiation among humans. In other words, the interaction or society constructed by each categorisation of selves is full of conflict and negotiation (cf. Katagiri, 2006, chs 1, 2).

3.3.3 Relationship between Coeckelbergh's relational-phenomenological approach and interpretive sociology and cognitive sociology

In this section, we explore human–robot interaction from the viewpoints of interpretive sociology and cognitive sociology. In today's society, for long-term care purposes, many social robots are being introduced in hospitals and facilities for the elderly. These robots are categorised differently, such as healing robots, companions, tools and so on. At home, we might seek entertainment or healing from social robots. Alternatively, we might consider them merely as tools that help us with household chores. In factories, various industrial robots are being introduced, independent of the degree of technology. There, robots may be regarded as mere tools for increasing productivity, or they may be recognised as colleagues who work alongside humans. Thus, the way humans perceive robots depends on their interpretations, which in turn depend on their interests and motivations. Further, our interpretations also change according to the naming, typification and categorisation of the robots. Hence, the way and under what circumstances robots are categorised corresponds to a problem of relevance.

As discussed above, interpretive and cognitive sociological viewpoints focus on the situations in which humans and robots interact and the definition of robots by exploring the categorisation. Therefore, these two standpoints are compatible with Coeckelbergh's relational-phenomenological approach. We can also say that Coeckelbergh's relational-phenomenological approach has more in common with cognitive sociology in that both emphasise the linguistic function of human–robot interactions by analysing categorisation. On the other hand, Coeckelbergh also focuses on the function of technology independent of humans. For example, when interacting with a robot, humans are forced to consider whether they are a robot user or a recipient of some benefit from a robot. Generally speaking, such questions do not occur in a one-to-one relationship between humans themselves. As interpretive sociology and cognitive sociology do not clearly focus on the unique function of technology, two theoretical standpoints, and Coeckelbergh's relational-phenomenological approach are different in this sense.

3.4 The Possibility of Interaction Between Humans and AI/Robots From a Relational Approach

To explore the possibility of interaction between AI/robots and humans, we have focused on the relationship humans develop with them. In doing so, we discussed Coeckelbergh's relational-phenomenological approach, pointing out the compatibility of his argument with cognitive sociology. In this section, we would like to consider the Turing test from a relational approach, as this test has been considered a classic study of the interaction abilities of robots with humans.

We will discuss the Turing test through the lens of E. Goffman's interaction theory 1963, one of the frameworks in interpretive sociologies; influenced by symbolic interactionism and Schutz's phenomenological sociology. Goffman's interaction theory has been used in previous research to explain the interaction between AI/robots and humans. For example, using Goffman's concepts of 'front region and back region', A. Elliott (2019) has explained digital communication. Earlier, American sociologist, R. Collins (1992) discussed whether sociology can explain AI based on the theories of interactions such as Goffman's concepts and ethnomethodology. He discussed some conditions that enable AI to interact with humans. For example, AI must understand the rules of turn taking and interaction rituals in conversation. AI must acquire 'cultural capital' to make conversations more attractive. AI needs to have an 'internal' motivation to continue conversations with humans. AI must be able to position the current conversation in the context of planning for the future. AI also needs to employ creativity in conversations with others. Collins's findings consist of 13 rules including the above examples. It is doubtful that ordinary people abide by such advanced conversational rules. However, Collins insists that AI will only be able to interact like humans if it can apply these rules successfully.

Collins's view, which applies Goffman's theory to explain the way AI interacts with humans, is based on the properties approach. Generally, the properties approach posits that it is possible for AI/robots to interact with humans if AI/robots have the interaction abilities (or properties) that Goffman (1963) and others point out. On the other hand, in the next section we will apply Goffman's concept to the relational approach, by regarding the Turing test as one form of passing.

3.4.1 Turing and Goffman

3.4.1.1 Turing and the Turing test

Alan Turing was a British mathematician known for his contribution to the development of the Turing machine and the Turing test. His life was full of ups and downs. During World War II, Turing succeeded in deciphering a highly important code by elucidating the structure of the German cryptographic machine called Enigma. Because it was a state secret, Turing's accomplishment was hidden from the British public and his efforts were not fully rewarded. He died in 1954, and it is suspected that Turing might have committed suicide due to his sexuality. Being homosexual was classified as a crime in Britain in the 1950s, and Turing was mentally burdened by his sexuality. He even faced trial and imprisonment for his sexuality. Considering his personal history, it is certainly possible that Turing committed suicide by experiencing accusation and criminalisation in his life (Copeland, 2012, p. 223).

Turing is now famous more as the developer of the Turing machine and Turing test than as a cryptanalyst. The term 'artificial intelligence' is said to have been coined publicly in 1956 by cognitive scientist Marvin Minsky and others at the Dartmouth Conference held at Dartmouth College in the United States. However, Turing is considered to have invented the basic idea of AI more than 10 years prior to the development of the Turing machine. While the Turing machine is a great achievement, what we focus on here is the Turing test.

As is well known, the Turing test involves a computer (AI) and a person, as well as another person who acts as a referee. When the referee asks questions, the person and the computer respond and submit their answers. If the person's response and the computer's response cannot be distinguished by the referee, the computer is considered to be the same as a human. Our question here is why Turing tried to devise such a test.

B. J. Copeland, an author of Turing's biography, explains that Turing wanted to judge whether computers can think. It is widely considered that computers do not perform the same processes as the human brain, nor do computers convert biological nutrition into energy like humans do. Therefore, it is difficult to conclude that computers have the same interaction abilities or properties as humans. So, Turing did not compare the interiors of humans and computers but rather assessed computers' outer processes of behaviour. If computers are indistinguishable from humans in behaviour and response, as in the Turing test, the computer is considered as 'thinking' like a human (Copeland, 2012, pp. 205–206). Although, surely, the Turing test explores the external process of behaviour rather than examining internal mechanisms such as the brain, judging the Turing test as a product of mere behaviourism is too simplistic.

How does this test relate to Goffman? The answer lies in Turing's sexuality and the consequent experience of criminalisation. Harari makes a very interesting point about Turing as follows:

> The Turing test is simply a replication of a mundane test every gay man had to undergo in 1950s Britain: can you pass for a straight man? Turing knew from personal experiences that it did not matter who you really are—it mattered only what others thought about you. According to Turing, in the future computers would be just like gay men in the 1950s. (Harari, 2017, p. 121)

It might be not possible to judge whether Harari's interpretation is correct; however, it is certain that Turing attempted to pass as a straight man in his society. In any case, it is compelling for Turing to connect his own real experience with the invention of his Turing test. As we will discuss in the following, Turing's experience concerning how to pass as a straight man is exactly the same problem as the practice of 'passing' that Goffman proposed.

3.4.1.2 Goffman and passing

The concept of passing is closely related to the concept of stigma. Stigma refers to, in Goffman's terms, properties or attributes seen as abnormal or undesirable that fall outside of the realm of the normal. Goffman (1963) classifies stigmas into three categories: physical stigmas (e.g., physical disabilities), stigma of personality (e.g., a weak will or other personality 'defects') and tribal stigma (e.g., racial, ethnic or religious stigmas). Homosexuality, along with mental illness and alcohol and drug addiction, is classified as one of the stigmas of personality.

There are two points to note in relation to this definition of stigma. First, stigma is considered an attribute associated with negative images that harm the person who carries the stigma. However, whether the attribute is stigmatised or not depends on the person's relationship with other people. For example, the attribute of being a college graduate can be a stigma in certain contexts, for example in a less desirable profession it might be considered superfluous (Goffman, 1963, pp. 14–15). In this way, anyone has the possibility to be stigmatised depending on the context. Secondly, stigma is a category, which enables us to understand some actions. Defining a person by stigma categorises and identifies the person and creates interactions. Given these assumptions, it is not important whether stigma is a real fact. Stigma is a category that defines identity and creates interaction with others.

For example, people with visual disabilities might behave as if they can see things when this act allows them to escape stigmatisation. On the other hand, some people with partial vision impairment might pretend to be completely blind and accept the category of total blindness, meeting the 'expectations' of those around them. They choose to do so to avoid explaining their partial blindness time after time (Goffman, 1963, p. 110). This idea of Goffman is similar to the argument of cognitive sociology and Coeckelbergh's relational-phenomenological approach in that all discuss the role of categorisation in defining identity and creating interaction.

In summary, stigma is both contextual and relational, functioning as a category of identity and thereby creating interaction. Next, let us move on to the concept of passing. Goffman defines passing as the manipulation of information about the self by concealing it. The exposure of the information results in a loss of trust in the individual. When we combine the concept of passing with the concept of stigma, the manipulated information is the stigma itself (Goffman, 1963, p. 42). As discussed earlier, the Turing test was founded on the idea that if a computer responds the same way as a human, it can be regarded as a human. In other words, the question is not whether the computer can actually think like a human, but whether it can respond like a human.

According to Harari, Turing struggled to pass as heterosexual in his daily life. As already discussed, the important points in Goffman's view on stigma and passing are twofold. If we apply these meanings to the case of homosexuality, stigma of the homosexuality exists in a relationship with others. Secondly, stigma of the homosexuality as a category identifies a person's identity and creates interaction. Passing is one of the ways to escape the oppression of stigma. Passing to escape stigma is associated with Goffman's idea that stigma is a product of relationships and that its factual properties are not necessarily important. This leads to the idea that interactions in general are not necessarily based on properties or attributes of the people but on the process of categorisation. In this respect, Turing and Goffman have the same position.

3.4.2 Passing as a kind of role taking

The concepts of stigma and passing are closely linked to the concept of role taking, theorised by G. H. Mead (1934) and symbolic interactionism. Taking a role is not the same as keeping a role as a norm. Role taking is taking of others' attitudes, which is the basic process of interaction (Mead, 1934, part III; Shibutani, 1986, ch. 3). For example, when we greet someone, we must predict whether the person will respond to our greeting. If we think that he/she will respond, we greet him/her. On the other hand, if we predict that he/she will not greet us back, we suspend greeting him/her. The prediction is only possible from the viewpoint of others—in other words, by taking the attitudes of others. This process is role taking. In cognitive sociology, how to categorise oneself and others is considered an important aspect to smoothly facilitate role taking. Furthermore, the way we categorise others as having stigma defines the way we take roles with others.

Such role taking is not limited to interactions among humans. Humans can interact with robots because they regard robots as humans and predict their reactions accordingly. In fact, Mead has pointed out the fascinating idea that role taking also exists between humans and physical objects. For example, let us suppose that we hit a desk. Then, the desk as an object reacts to our collision. The reaction of the desk is different depending on whether the desk is made of wood or metal such as very hard steel. Even if we hit hard a wooden desk, we would not feel so hurt, but if we hit hard a desk made of metal, we may feel hurt, or we may hesitate to hit so hard a desk made of metal. We can say that we hit physical things according to the reaction of those things. In other words, we feel as if physical things themselves react to us.

Furthermore, we can predict how the desk would react before we hit the desk. According to Mead, role taking with physical things is divided into two: 'manipulatory experience' and 'distance experience'. Role taking by distance experience is created not only by the accumulation of direct manipulatory experience in the past but also by the accumulation of knowledge not based on direct manipulatory experience (Mead, 1932, part II). These processes, including manipulatory and distance experience, indicate role taking between humans and physical things. This is because humans can stand in the position of physical things and, assessing them, predict the reactions of physical things and coordinate their actions to those reactions. And, in the case of hitting a desk, from the viewpoint of cognitive sociology, to categorise the desk as wood or metal facilitates role taking with the desk more smoothly. When thinking about role taking with physical things, it is important to understand that physical things themselves do not have the ability to role take, but humans possibly read that ability into physical things.

Mead's discussion of role taking shows important implications in the potential interaction between AI/robots and humans. Even if AI/robots do not have the ability to role take with humans, if humans can predict the reactions of AI/robots, role taking will be possible in the same way as interaction between humans. As in the Turing test, whether the computer (AI) actually thinks like humans is not important. What is important is whether the computer is perceived as human. In other words, the relational approach is more important than the properties approach when considering potential interaction between AI/robots and humans. Further, according to the standpoint of the relational approach, to interact with humans, it is not necessary for AI/robots to overcome the frame problem because the interaction between humans and AI/robots is possible if humans can read intentions and emotions into AI/robots.

From the viewpoint of cognitive sociology, the following points can be discussed. In the Turing test, it is not obvious whether a respondent is a human or a computer, meaning that there is room to categorise the respondent as either. Needless to say, the Turing test becomes meaningless if the categorisation is taken for granted from the beginning. On the other hand, as inferred from the response of humans to social robots, even if the respondent is categorised as a computer, the possibility for humans to build intimate relationships with computers cannot be denied.

3.5 Summary and Conclusion: What Does the Relational Approach Tell Us About AI/Robots?

3.5.1 Paying attention to the relational approach

The properties approach has investigated the question of whether robots have the ability to interact with humans. For example, it has been assumed that if the frame problem is solved, robots can interact like humans. In contrast, instead of questioning whether robots have the ability to interact like humans, the relational approach focuses on how humans interact with robots by exploring how humans perceive robots. Coeckelbergh has called this standpoint a relational/phenomenological approach, noting that humans form relationships with robots when they are in front of them. We perceive robots' appearance not externally or objectively but subjectively. In other words, the relationship between robots and humans is subjectively constructed by humans, and it is defined by the situation and the system of language. Coeckelbergh criticised the properties approach because, due to the limited AI technology of properties, the approach does not fully explain the potential interaction between humans and robots. Whether AI can overcome the frame problem remains a big question in contemporary AI theory where deep-learning technology has been developed as in the case of Chat GPT. Therefore, if it is concluded that robots do not have the same interaction abilities as humans, the discussion of the possibility of interaction between humans and AI /robots ends. On the other hand, humans have already formed relationships with various robots, especially social robots, in various situations in everyday life. According to the relational approach, it is not considered necessary for AI/robots to overcome the frame problem in order to interact with humans. Interaction between humans and AI/robots is possible because humans can read intentions, motivations or emotions into AI/robots depending on the situation.

To reiterate, AI/robots present themselves to humans regardless of their properties or abilities to interact, and humans react subjectively to them (Coeckelbergh, 2020, p. 161). In other words, humans read intentions and emotions into AI/robots just as they do with other humans. Such human–AI/robot interaction also depends on the situation in which AI/robots and humans are placed and the linguistic categorisation of AI/robots. In this sense, Coeckelbergh's relational-phenomenological approach is very compatible with the viewpoints of interpretive sociology and particularly cognitive sociology.

3.5.2 Diversity and context dependence of interaction

Because humans read intentions and emotions into robots, the way to read them is situationally dependent and diverse. Coeckelbergh has pointed out that humans interact with robots differently depending on whether robots are in a factory or at home. As mentioned aove, the way of dealing with robots differs between Western and Japanese society. In Western society, robots are treated as mere tools or machines, not humans. In Japanese society, as seen in the active development of entertainment robots, people regard robots as social and intimate companions. A typical example is the existence of memorial services for robots in Japan. This practice involves memorialising 'dead'/broken robots in a manner similar to how humans are remembered. It is also cited as an example of the animistic attitudes of Japanese people towards robots. As seen in the development and research of AI/robots by Ishiguro and Okada, the relational approach to robots may affirm the way Japanese people already have contact with robots. Although the substantive and fixed cultural comparison between the West and Japan should be avoided (cf. Kaplan, 2005, chs 9, 10), the diversity in concrete and empirical situations should be given attention, while avoiding stereotypes (cf. Hsu et al., 2020).

The viewpoint that humans read intentions and emotions into AI/robots does not ask whether AI/robots actually have intentions and emotions. Instead, it is based on a relational approach that asks how humans interact with or have relationships with AI/robots. The inherent subjectivity of this process is the reason why reading intentions and emotions into AI/robots varies depending on the context of each interaction. This viewpoint is also important for exploring the ways humans interact with each other. As seen from the standpoint of interpretive sociology, human interaction varies from situation to situation, as humans pursue their own interests and motivations in each interaction, as the concepts of 'perspective' and 'relevance' suggest. In conclusion, by examining the nature of interactions between humans and AI/robots, we can also explore more deeply how humans themselves interact with one another.

References

Bartneck, C. et al. (2020) Human-robot interaction: *An introduction, Cambridge* University Press.
Braidotti, R. (2013). *The posthuman*. Polity Press.
Coeckelbergh, M. (2011). Humans, animals, and robots: A phenomenological approach to human–robot relations. *International Journal of Social Robotics*, *3*, 197–204.
Coeckelbergh, M. (2018). Why care about robots? Empathy, moral standing, and the language of suffering. *Kairos: Journal of Philosophy & Science*, *20*, 141–158.
Coeckelbergh, M. (2020). *Introduction to philosophy of technology*. Oxford University Press.
Coeckelbergh, M., and Gunkel, D. J. (2014). Facing animals: A relational, other-oriented approach to moral standing. *Journal of Agricultural and Environmental Ethics*, *27*, 715–733.

Collins, R. (1992). *Sociological insight: An introduction to non-obvious sociology* (2nd ed.). Oxford University Press.
Copeland, B. J. (2012). *Turing: Pioneer of the information age*. Oxford University Press.
Dennett, D. C. (1984). Cognitive wheels: The frame problems of AI. In C. Hookway (ed.), *Mind, machine and evolution: Philosophical studies* (pp. 129–145). Cambridge University Press.
Dreyfus, H. L. (2007). Why Heideggerian AI failed and how fixing it would require making it more Heideggerian. *Philosophical Psychology, 20*(2), 247–268.
Elliott, A. (2019). *The culture of AI*. Routledge.
Goffman, E. (1963). *Stigma*. Prentice-Hall.
Harari, Y. N. (2011). *Sapiens: A brief history of humankind*. Harper.
Harari, Y. N. (2017). *Homo Deus: A brief history of tomorrow*. Harper.
Hsu, E., Elliott, A., Ishii, Y., Sawai, A., and Katagiri, M. (2020). The development of aged care robots in Japan as a varied process. *Technology in Society, 63*. doi: 10.1016/j.techsoc.2020.101366.
Ishiguro, H. (2009). *Robotto towa nanika: Hito no kokoro wo utsusu kagami* [What is a robot? The mirror which reflects the human mind]. Kodan-Sha.
Kaplan, F. (2005). *Les machines apprivoisées: Comprendre les robots de loisir*. Vuibert.
Katagiri, M. (1993). *Schuttsu no shakaigaku* [Sociology of Schutz]. Inaho-Shobo.
Katagiri, M. (2006). *Ninchi shakaigaku no koso: Kategori, jiko, shakai* [The idea of cognitive sociology: Category, self, and society]. Sekai-Shiso-Sha.
Katagiri, M. (2022). *Ningen, AI, Dobutsu: Posuto hyuman no shakaigaku* [Human, AI, and animal: Sociology of the post-human]. Maruzen-Shuppan.
Kurzweil, R. (2005). *The singularity is near: When humans transcend biology*. Penguin Books.
Mead, G. H. (1932). The physical thing. In *The philosophy of the present (Supplementary Essays II)* (pp. 119–139). University of Chicago Press.
Mead, G. H. (1934). *Mind, self, and society from the standpoint of a social behaviourist* (C. W. Morris, ed.). University of Chicago Press.
Okada, M. (2017). *'Yowai robotto' no shiko: Watashi, shintai, komyunikeishon* [The thinking of the 'weak robots': I, body, and communication]. Kodan-Sha.
Sato, S. (2021) Modernity, postmodernity, and late modernity: A review of sociological theories in contemporary Japan, *International Sociology Reviews, 36*(2), 148–159.
Schutz, A. (1964). *Collected papers II*. Nijhoff.
Schutz, A., and Luckmann, T. (1984). *Strukturen der lebenswelt*. Suhrkamp.
Shibutani, T. (1955). Reference groups as perspectives. *American Journal of Sociology, 60*(5), 562–569.
Shibutani, T. (1986). *Social processes: An introduction to sociology*. University of California Press.

Ludovica Paseri

4 Science and Technology Studies, AI and the Research Sector: Questions of Identity

Abstract: In various normative contexts, individuals take on diverse roles, such as 'patient', 'consumer' or 'citizen'. Social sciences contribute significantly to shaping these representations, often with normative consequences. The rise of digital technologies has profoundly impacted scientific research, influencing data collection, processing and analytics, and thereby impinging on identity construction. Computational models, particularly artificial intelligence and machine-learning systems, play a pivotal role in reshaping individuals' identities, leading to issues of discrimination and inequality. The prevalence of bias in data processed by computational systems contributes to these challenges. Addressing this issue becomes crucial, prompting varied responses explored in extensive studies. Against this backdrop, EU institutions' policies advocating Open Science and Fair Research Data emerge as potential solutions to mitigate bias, discrimination and inequality in scientific research. The chapter intends to shed light on the relationship between Open Science policies, scientific research and the construction of individuals' identities, investigating the implications concerning the engendering of bias and the emergence of forms of discrimination.

Keywords: scientific research, artificial intelligence, AI, research data, identity, institutions, STS

4.1 Introduction

Several representations of individuals emerge in different normative contexts: "identity is more than just ideas, as well as more than the realm of ordinary experience. Identity is also, fundamentally, political. The political currents which dominate the global agenda in the early twenty-first century—including environmental politics, sexual politics and ethnic struggle—place identity at their center" (Elliott, 2012, p. xxi). For this reason, the normative dimension which follows the political stance is extremely relevant. For example, the individual may be a 'patient' in the medical context, a 'consumer' in the commercial sphere, a 'citizen' in the public sector, and so on.

A multiplicity of possible representations of the individual from a legal point of view have an established origin, being already discernible in the legal system of the ancient Roman age. Consider, for instance, the *summa divisio* between things (*res*) and

Ludovica Paseri, University of Turin, Italy

https://doi.org/10.1515/9783110721751-004

individuals (*personae*), and the further problematic legal framing of slaves (Gaius, n. d., pp. 9–12). From the ancient to contemporary times a multiplication of subjective figures of law has taken place: think, for example, of the distinction between natural person and legal person; or the notion of subjective right as opposed to collective right, in which the individual identifies her or his identity in relation to being part of a qualified group. It is worth stressing that in many instances, the notion of the individual is being declined differently from a normative point of view: individual as citizen; individual as user; individual as consumer; but also, individual as 'data subject' in the processing of personal data, within the framework of the General Data Protection Regulation, that is, the European data protection discipline. These different articulations and representations of the individual as a rights holder stand in line with certain and specific configurations of society (e.g., more collectivist or more individualist, see Elliott, 2015, pp. 1–29).

Nowadays, social sciences in particular contribute to convey representations of individuals. These representations, which are mostly meant to be descriptions or accounts of the individual, may have, and often do have, normative consequences. A 'consumer', for instance, may be further described and profiled as either a 'strong' or 'weak' consumer, and therefore be offered or denied opportunities (in terms of price, goods, services, etc.).

Against this backdrop, scientific research has been impacted by digital technologies, as they are deeply involved in data collection, processing and analytics. This also has significant repercussions for identity construction. As has been widely remarked, computational models, artificial intelligence (AI) and machine-learning systems play a considerable role in the reshaping of individuals' identity: 'For the global communications revolution has impacted profoundly on modern society, and perhaps nowhere more so than in the everyday routines through which individuals forge and sustain a sense of self' (Elliott, 2013, p. 2).

In this perspective, it may be remarked that AI systems

> have the resources (in terms of the amount of memory, sharing and availability of data) and are programmed in such a way (based on increasingly sophisticated computational models able to learn from experience) that they reduce the representation of reality to a series of elements, operations, processes and data that can be processed syntactically, so as to adapt it to their mode of operation. (Durante, 2021, p. 72)

As a result, digital technologies, and in particular the AI systems, to some extent impose a representation of reality which fits their distinctive syntactic capabilities, as opposed to the semantics typical of humans.

But digital technologies and AI systems are not limited to proposing representations of reality, exerting an impact on the identity of individuals as well: 'Robotics and AI have spread throughout social life, reorganising the dynamics of identity and the self' (Elliott, 2019, p. 83). The philosopher Luciano Floridi, adopting the notion of Michel Foucault (1988), describes information and communication technologies (ICTs) as 'the most powerful *technologies of the self* to which we have ever been exposed', emphasising the relevance of the matter: 'Clearly, we should handle them carefully, as they are significantly modifying the contexts and the practices through which we shape ourselves' (Floridi, 2014, p. 106). The impact of ICTs and digital technologies on the construction of identity is far reaching (Floridi, 2014):

> Change the social conditions in which you live, modify the network of relations and the flows of information you enjoy, reshape the nature and scope of the constraints and affordances that regulate your presentation of yourself to the world and indirectly to yourself, and then your social self may be radically updated, feeding back into your self-conception, which ends up shaping your personal identity. (p. 107)

The philosopher Michel Serres has illustrated the impact of technology on identity in terms of 'hominescence', as a sort of commencement of a new human and, to some extent, of a new identity (Serres, 2001).

Throughout the evolution of thought, as well as in the history of politics, whenever an attempt has been made to delineate a notion, however comprehensive, of identity, the risk has been that of generating a relationship of inclusion and exclusion: any definition of identity (any construction of an identity process, which passes through the identification of certain characteristics) produces or is likely to produce an implicit subdivision or classification that both integrates and excludes or discriminates. This may be more striking today, when the enormous availability of data and personal information allows for the incessant construction and reconstruction of profiles, categories and clusters within which individuals find themselves inserted and subsumed.

In this context, one of the most striking aspects of the identity issues raised by digital technologies is the occurrence of forms of discrimination and inequality. Such forms of discrimination and inequality are often due to bias related to the data processed or trained upon by computational systems.

How to face this issue? Can bias be limited, mitigated or even eliminated? There can be diverse answers to these questions, as illustrated by the great number of studies in this field.

Social sciences are thus put under pressure. The strand of research of Science and Technology Studies (STS), since the 1980s has been concerned with the interaction of technology with many and various fields of knowledge, investigating 'from a broad social-scientific perspective how social, political, and cultural values affect scientific research and technological innovation, and how these in turn affect society, politics, and culture' (Franssen, Lokhorst and van de Poel, 2018). The most compelling angle, for the purposes of this chapter, is the representation of the relationship between science and technology, according to the STS field. As illustrated by Sheila Jasanoff, 'STS scholarship acknowledges that science and technology do not unidirectionally shape our values and norms. Rather, and symmetrically, our sense of how we ought to organise and govern ourselves profoundly influences what we make of nature, society, and the "real world"' (Jasanoff, 2015, p. 3). In this way, the normative dimension of science and technology is, to some extent, encompassed by the STS field itself.

The aim of this chapter is to draw attention to the role that scientific research plays in the current scenario of ongoing digitisation, taking into account the perspective of STS. In particular, in the field of STS, co-production, understood as 'the simultaneous production of knowledge and social order' (Jasanoff, 1996, p. 393), represents the conceptual framework needed to define the interplay between science and society.

In this context, the next section delves into the role of scientific research in identity construction. After that, section 4.3 focuses on the emerging paradigm of scientific research, the so-called Open Science paradigm, which represents a new way of doing science triggered precisely by the digital revolution, which amis to the openness of every phase of the scientific research cycle, which may well illustrate some aspects of the STS approach. Having clarified the terms of the issue, the relationship between technology and scientific research, and the role of institutions in supporting such a profound transformation, section 4.4 investigates how openness in scientific research can represent a means to help mitigate biases and therefore inequalities and discriminations arising from the use of new technologies. Finally, section 4.5 integrates the different aspects of the discussion, clarifying the relationship between science, technology and identity.

4.2 The Role of Scientific Research in Shaping Personal Identity

If, just a few years ago, it was necessary to stress the reasons for investigating the impact of new and digital technologies on various aspects of our society, today this impact is so evident that exploring its consequences is a real duty for the contemporary researcher.

The presence of AI is increasingly pervasive in many areas of our societies. First, an understanding of what is meant by the expression 'artificial intelligence' is needed, since it is has been increasingly evoked, from 1956, when John McCarthy coined the term (McCarthy et al. 2006), up to the present day. In addition, the debate in recent months has specifically revolved around what is meant by AI in the European institutions' new proposal for a Regulation, the so-called Artificial Intelligence Act, which is now in the final stages of approval.

According to Pagallo and Barfield (2020): "AI refers to a program that is able to mimic or re-create the thought processes demonstrated by the human brain. This usually involves making observations or receiving input for use in a problem-solving situation, and the ability to categorize and identify different objects and the properties associated with those objects" (p. 1). Generally, the inputs mentioned by the authors are represented by large amounts of data. In other words, "AI programming allows to infer complex, nonlinear relationship from the data which the software analyses. It does so to achieve a specific goal by identifying and, if necessary, automatically refining (or prompting the refinement of) a system's operations' (Hofmann, 2021, pp. 2–3). AI systems designed by programmers, eventually capable of self-learning and inferring complex relationships, therefore, entail that "AI-based technological creativity and innovation are co-produced by humans and machines" (Holton and Boyd, 2021, p. 187).

When policymakers and institutions, whether international, national or local, in the field of scientific research (also including universities and research centres), approach the matter, these features of AI should necessarily be taken into account. This is essential in order to put in place a framework of provisions, in terms of inspiring both principles and detailed rules, that are aimed at minimising the risks associated with their use and maximising their benefits.

Since the pervasiveness of digital technologies and AI systems is clear-cut in every sector of our lives and in every aspect of our society, neither does scientific research sidestep this dynamic. However, this statement, that may appear at first sight straightforward, requires some clarification: first a definitional clarification, and then a functional one.

The definitional clarification concerns the very notion of scientific research. Far from engaging in the debates that characterise the field of philosophy of science, epistemology and gnoseology, it is necessary to agree on what is meant by scientific research. In this chapter, the term 'scientific research' is used to refer to the process that develops through the scientific method, relating to the work of the scientific community, broadly understood, that is, including both the so-called STEM or hard sciences, such as science, technology, engineering and medicine, and the social sciences and humanities. Specifically, furthermore, it is intended to refer primarily to publicly funded scientific research, that is, scientific research funded by national bodies or international and European institutions, in various ways (*in primis*, e.g., consider the new EU funding programme for science and innovation, the 'Horizone Europe' programme: see Regulation (EU) 2021/695).

The other clarification, which is defined as functional, concerns precisely the function that may characterise a given AI system in the context of scientific research. In fact, AI systems can have a twofold function in relation to the field of science: AI systems can be either the tool or the result of a scientific research project.

First, such systems can be tools used by the researchers of the team to implement a specific research project. In this circumstance, AI is instrumental in obtaining the results of the research project: the AI system is an instrument—a very powerful one —and, also, an excellent syntactic engine, for the benefit of researchers, that enables the achievement of results, generally previously unseen before the digital revolution. An excellent example of the use of AI as a means through which to carry out a research project is provided by the Institute of Photonics and Nanotechnology of the Physics Department of the University of Milan, which precisely by applying an AI system has developed a quantum compiler capable of programming algorithms applicable to any quantum computer, through the use of logic gates (Moro et al., 2021). Consider that it is often mistakenly believed that only the so-called 'hard sciences' make use of the most advanced AI systems as tools to conduct research projects. Recent studies prove otherwise. For instance, an analysis conducted by the University of Luxembourg, concerning the use of the computational power derived from the High Performance Computing platform held by the university itself, for scientific research purposes, showed the opposite: during the year 2020, 13.5% of the available CPUs were used for projects in the life sciences, 1.8% for projects in the field of history and social sciences, and 2.7% by legal scholars and economists (Paseri, Varrette and Bouvry, 2021, p. 127).

Secondly, an AI system can also represent the result of a research project: the objective of the research project itself may be the design and development of an AI system aimed at some specific further use. The idea, then, is to have an AI system developed in the academic sphere, by public researchers, which can later become the object of use by third parties, both academic and non-academic, as well as possibly even commercialised.

While there are many examples of the use of AI systems as a tool for research projects, it is more difficult to identify AI systems developed as a result of research projects financed with public money and then, only afterwards, commercialised (Ebers et al., 2021, p. 591). Frequently, in this type of scenario there is, anyhow, the presence, support and participation of private actors in the research project (Lior, 2023). There may be several reasons why AI products are seldom developed as an outcome of a public research project. But the reason will primarily be economic, since these types of operations require substantial funding in terms of both development and commercialisation; the causes could also be legal, for instance related to the ownership of the results of the project.

Thus, having clarified these fundamental preliminary aspects, what is the impact of the AI systems in scientific research on the shaping of identity? One of the most marked and troubling aspects of identity issues raised by digital technologies is the increasing emergence of forms of discrimination and inequality. Most forms of discrimination and inequality arising from the use of AI systems are due to the presence of biases related to the data that is processed by the systems themselves. The concept of bias is defined as 'the lack of internal validity or incorrect assessment of the association between an exposure and an effect in the target population' (Delgado-Rodríguez and Llorca, 2004, p. 635). A biased AI system may generate highly impactful consequences on the human and fundamental rights of individuals, especially in terms of discrimination and inequalities. Consider that the right to equality and non-discrimination is enshrined in Article 2 of the Universal Declaration of Human Rights (UDHR); the principle of non-discrimination is also protected under Article 14 of the Convention for the Protection of Human Rights and Fundamental Freedoms (ECHR) and Article 21 of the Charter of Fundamental Rights of the European Union; whereas the principle of equality is also guaranteed by Article 2 of the Treaty on European Union, representing one of the pillars of the entire European system.

An emblematic example of how biased AI systems can affect the rights of individuals, and consequently their identity, is provided in a prominent 2016 study which showed how biased algorithms can easily convey gender discrimination in the identification of online job advertisements (Knight, 2016).

In both cases, whether AI represents a tool of scientific research or the result of the project itself, the impact on the shaping of the identity of individuals is considerable and carries with it relevant normative consequences.

The impact of technological developments on individual identity and on the human condition has deeper roots. Michel Serres speaks of 'hominescence' insofar as the human condition has undergone a radical transformation due to the extraordinary power of life and death permitted and generated by the enormous scientific-technological progress, going so far as to declare that "nous accouchons peut-être d'une autre humanité"[1] (Serres, 2001, p. 15).

The stance of the human and the possible representations of identity have been variously investigated in relation to technical and technological developments. In particular, in the development of AI, according to a socio-technical perspective, a co-production results from the interaction of social and technical rules (Holton and Boyd, 2021, p. 180). It has been argued by some scholars that, in this scenario, the human actors lie in a substantial asymmetrical position (Holton and Boyd, 2021, p. 179). In some ways, the rupture between subject and object, according to Serres's (2009) philosophy, is found, from a socio-technical point of view, in the co-presence of human and machine actors in a communicative and cognitive sense (Holton and Boyd, 2021, p. 182). The concept of co-production, a fundamental pillar of the STS field, stands in opposition to the mere determinism of scientific-technical progress: "Making knowledge is thus never 'innocent' activity; nothing can be regarded as 'natural' or 'simply given'. Rather, knowledge plays a central role in the way we understand ourselves, people around us, and our societies. At the same time, people and societies as well as the values, aspirations, and imaginaries they develop, circulate, and share, also shape the making of knowledge" (Felt, 2016, p. 253).

Therefore, researchers should proceed in their work keeping in mind that scientific research—*a fortiori* if it involves AI systems—takes part in shaping the identity of individuals. But how can the profound transformation that ICTs and digital technologies are exerting be interpreted? Besides, this impact is exercised both in the way of doing science and in the way of wondering about science itself. The next section intends to provide an answer to this question by illustrating the approach of Open Science.

1 "We may be generating another humanity".

4.3 The Emergence of the Open Science Paradigm

The impact of new technologies, ever-increasing computational power and enormous amounts of data in the field of scientific research can be interpreted in many different manners. This chapter aims to represent this profound transformation in terms of a paradigm shift, evoking the terminology of the philosopher of science Thomas Kuhn (Kuhn, 1962), represented by the emergence of the paradigm of Open Science. This paradigm shift identifies a profound upheaval in the way science operates, which affects the entire cycle of scientific research: the goal becomes to unlock every phase of the research process. All steps of the scientific research process are involved and touched by this profound transformation: the collection or creation of data; the selection of the most suitable methodologies adopted to carry out the project; the choice of tools to process the vast amounts of data collected or created; up to including all aspects related to the dissemination of the results of the scientific research project, such as the adoption of open educational practices (i.e., 'open education') or open access to scientific publications.

The approach to openness is, in fact, the scientific community's response to the unfolding of the enormous opportunities provided by the application of new and digital technologies to the field of scientific research. What is now commonly referred to as the Open Science paradigm actually originated from a bottom-up impulse, limited to the aspect of free access to scientific literature. In fact, in the early 2000s a part of the scientific community was been able to grasp the potential of digital technology as a means to expand access to scientific literature on a very wide scale: scientific publications, thanks to ICTs, may be consulted directly online, without physically going to a library. In addition, the early pioneers of Open Science were originally able to transform these opportunities related to the emergence of new digital technologies into real instances aimed to change the business model that was imposing a payment to scientific publishers in order to benefit from scientific literature.[2] From the first instances, represented by the well-known Budapest,[3] Bethesda[4] and Berlin[5] Declarations, the Open Science movement has been developed into a true paradigm shift, involving almost every aspect of research.

The expression 'Open Science', now, is intended to indicate 'good' science (Budroni, Burgelman and Schouppe, 2019, p. 136), based on communication and cooperation (as opposed to thrusting competition), being transparent, replicable and responsive

---

2 However, this business model has not been completely overturned—it still persists today. Nevertheless, the great merit of the first instances attempting to achieve open access to the scientific literature was precisely to bring the debate to the forefront and start the process of change.

3 Budapest Declaration, 2002, https://www.budapestopenaccessinitiative.org/read.

4 Bethesda Declaration, 2003, https://osc.universityofcalifornia.edu/2003/04/bethesda-statement-on-open-access-publishing/.

5 Berlin Declaration, 2003, https://openaccess.mpg.de/Berlin-Declaration.

(i.e., able to take full advantage of the use of new ICTs and digital technologies) but also inclusive and pluralistic (Leonelli, 2023).

The aspect of replicability has become central. Science is facing the so-called 'replicability' or 'reproducibility crisis' (Baker, 2016; Fanelli, 2018; Mede et al., 2021), since scientific experiments often cannot be replicated by other scientists, or sometimes even by the very scientists who performed the experiment in the first place, despite the fact that the results of those experiments may have been published in the most famous scientific journals. The reasons for the replication crisis are manifold: the lack of raw data, or the lack of sharing the source code of the algorithm used to perform the search, and so on (Miyakawa, 2020; Hutson, 2018). Some causes of the replication crisis, then, are not necessarily liked to the technologies involved. For instance, think about the field of quantum physics, which is currently facing a considerable problem concerning not only the verifiability, but also the falsifiability, of string theories (Ritson and Camilleri, 2015; Greene, 1999). Without claiming to investigate the many causes of this crisis of replicability or reproducibility, it is worth pointing out just a few consequences of the crisis. If the scientific results are not replicable, then these findings do not stand up to peer scrutiny and therefore the foundations of science itself are undermined, both from a practical and an ethical point of view. From a practical point of view, the scientific method developed from Galileo Galilei onwards is undermined. From the ethical point of view, these unreplicable scientific results unhinge the pillars of the *ethos* of science according to Robert Merton: (i) disinterestedness; (ii) universalism; (iii) communality (instead of 'communism', adopting the interpretation of Anderson et al., 2010); and (iv) organised scepticism (Merton, 1973). As described by Stuart Ritchie: 'science is crucial to our society, and we mustn't let any of it be compromised by low-quality, unreplicable studies. If we let standards slip in any part of science, we risk tarnishing the reputation of the scientific enterprise more generally' (Ritchie, 2020, p. 36).

This issue is closely linked to a matter very relevant to the STS field: 'How many times have politicians made laws or policies that directly impact people's lives, citing science that won't stand up to scrutiny?' (Ritchie, 2020, p. 9). In light of these considerations, therefore, the principle of transparency in scientific research, one of the pillars of the Open Science framework, acquires fundamental relevance: Open Science also indicates a kind of renewal or rethinking of the traditional mechanisms of scientific research, as well as a way of responding to the current challenges it faces.

A change of mind and a rethinking are required, above all, in relation to the management of research data: one of the most important dimensions of the Open Science paradigm is the openness of scientific research data. The concept of 'openness', when related to data, must be understood in terms of sharing and re-use of data by third parties who did not participate in the original collection or creation of that data. The purpose of Open Science, however, even before that of openness, is that of good management of research data. To achieve this goal, the so-called FAIR Data Principles have been developed, that is, a set of guidelines which aim to have data that is 'well constructed' from a technical point of view (Wilkinson et al., 2016). FAIR is an acronym that represents the characteristics that data must have in order to be considered robust and technically flawless: Findability, Accessibility, Interoperability and Reusability. Findability represents the set of operations that must be put in place in order to guarantee the effective retrievability of data (such as the indication of a unique and persistent identifier; or rich metadata, etc.); accessibility indicates the—potential—availability of data, in case there is both the need and the authority to access it (e.g., a qualified role, such as a reviewer of a scientific paper submitted for scrutiny); interoperability represents the set of technical features that allow data developed by a specific research group to be technically compatible with other technologies, and potentially usable by third parties intending to re-use such data (e.g., a common vocabulary or formal and shared rules for the representation of knowledge); and reusability refers, in particular, to the legal conditions that allow legitimate sharing and reuse of data (e.g., an accurate description of the provenance of the data; or the correct licence for reuse).

After a bottom-up origin of the phenomenon, a strong endorsement of Open Science by both international and European institutions can now be claimed.

At the European level, the most relevant manifestation of the intention of institutions to embrace this paradigm shift is represented by Article 14 of Regulation (EU) 2021/695 establishing the new framework programme for research and innovation, the so-called 'Horizon Europe' that replaces the previous 'Horizon 2020'. This article, titled 'Open Science', states that

> [T]he Programme shall encourage open science as an approach to the scientific process based on cooperative work and diffusing knowledge, in particular in accordance with the following elements: . . . a) open access to scientific publications resulting from the research funded under the programme; b) open access to research data, including those underlying scientific publications, in accordance with the principle 'as open as possible, closed as necessary'.

This article is significant because it strongly anchors the development of European research projects over the next four years to openness, embracing the paradigm shift represented by Open Science. In addition, this article enshrines in hard law the fundamental formula underlying the concept of openness in scientific research, namely the principle of 'as open as possible, as closed as necessary'. This principle precisely identifies the limits of this openness: Open Science is not designed to open up the various stages of the scientific research process indiscriminately and blindly; on the contrary, the formula shows the intention to guarantee the limits that hinder openness, as the outcome of a process of balancing with any opposing interests.

For several years, however, the European Union seems to have embraced the paradigm of Open Science. In fact, EU Open Science policies may be interpreted twofold. First, they can be considered the embodiment of the so-called 'fifth European freedom', the freedom of circulation of knowledge, introduced by the Council of Europe held in Ljubljana in 2008. This freedom of movement of knowledge underpins the establishment of the European Research Area, an area of free circulation of knowledge and researchers, which aims to develop a further dimension of the European process of integration that goes beyond the purely economic sector. Secondly, European Open Science policies are also a key part of the European Union's digital strategy. As shown in 2015, in establishing the Digital Single Market and in 2020, in outlining the 'European Strategy for data' (European Commission, 2015, 2020), as well as later proposing a Regulation about the European Health Data Space (European Commission, 2022), the European institutions have devoted particular attention to the research sector. Consider, *in primis*, the European Open Science Cloud project. This project represents a trusted space of federated resources, conceived and developed as an asset for European researchers, so that they can take advantage of data, infrastructures and services, and thus be able to benefit as much as possible from technological developments.

At the international level, the most significant recent event has been the adoption of the first UNESCO Recommendation on Open Science on 23 November 2021. The adoption of this Recommendation is the outcome of a process that started in 2020, during the second wave of the COVID-19 pandemic, with the adoption of a joint appeal by UNESCO, WHO, CERN and the Office of the UN High Commissioner for Human Rights, specifically on Open Science, acknowledging it as the way forward for 'the urgency of addressing complex and interconnected environmental, social and economic challenges for the people and the planet, including poverty, health issues, access to education, rising inequalities and disparities of opportunity, increasing science, technology and innovation gaps, natural resource depletion, loss of biodiversity, land degradation, climate change, natural and human-made disasters, spiralling conflicts and related humanitarian crises' (UNESCO, 2021a, p. 2). Paying attention to this list of grand challenges identified by UNESCO, what is significant is that the digital revolution—and the consequent impact of technology and the pervasiveness of AI—appears across every field: acquiring this awareness becomes the prerequisite for understanding the change that is affecting not only scientific research, but also society more generally. Significantly, on 23 November 2021, alongside the Recommendation on Open Science, the first draft of the UNESCO Recommendation on the Ethics of Artificial Intelligence was also released (UNESCO, 2021b). Once again, the link between the two topics, considered priorities by international institutions, clearly emerges.

The interest developed at the international level in Open Science is closely connected to a remarkable framework of human and fundamental rights. The first and fundamental reference is Article 27 of the UDHR. The first paragraph states: '[E]veryone has the right freely to participate in the cultural life of the community, to enjoy the arts and to share in scientific advancement and its benefits', which becomes an expression of the so-called 'right to science' (Porsdam and Porsdam Mann, 2021). This article, which for a long time was merely an inspirational principle, has recently, also in light of the debate generated by the COVID-19 pandemic, acquired new relevance and strength. In addition, the reference to the right of individuals to 'take part' in scientific development refers to another essential dimension of Open Science, namely that connected to the active participation of individuals in scientific research projects. Such participation, identified by the expression 'Citizen Science', indicates an engagement of civil society in the operations of science, either directly, by performing small tasks within specific research projects, or indirectly, perhaps through micro-research grants represented by projects of crowdfunding of science. Thereby, Open Science, which represents the paradigm of science of the digital revolution era, also encompasses a participatory science that becomes a 'joint commitment', as conceived by the philosopher Margaret Gilbert (Gilbert, 2013, pp. 37–57).

Alongside the UDHR, Article 15 of the International Covenant on Economic, Social and Cultural Rights, of 1966, in its first paragraph reaffirms the content of Article 27 of the UDHR, while its second paragraph identifies a real proactive role of states aimed at fostering the exercise of this right: 'The steps to be taken by the States Parties to the present Covenant to achieve the full realisation of this right shall include those necessary for the conservation, the development and the diffusion of science and culture'.

The ECHR, on the other hand, while making no explicit reference to science *per se*, does, however, enshrine the right to communicate and receive information and ideas, within the framework of freedom of expression (Article 10 ECHR). The benefits of scientific progress mentioned in the UDHR, as well as the practices of dissemination of science referred to in the International Covenant, can easily be represented in terms of the communication and sharing of scientific information, that is, the outcome of a scientific research process. More significantly, the ECHR emphasises a twofold condition to this right of communication: (i) the absence of interference by public authorities; and (ii) the absence of boundary limits. These conditions are also entirely consistent with the requirements of science: non-interference by public authorities, specifically, relates to the academic freedom enshrined in Article 13 of the Charter of Fundamental Rights of the European Union; while the absence of border limits is linked with the fifth European freedom, that is, the freedom of circulation of knowledge, mentioned above.

The context is, thus, represented by profound changes and by the consequent strategies of the international and European institutions, as a reaction. Against this backdrop, focusing on the impact of new technologies on science and the further consequences on identity construction appears crucial. The next section argues that Open Science policies may become the means by which to help address inequalities and discrimination generated by the use of AI.

4.4 Open Science, Identity and Social Justice

Most forms of discrimination and inequality resulting from the use of AI systems are due to the biases that characterise the data processed by these systems. As previously argued, a biased AI system may generate highly impactful consequences on the construction of the individual's identity, especially in terms of discrimination and inequality.

How can any bias in the generation of an AI system be limited, mitigated or even completely eliminated? There may be many answers and possible solutions, also considering the fact that, nowadays, this area of research is becoming widely investigated.

Chiefly, it is essential to specify that '[B]iases are an unavoidable part of human nature and it's naïve to think they could ever be eradicated from anything that we do' (Ritchie, 2020, p. 121), even when we design and develop an AI system. Nonetheless, '[W]e do, however, have tools that are supposed to bring a little more objectivity to the table' (Ritchie, 2020, p. 121). The following argument starts from this assumption.

Without claiming to provide an exhaustive analysis on a subject as broad as that of the study of bias in AI systems, transparency is certainly a fundamental element in facing these drifts. Greater transparency connected to the datasets underlying the AI system has been indicated by researchers as a means of limiting the proliferation of bias, and hence of forms of discrimination and inequality (Daneshjou et al., 2021).

The full relevance of Open Science emerges precisely in relation to the principle of transparency. Transparency is one of the cardinal principles of this emerging paradigm. Besides being historically an inspirational and pivotal principle in the progress of scientific investigation, today the principle of transparency also becomes the means to cope with the already mentioned replication crisis of scientific results.

Transparency, within the paradigm of Open Science, can take on various meanings. Transparency in the management of public funds aimed at financing scientific research projects. Transparency in the evaluation of the validity of the projects. Transparency in the identification of criteria aimed at evaluating the advancement of the researchers' careers. Above all, transparency of data, means and results of research projects, which aim to generate scientific knowledge according to a specific scientific community of reference.

However, recently, in relation to data, the issue of transparency has become extremely problematic, connected with the aforementioned replicability crisis. As illustrated above, the causes of this replicability crisis are many. However, the low inclination to share datasets (even non-publicly, but only with reviewers to verify the validity of a scientific investigation for the purpose of publication, Miyakawa, 2020, pp. 13–24) is certainly one of the most important. Open Science, therefore, aims precisely to strengthen transparency, in order to overcome the crisis of replication of scientific results, returning to adhere to the *ethos* of science as outlined in the early twentieth century by Robert Merton. Furthermore, Open Science, as a further guarantee of transparency, implies the adoption of the FAIR Data Principles as a possible approach in handling, operating and processing research data. These Principles are a set of rules that guarantee good data management from a technical point of view and may represent an additional guarantee on the management of the dataset underlying the AI system.

It should be borne in mind that the necessary transparency addressed to limit the formation of bias in the AI systems is often not practically feasible, insofar as a private company is involved in the development and design of the AI system. Private actors generally have legitimately conflicting interests in openness and sharing. However, just above it was analysed how AI systems, in the sector of public scientific research, generally represent a *tool*, and are rarely a *result* of the project itself.

Thus, if one of the goals of the institutions is to foster AI as a means to face inequality and discrimination, then in parallel institutions should concretely implement solid national Open Science strategies. These strategies promote the realisation of AI systems, developed by publicly funded scientific research, in which AI does not represent just a tool, but is the real outcome of the project, respecting the FAIR Data criteria and the principle of transparency.

In this regard, concerning the role of institutions, it is indeed necessary to emphasise the twofold objective that any intervention in the field of AI should take into consideration. On the one hand, the aim of the intervention by institutions should be to put in place a set of provisions that avoid the emergence of inequalities and discrimination that can be generated or perpetuated by the use of AI systems. On the other hand, however, it is crucial that the institutional initiatives are not limited to this aspect: reducing legislative intervention to a means through which to protect oneself from any risks posed by AI systems is symptomatic of a vision of technology as an enemy, implying only risks from which to protect individuals. Such an interpretation is neither sharable nor empirically proven. An essential element that institutions and lawmakers involved in this field should bear in mind is that AI systems can become a means through which to tackle pre-existing inequalities and discrimination. Thus, the second objective that institutions are called upon to pursue is to outline a legal framework that does not represent a limitation but is capable of promoting the use of AI (Pagallo, 2022), aimed to protect fundamental principles such as the principle of equality, as well as the principle of non-discrimination.

In other words, supporting Open Science becomes a means of creating a suitable ecosystem for the design, development and application of AI systems in publicly funded research, as a *result* and not merely as a *tool*.

This approach that the institutions are suggested to adopt is, moreover, in line with the strategies developed in the STS sector, in order to 'counteract some of the anti-democratic tendencies observable in recent decades' (Jasanoff, 2015, p. 621) related to the governance of science and technology. If scientific research is capable of proposing and conveying specific representations of the identity of individuals, consider the impact that scientific knowledge may have when it becomes the basis for political action. In a recent study, "Assessing National Institutional Capacity for Evidence-Informed Policymaking" (Oliver, Krieger and Melchor, 2022), the 'science-for-policy system' was investigated. According to the authors, the science-for-policy system represents '[T]he set of actors and connections through which scientific knowledge is acquired, synthesised, translated, presented for use, and applied in the policymaking process' (p. 6). The science-for-policy system represents concretely the scenario in which specific representations of the identity of individuals can be conveyed into the space of politics and generate considerable consequences.

For instance, consider scientific knowledge developed with biased AI systems. The consequences could have a thrust that goes far beyond the field of scientific research, to have a direct impact on democracy. Accordingly, in the aforementioned study on the science-for-policy system, transparency is identified as one of the key features:

> When research is reproducible and a clear line can be traced between research question, methods, results and conclusions, it is easier for others to assess the validity of claims made. The same principle applies to the mobilisation and use of evidence. Transparent systems to select, prioritise, and interpret research, and to offer recommendations for action enable observers to understand where policies and practices come from. (Oliver, Krieger and Melchor, 2022, p. 8)

Once again, therefore, the Open Science paradigm seems to point the direction to follow.

As a result, the policies promoted by the EU institutions in the field of science and designed to foster greater openness and transparency in scientific research may help address bias and hence discrimination and inequality. Scientific research, understood in terms of the emerging paradigm of Open Science, may represent a means by which to support the protection the construction of individuals' identity, and to limit how a growing range of human subjectivities (weak, vulnerable, diverse and excluded persons) may be discriminated and set apart from the community. In other words, European Open Science policies can be interpreted as the expression of the aforementioned relationship between science and technology on the one hand, and social values and norms on the other, that is not unidirectional, but rather of mutual exchange. In light of the foregoing, the role of institutions, especially national and local ones (i.e., universities and research centres), is to implement Open Science policies without distorting its principles.

4.5 Conclusions

This chapter claimed that EU Open Science policies may help mitigate the design and development of biased AI systems and, as a result, may support the right to equality and non-discrimination. The paradigm of science of the digital revolution era, described in terms of Open Science, may represent a means by which to participate in the protection of individuals' identity construction, limiting a multitude of human subjectivities (such as weak, vulnerable, diverse and excluded persons), which may embody forms of discrimination, within the society.

The chapter initially analysed the impact of AI systems on identity, focusing on the distortions that can occur in terms of forms of discrimination and infringements of the principle of equality: AI systems may generate representations of individuals that are potentially detrimental in certain contexts. However, as the field of STS shows, science and technology do not unilaterally shape norms and values. Therefore, there is a leeway in this relationship that takes on a normative dimension.

The impact of scientific research in shaping personal identity, particularly where it makes use of AI, was then explored. In so doing, it was first essential to understand what role an AI system can assume within a scientific research context. According to the proposed analysis, in light of the characteristics of AI systems, they can either represent the *tool* through which a scientific project is carried out, or the *result* of the project itself. Concerning the impact on identity, on the one hand technology, and in particular AI, is profoundly impacting, so much so that Serres speaks of 'another humanity (Serres, 2001)'. On the other hand, as emphasised by the STS sector, the formation of scientific knowledge is never a neutral activity, but rather is always profoundly interconnected and shaped by social values and those who participate in it, within a mechanism of co-production.

Thus, the representations of the identity of individuals conveyed through scientific research—and even more so if the scientific research is implemented through an AI system—emerges to be highly impactful on the human and fundamental rights of individuals.

The profound transformation that digital technologies are exerting on the field of scientific research has subsequently been interpreted in terms of Open Science.

Since most of the forms of discrimination and inequality that can be generated through AI systems are due to the presence of bias related to the underlying dataset, proper and—above all—transparent data management can be the means to address this issue. The principle of transparency is a pillar of the Open Science paradigm. Transparency related to research data is specifically relevant here. The adoption of the FAIR Data Principles, that is, the set of guidelines that guarantee good data management from a technical point of view, goes in this direction. For this reason, it has been argued that supporting Open Science can take part in mitigating the formation of bias and, consequently, limiting forms of discrimination and inequality connected to the use of AI systems. Given the implications that arise from the relationship between scientific knowledge and democracy, especially when it underpins political action, the adoption of Open Science strategies at the national level becomes essential. In so doing, the realisation of AI systems as a *result* of publicly funded scientific research should be promoted, respecting the Open Science Principles.

References

Anderson, M. S., Ronning, E. A., DeVries, R., and Martinson, B. C. (2010). Extending the Mertonian norms: Scientists' subscription to norms of research. *Journal of Higher Education*, *81*(3), 366–393.

Baker, M. (2016). Reproducibility crisis. *Nature*, *533*(26), 353–366.

Budroni, P., Burgelman, J.-C., and Schouppe, M. (2019). Architectures of knowledge: The European Open Science Cloud. *ABI Technik*, *39*(2), 130–141.

Daneshjou, R., Smith, M. P., Sun, M. D., Rotemberg, V., and Zou, J. (2021). Lack of transparency and potential bias in artificial intelligence data sets and algorithms: A scoping review. *JAMA Dermatology, 157*(11), 1362–1369.

Delgado-Rodríguez, M., and Llorca, J. (2004). Bias. *Journal of Epidemiology & Community Health, 58*(8). doi: 10.1136/jech.2003.008466.

Durante, M. (2021). *Computational power: The impact of ICT on law, society, and knowledge*. Routledge.

Ebers, M., Hoch, V. R. S., Rosenkranz, F., Ruschemeier, H., and Steinrötter, B. (2021). The European Commission's proposal for an Artificial Intelligence Act—A critical assessment by members of the Robotics and AI Law Society (RAILS). *J, 4*(4), 589–603.

Elliott, A. (2012). *Routledge handbook of identity studies*. Taylor & Francis.

Elliott, A. (2013). *Concepts of the self*. Polity Press.

Elliott, A. (2015). *Identity troubles*. Taylor & Francis.

Elliott, A. (2019). *The culture of AI: Everyday life and the digital revolution*. Routledge.

European Commission (2015). A digital single market strategy for Europe. COM/2015/192 final. https://eur-lex.europa.eu/legal-content/EN/TXT/?uri=CELEX:52015DC0192.

European Commission (2020). A European strategy for data. COM/2020/66 final. https://eur-lex.europa.eu/legal-content/EN/TXT/?uri=CELEX%3A52020DC0066.

European Commission (2022). Proposal for a Regulation of the European Parliament and of the Council on the European Health Data Spaces. COM/2022/197 final. https://eur-lex.europa.eu/legal-content/EN/TXT/?uri=celex%3A52022PC0197.

Fanelli, D. (2018). Opinion: Is science really facing a reproducibility crisis, and do we need it to? *Proceedings of the National Academy of Sciences, 115*(11), 2628–2631.

Felt, U. (2016). Making knowledge, people, and societies. In U. Felt, R. Fouché, C. A. Miller and L. Smith-Doerr (eds), *The handbook of science and technology studies* (4th ed.). MIT Press.

Floridi, L. (2014). *The fourth revolution: How the infosphere is reshaping human reality*. Oxford University Press.

Foucault, M. (1988). Technologies of the self. In *Technologies of the self: A seminar with Michel Foucault* (Vol. 18).

Franssen, M., Lokhorst, G., and van de Poel, I. (2018). Philosophy of technology. In E. N. Zalta (ed.), *The Stanford Encyclopedia of Philosophy*. https://plato.stanford.edu/archives/fall2018/entries/technology/.

Gaius (n. d.). *Institutiones*. (I), 9–12.

Gilbert, M. (2013). *Joint commitment: How we make the social world*. Oxford University Press.

Greene, B. (1999). *The elegant universe: Superstrings, hidden dimensions, and the quest for the ultimate theory*. W. W. Norton & Company.

Hofmann, H. C. H. (2021). An introduction to automated decision-making (ADM) and cyber-delegation in the scope of EU public law. *University of Luxembourg Law Working Paper 2021–008*.

Holton, R., and Boyd, R. (2021). "Where are the people? What are they doing? Why are they doing it?" (Mindell) Situating artificial intelligence within a socio-technical framework. *Journal of Sociology, 57*(2), 179–195.

Hutson, M. (2018). Artificial intelligence faces reproducibility crisis. *Science, 359*(6377), 725–726.

Jasanoff, S. (1996). Beyond epistemology: Relativism and engagement in the politics of science. *Social Studies of Science, 26*(2), 393–418.

Jasanoff, S. (2015). Future imperfect: Science, technology, and the imaginations of modernity. In S. Jasanoff and S.-H. Kim (eds), *Dreamscapes of modernity: Sociotechnical imaginaries and the fabrication of power*. University of Chicago Press.

Knight, W. (2016). How to fix Silicon Valley's sexist algorithms. *MIT Technology Review, 23*.

Kuhn, T. S. (1962). *The structure of scientific revolutions*. University of Chicago Press.

Leonelli, S. (2023). *Philosophy of open science*. Cambridge University Press.

Lior, A. (2023). Private and academic AI collaboration: Opportunities and challenges to open science in the US. In L. Paseri (ed.), *Open science and data protection* (pp. 1–14). *Journal of Open Access to Law, 11*(2).
McCarthy, J., et al. (2006). A proposal for the dartmouth summer research project on artificial intelligence, august 31, 1955. AI magazine *27*(4), 12–12.
Mede, N. G., Schäfer, M. S., Ziegler, R., and Weißkopf, M. (2021). The 'replication crisis' in the public eye: Germans' awareness and perceptions of the (ir)reproducibility of scientific research. *Public Understanding of Science, 30*(1), 91–102.
Merton, R. K. (1973). *The sociology of science: Theoretical and empirical investigations*. University of Chicago Press.
Miyakawa, T. (2020). No raw data, no science: Another possible source of the reproducibility crisis. *Molecular Brain, 13*, 24. doi: 10.1186/s13041-020-0552-2.
Moro, L., Paris, M. G. A., Restelli, M., and Prati, E. (2021). Quantum compiling by deep reinforcement learning. *Communication Physics, 4*, 178. doi: 10.1038/s42005-021-00684-3.
Oliver, K., Krieger, K., and Melchor, L. (2022). Assessing national institutional capacity for evidence-informed policymaking: The role of a science-for-policy system. Publications Office of the European Union. https://data.europa.eu/doi/10.2760/951556.
Pagallo, U. (2022). *Il dovere alla salute: Sul rischio di sottoutilizzo dell'intelligenza artificiale in ambito sanitario*. Mimesis.
Pagallo, U., and Barfield, W. (2020). *Advanced introduction to law and artificial intelligence*. Edward Elgar Publishing.
Paseri, L., Varrette, S., and Bouvry, P. (2021). Protection of personal data in high performance computing platform for scientific research purposes. In N. Gruschka, L. F. C. Antunes, K. Rannenberg and P. Drogkaris (eds), *Privacy technologies and policy*. Lecture Notes in Computer Science, vol. 12703. Springer.
Porsdam, H., and Porsdam Mann, S. (2021). *The right to science: Then and now*. Cambridge University Press.
Regulation (EU) 2021/695 of the European Parliament and of the Council of 28 April 2021, *establishing Horizon Europe—The Framework Programme for Research and Innovation, laying down its rules for participation and dissemination, and repealing Regulations (EU) No 1290/2013 and (EU) No 1291/2013*, OJ L 170, 12.5.2021. http://data.europa.eu/eli/reg/2021/695/oj.
Ritchie, S. (2020). *Science fictions: Exposing fraud, bias, negligence and hype in science*. Penguin Books.
Ritson, S., and Camilleri, K. (2015). Contested boundaries: The string theory debates and ideologies of science. *Perspectives on Science, 23*(2), 192–227.
Serres, M. (2001). *Hominescence*. Le Pommier.
Serres, M. (2009). *Temps des crises*. Le Pommier.
UNESCO (2021a). UNESCO recommendation on open science. https://unesdoc.unesco.org/ark:/48223/pf0000379949.locale=en.
UNESCO (2021b). Recommendation on the ethics of artificial intelligence. https://en.unesco.org/artificial-intelligence/ethics.
Wilkinson, M. D., et al. (2016). The FAIR guiding principles for scientific data management and stewardship. *Scientific Data, 3*(1), 160018. doi: 10.1038/sdata.2016.18.

J. Michael Innes and Ben W. Morrison

5 Social Identities and the Roles of Artificial Intelligence

Abstract: We examine a key social concept in the field of social psychology—shared social identity, namely how a person identifies their membership of social groups to provide a sense of purpose and position in society—and relate that to the development of artificial intelligence systems that increasingly influence the nature of work and employment and knowledge of social and political systems. The presence and influence of other people, and the effects of exposure to social information mediated by institutions, government and organisations, affects the sense of being similar to and also being able to differentiate oneself from other people. We examine developments in the digital world impacting upon the social and political values and actions of the individual, especially regarding the sharing of information and also the restriction of access to information for members of a group. We look at the roles of membership and leadership in forming and changing social identity.

Keywords: artificial intelligence, social identity, media, leadership

5.1 Introduction

In this chapter we examine a key social psychological concept—shared social identity—and relate that to the development of the digitisation of information and media and of artificial intelligence (AI) systems that impact upon everyday lives and influence the nature of work and employment. The concept that a person identifies their membership of a group or groups in order to provide a sense of identity, purpose and agency is central to understanding the human being.

We can place the concept of identity into a broad context by considering the importance of the idea in literature and the ways that the novelist constructs characters to demonstrate the centrality of human identity and its role in coping with adversity. In the work of Nobel Laureate Kazuo Ishiguro, his main characters, for example the butler, Stevens, in *The Remains of the Day*, develop identity to deal with trauma, and Ishiguro expresses this through an exploration of psychological defences. In his analysis of the character, Ishiguro introduces ideas commensurate with psychoanalysis, with the demonstration of how low affect and a weak ability thereby to form relationships evolve from the maintenance of the defences. It is interesting in the present con-

J. Michael Innes, University of South Australia, Australia
Ben W. Morrison, Macquarie University, Australia

https://doi.org/10.1515/9783110721751-005

text to note that Ishiguro, in his evolving opera of work, *Klara and the Sun*, has explored the identity and relationships of an artificial person. We will provide a social psychological analysis of social identity in relation to the modern technological and social environment.

The social environment of a person, the presence and direct influence of other persons, and the mediated effects of exposure to social information, affect the growth of social identity, the sense of being similar to and sharing the stereotype of being a member of a group, and also being dissimilar to others. We shall extend the analysis of the social environment to examine developments in the digital world, with resultant impact upon the social and political acts of the individual. We shall also explore the value of sharing information and restricting access to information for members of a group and how that is affected by the digitisation of information and the explosion of information in social media.

The major intellectual influence on our understanding of identity in this chapter stems from the discipline of social psychology. In this process we adopt a cognitive approach to an understanding of social identity, to relate the social to the artificial mechanical world. However, reference will be made to political science and sociology, to understand influences on social and political behaviour and how that is related to the understanding that people have about artificially intelligent systems. First, we shall examine a philosophical context that will enable us to relate two separate traditions, the social psychology of identity and the psychology of artificial intelligence.

5.2 A Philosophical Context

In 1942 the philosopher S. C. Pepper wrote a book on *World Hypotheses* (Pepper, 1942). In this he presented an analysis of four hypotheses, or *World Views*, that have developed in human thought over time in attempts to understand the world and universe. We shall emphasise two of these hypotheses to understand the evolving relationship between humans and artificially intelligent systems. These hypotheses derive from 'root metaphors', common sense conceptualisations of events which in turn provide the structure for an understanding of technical analyses of those events.

The four hypotheses of Pepper include *Formism*, whereby plants, animals and physical structure can be understood in terms of the similarities and differences in form and shape that exist in the observation of them. Another hypothesis is *Organicism*, namely that understanding of actions can be understood via the metaphor of a total living system where all parts of the system interact and feed upon each other. The understanding of the whole is necessary to understand the parts; the parts do not have meaning without being part of that whole.

We shall concentrate on two other hypotheses for our present purposes. The first, *Mechanism*, embodies the view that living things can be understood by analogy with our understanding of the actions of machines. In a basic understanding of a machine, the linkages between separate parts of the machine act to transmit energy from one part of the machine to another, eventually to provide an outcome which affects the world external to the machine. While we may in our present civilisation conceive of complex machines such as aircraft, internal combustion engines and now robots, the concept of a machine goes back millennia to the creation of systems to provide transport (the wheel), to conduct water over long distances and shoot missiles from siege machines.

The second hypothesis, *Contextualism*, proposes that we understand actions through an analysis of the texture and context of persons, acts and events that surround the person or the event that is being observed. This takes a historical stance; at one point in time a confluence of events may occur to produce an effect that is unique, different from the confluence of the same group of events which may occur at a later time. The hypothesis requires an understanding of the interaction of many events to understand a cause-and-effect relationship. This hypothesis is clearly exemplified in the work of Mackie (1974), where he considers the need to understand the framing of clusters of events, which are *sufficient* to cause an outcome, but in which no cluster is *necessary*. Framing, or the identification of context, is essential in understanding how we identify hypothetical cause-and-effect relationships.

Each of these *root metaphors* has risen and declined in influence upon thought over time. For our current concerns, the mechanistic metaphor has in the past been dominant in the discipline of psychology. The rise of stimulus–response learning theory in behaviourist terms (Koch, 1964), the cognitivist approach—where models of cause and effect are explained in terms of modules of actions and pathways between modules to understand systems of memory, the solving of problems and the making of judgements and decisions (Broadbent, 1958)—and also the development of psychoanalytic thinking—in which a metaphor was derived from plumbing, to understand the homeostasis of motivational systems—all relied on a basic metaphor derived from an understanding of machines.

In recent years the contextual metaphor has been dominant in the field of social psychology, to enable an understanding of the complexity of causal influences on human and animal social behaviour (Rosnow and Georgoudi, 1986; see also Guerin, 2016). Any event which occurs which may influence a human thought or action may in turn be tempered by the occurrence of another event or events which moderate the action. In thinking, as a social psychologist, about cause and effect in human behaviour, there is always the need to add the proviso that ‘it depends’. Other variables must be accounted for in understanding how events interact to produce outcomes. This approach has also been emphasised in the work of McGuire (1999) in his perspectivist approach, both to the construction of theory and the planning and execution of experimentation.

This metaphor has also been influential in other ways in psychology, namely in the analysis of behaviour, associated with the work of B. F. Skinner (Skinner, 1963).

While this sub-discipline is overtly similar to the study of stimulus–response learning theory, it has had a separate and very influential impact within psychology and now has experienced a resurgence in the analysis of artificial intelligence systems. The concept of the operant, fundamental to behavioural analysis, is defined as the relation between the behaviour of an organism, events in the environment which provide a context for the enactment of an action, and the consequences of that action for the organism that together create the behavioural act and the effects on the environment which is changed by the action. These events must be understood as a system and therefore carry the sense of contextualism. This extension of Pepperian thinking to behaviour analysis is fully developed in Hayes, Hayes and Reese (1988). What is of interest, in our present thinking, is that the contemporary rise of artificially intelligent systems, based upon deep learning processes, as we shall explore later in this chapter, are not based upon cognitivist/mechanistic systems as one might expect, but upon systems of stimulus and reward as developed in Skinnerian thinking (Sejnowski, 2018).

For our analysis, the combination of the analysis of social identity and the role of artificial intelligence systems, both developed within a contextualistic metaphor, allows a comparison to be made between systems, which Pepper (1942) proposes is not possible across metaphors. This idea of incommensurability of concepts and how to overcome it arises also in Kuhn's thinking about the role of paradigms in the process of 'normal' science (Kuhn, 1962). Pepper's analysis, however, delves more deeply into the nature of the metaphors which form the basis of understanding the structure of events and the methods and theories that are constructed to study them.

We are aware that the *organic* metaphor is powerful in the biological sciences and is currently being developed further within thinking about global macro systems such as the study of meteorological events and climate change. Lovelock (2019), whose work on the Gaia Hypothesis, which conceives biological and geological systems and their mutual evolution to impact upon the creation and survival of human life, is fundamentally an organicist. He has extended the concept to consider the development of the *Novacene*, with the evolution of hyperintelligent systems, extending human behaviour and existence beyond the Anthropocene. The idea that hyperintelligence may be organic, with the parts only understandable within the context of the whole, is clearly an exercise worthy of further development. We do not extend to an analysis of that metaphor in the present chapter with limited length.

We shall first consider contemporary views of artificial intelligence systems from the point of view of the discipline of psychology and how they are influencing ways of thinking about employment and everyday life. We shall then examine features of the social psychological concept of social identity, its historical development and the processes which are hypothesised to lead to the development of identity in a social context, and link this contextualist view of understanding of social identity to the development of AI. We shall then consider the synergistic effects that occur with the interaction of the two systems related conceptually to the same root metaphor.

5.3 The Modern Development of Artificial Intelligence

In this analysis we adopt our analysis of the relationship between the development of AI and the broad discipline of psychology presented earlier (Innes and Morrison, 2017, 2020; Innes, Innes and Morrison, 2022; Morrison and Innes, 2022). The modern era of AI is concerned with a technology of 'machine learning'. Machine-learning systems are presented with a series of examples of the things/objects/concept that they are trying to learn. This allies the machines with the earlier concepts of human psychology in learning theory, not with the stimulus–response theories, but rather with the work of radical behaviourists, such as Skinner (1963), to whom we have referred.

How can the machine/organism learn to recognise an object or concept? The idea of 'reinforcement' presents the system with feedback that a decision was incorrect or correct, to increase the probability of making the correct decision in the future. This is the idea of contingencies of reinforcement and the experimental demonstration that complex action sequences can be explained by exposure of the organism, or learning mechanism, to varied structured sequences of consequences (Ferster and Skinner, 1957). The processes of contemporary AI systems, based upon principles of learning and contingent reinforcement, are based on behaviouristic views of operation. Sejnowski (2018) notes the explicit reference to the work of Skinner.

Machines, given basic principles, are exposed to many diverse stimuli on multiple occasions and can develop outcomes that both exceed human capacity and yet provide no insight into how they arrived at the outcomes. A particularly relevant example comes from the study of the chess-playing program 'AlphaZero' (Sadler and Regan, 2019). The program, which learns to play chess given only the rules of the game and guided by reinforcement-based learning (bolstered significantly from unsupervised 'Self-Play') demonstrates human-like tactics in playing the game. For instance, the program plays sacrifices and employs very unusual moves, more typically associated with grandmasters, than the knowledge-based programs. Thus, an emergent system has developed beyond the constraints of the initial program. Indeed, AlphaZero is said to have achieved a 'superhuman' level of play within 24 hours of training.

The development of such systems appears to enable the systems to go beyond the limits of knowledge-based systems as suggested by, for example, Rasskin-Gutman (2009). Sejnowski (2018) and Rahwan et al. (2019) give extensive accounts of machine behaviour in their recent reviews. The success of machine learning depends upon the opportunity to present vast numbers of examples to the systems and construct contingent rewards for the execution of selected actions. For instance, in the first nine hours of its life, the AlphaZero program had played itself around 44 million times.

It should be noted that a model based upon a contextual analysis to explain creative thought, the 'blind variation and selective retention' of thoughts and actions, was proposed by Campbell (1960) and has received significant empirical support (e.g., Simonton, 2012). The relationships between creativity and associative processers have also been noted by Mednick (1962), and have been validated in studies of AI systems (Olteteanu and Zunjani, 2020). Creative thought can emerge from such systems of actions and consequences.

The development of reinforced-based, 'deep' learning systems is highly advanced mathematically and involves the analysis of sequences of acts which are contingent not only upon the external reinforcers, but also on the continually altered states of the operator (human or machine) which have been affected by prior reinforcers, Markov Decision Processes (MDPs) (cf., Graesser and King, 2020). The systems also take into account another feature of the operator within the system, namely the difference between the external totally observable influences (by an external, wholly knowledgeable observer) and the partially observable (PO) process available to the operator itself within the system (POMDPs). This is a mathematically formulated image of a person by system (PxS) interaction in psychological research that has driven experimental research for decades.

5.4 Complexity, Opacity and Bias of Algorithms: The Question of Trust

A feature which emerges from such developments is that the outcomes of the programs surprise the people who created them. This is the notion of 'explainable artificial intelligence' (XAI), that in understanding the outcomes of the AI systems there needs to be a means whereby humans can excise and examine the processes which lead to the outcomes. Bridle (2018) suggests that the 'fourth law of robotics', an extension of Isaac Asimov's three laws of robotics, should be that the robot must be open to examination by a human. The systems need to be transparent and open to explanation, to allow trust to evolve by humans in the operation of the automated systems. For instance, the uptake of algorithm-generated advice has been widely found to be determined by its relative degree of transparency in formulating a response (Dietvorst, Simmons and Massey, 2015; Dietvorst and Bharti, 2019; Seong and Bisantz, 2008; Sinha and Swearingen, 2002; Yaniv, 2004; Yaniv and Kleinberger, 2000; Yeomans et al., 2019). Indeed, Yeomans et al. (2019) found that in a task where humans would be expected to outperform artificial systems (predicting which jokes people find funny), such systems still outperformed humans. Aversion to the system was observed partly due to users' belief that they understood the human process better (Yeomans et al., 2019), even though the output was seemingly sophisticated.

We distinguish different senses of the term ‘opaque’ to use it in the sense of an emergent characteristic of machine learning and not something deliberately built into the systems (Burrell, 2016). And we follow a suggestion by Seaver (2017) to adopt an anthropological method, an ethnography of systems. This raises an analogy with the developments of behaviourism in psychology in the 1940s and 1950s when there was debate about ‘intervening variables’ and ‘hypothetical constructs’, unobserved constructs used to explain links between the external stimuli and the observed output (MacCorquodale and Meehl, 1948). The experimental methods used then may be reconsidered to interrogate the workings of the robotic mind. It is not too far a stretch to suggest that we, as investigators of the complexities of the robotic mind, should pay attention to the attempts by ethologists to understand the animal mind (De Wall, 2016). It may be that we may be better able to understand biologically based systems, derived from evolutionary pasts (cf., Campbell, 1994; Scerri, 2016), but we need to remember that the minds of robots capture the thoughts and insights of human beings, and hence an understanding of human insights may enable insights into the robot. The work of Campbell (1963) in his exposition of the nature of human dispositions, invoking the notion of the human homunculus, an idea which emulates the concept of the artificial person or robot simulating or representing behaviour and consciousness, deserves a re-evaluation in the present era. Robots acquire ‘behavioural dispositions’ and should be examined in that light just as are humans, about whom the investigator knows much less regarding history and environment than in the case of a robot. The work of Yeomans et al. (2019) referred to above suggests, however, that humans, in the context of interacting with a robot, apply different levels of proof in making a decision.

We should remember, also, that complex human behaviour at first may appear to be opaque to understanding, but that comprehension may be achieved by experiments which disentangle the complex interaction of variables (cf., Ferster and Skinner, 1957). Bem 1965 demonstrated that such experimental procedures may be conducted by the person upon themselves in order to ascertain their states of belief based upon their actions which may be executed within conscious awareness. The actor, while apparently introspecting, is acting as an observer would in examining the behaviour of another person and inferring the thoughts of that person.

The complex algorithms of AI systems are created by humans. The opening of them to scrutiny, therefore, may be achieved by experimental methods augmented by human understanding of human thought processes, not by self-reflection but by previous work with non-humans, namely animals. If the programs, written by humans, digitize the thought processes and the biases of those humans, it would be fruitful to explore the cognitive belief systems and socio-economic ideologies which are present.

There is clear evidence that the algorithms developed encapsulate the beliefs of the programmers (e.g., Broad, 2018; Muller, 2018; O’Neil, 2016; Russell, 2019; Smith, 2019). The exploration of the intrinsic and subtle biases of human experts, which are based upon 'common-sense' knowledge and socio-political beliefs, using technologies from the social sciences, including social psychology, could clarify where the programs 'come from'. We have known for years about the intrinsic biases of human 'experts' (e.g., Innes and Fraser, 1971; Kahneman, 2011) and, furthermore, about the opacity of human decision-making processes to conscious awareness (Guerin and Innes,1981; Nisbett and Wilson, 1977; Newell and Shanks, 2014). The same technical 'opacity' can be revealed, however, through experimental investigation and the relationship between the behaviour and the causal environmental factors demonstrated. There is a central role for psychology and the use of psychological methodology in revealing the processes embedded in the development of AI systems.

We note here the suggestion by Miller (2019) that the ability to explain algorithmic processes may be facilitated by recourse to social scientific research into the ways in which it is possible to question people to elucidate their explanations for their own behaviour and the behaviour of others. This is essentially an extension of the Turing test (i.e., a test of a machine's capacity to demonstrate intelligent behaviour indistinguishable from that of a human; Turing, 1950) to demonstrate the ways in which an experimental understanding of social behaviour can be used to establish the humanity of the agent. Miller's review does not account for the failure of people to introspect and reveal the processes whereby cognitions arise into consciousness, but his can be explored by experimental means (Nisbett and Wilson, 1977).

We need to recognise recent developments in psychology, however, where criticisms are presented about the ability of experiments to identify cause-and-effect relationships (Cook, 2018; Diener et al., 2022). Previous work by us on the methodology of experimentation in human–robot interaction points in a similar direction (Innes and Morrison, 2020). In attempting to understand the work on AI and how it relates to human behaviour, much of the experimental work in the field contains fundamental flaws in operation. Methodological lacunae include a failure to acknowledge the expectations and biases of the investigators in creating their experiments, the expectations (demand characteristics) of the participants in the experiments as to what they have to do and think in their interactions, and also the general and pervasive social and political thought that embraces all of the contemporary work on AI and what it can achieve (Innes and Morrison, 2020). The adoption of a more ethnographic approach to collecting data, with less emphasis upon artificial controlled experiments, may be needed to delve into the complexity of algorithmic systems, as proposed by Seaver (2017).

5.5 Psychology and Living with AI

The rapid development of AI brings with it increasing deployment of robots into places where human beings have previously been the sole or principal occupant. Interaction with robotic systems in the office, factory and home is becoming ubiquitous (Walsh, 2018) and humans will be expected to interact and cooperate with these systems in everyday life. There are many studies which are relevant, including those which refer to theories and methods in disciplines other than robotic studies (e.g., Young et al., 2009).

This interaction goes beyond traditional human–computer interaction, whereby humans are aided by the efficiency of the processing systems of the computer. The interaction will take the form of physical and mental interaction with a responsive 'other', which will add to the outcome of the interaction over and above the capacity of humans in the system. Indeed, the robot may make significantly different inputs, changing the direction of the course of the interaction. The quality of such interaction, however, will depend upon the ability of humans to adjust to interaction with non-human artefacts, and the ability of designers to manipulate the robot to behave in a manner that will be acceptable to the human.

The modern issue of machine–human interaction goes beyond the age-old relationship between humans and their tools. Robotic systems engage a new sense of relationship with the human: how does the human being represent the robot? It is not merely an attitude to the robot, a sense of liking or disliking the machine with an influence on behaviour. The representation of the robot represents more a collective sense of relationship, a relationship of a human being shared with other human beings' sense of what the robot is and what it means.

Commentators point out that the design of artificial systems starts as a human endeavour, embedding within them human values, biases, preconceptions and intentions (Broad, 2018). The initial interaction of the human with such systems, therefore, involves the interaction of a human with another human, mediated by the embodiment in the robot. However, the evolution of the system, as we have seen, will eventually result in an emergent organism/machine participating in the process. The question arises: how can we explore the nature of such emergent and developing interactions so that they may be beneficial and productive and continue to be acceptable to the human participant?

The AlphaZero program provides evidence of how changes may emerge in interaction with humans. But the rules of chess are transparent, simple, and can be defined in a program relatively easily. Social interaction, on the other hand, has no clear rules and depends heavily on very subtle cues and responses. How do designers acquire information about social interaction that enables them to program the relevant cues and responses to enable the interactions to develop? The decomposition of such intuitive processes, and the elicitation of valid cues, have proven challenging (Klein, 2015; Wiggins, 2021).

We consider the fundamental idea of 'common sense' in understanding of scientific and engineering change. There is the common sense of the public coming to understand what robots represent, essential to the acceptance of AI into business, commerce and education. We need to understand the common sense, or tacit knowledge (Collins and Evans, 2007 term this 'ubiquitous tacit knowledge'), of the experts who design the systems. This tacit knowledge is in addition to the explicit, technical knowledge which is used in the design and implementation of the systems. Engineers may be unaware of the influence that this may have in the design process. There are many who conceive of the role of common sense as fundamental to all knowledge systems (Campbell, 1975; Markova, 2014). Finally, there is the common sense of those who design the experiments to explore the factors which affect the interactions of robots with humans and cause the outcomes to provide the data that is entered into the algorithms which guide the machines.

Numerous experimental studies of interaction between humans and robot systems continue to demonstrate that humans can come to enter in trusting relationships with artificial intelligence systems, dependent upon the nature of the task involved (e.g., Jercic et al., 2018), the learning environment in which the interaction evolves (e.g., Stephan, 2015) and extent to which the robot during the interaction appears to demonstrate human-like characteristics (e.g., Kim and Duhachek, 2020). And further exploration of human behaviour using developing methodologies shows that artificial intelligence systems can be provided with finer-grained information that enables the systems to emulate human behaviour, such as in facial recognition systems (Cowen and Keltner, 2020). Humans adapt to the behaviour of the machine as well as the machine being programmed to adapt to the needs of the human, as shown in machine systems which can help humans learn to empathise (Shamay-Tsoory and Hertz, 2022).

5.6 Broader Social Effects Through Digitisation of Information

In considering the effects of artificial intelligence upon human behaviour, we broaden the scope of coverage beyond the possibilities of interaction between humans and individual robots or artificially intelligent devices. The ubiquity of the digitization of information affects humans across the entire range of activity, from the home, to leisure, to work. A discussion of the extent of these influences is beyond the scope of this chapter, but we can examine some features of the influence as they impact upon the focus of our analysis on the nature of social identity. Zuboff (2019) has described the 'fight for a human future' under the gaze of surveillance capitalism. Surveillance systems at work to monitor the activities of employees are having an impact on the well-being of those employees. There are documented effects of social media on the health and well-being of adolescents (Hamilton, Nesi and Choukas-Bradley, 2022). And the 'meme-ification' of politically oriented news and information has been cited in the rise of political extremity in Western liberal democracies (e.g., Khalil, 2022). The adoption in the Republic of China of ubiquitous methods of surveillance and the monitoring of social behaviour, either in real-life settings or electronically, are noted with concern for their use and the invasion of human rights and privacy (Babones, 2020). These effects can all be conceived to have an influence through the medium of social identity of the recipients and targets.

It should be noted at this point that the developments in electronic means of surveillance and monitoring are mimicked by human agents who use the same means of access to 'hack' human systems and use the means whereby people can persuade people and influence their behaviour to subvert systems designed to protect humans from machine interference. The human agent uses human agency to subvert machine intelligence (Cialdini, 2021; Hadnagy, 2018).

The 'tsunami' of information through digitization and the development of social media (Facebook, Tik-Tok, Twitter) has been credited with changes, anxiety, depression, conduct disorder and other psychological states, especially in adolescents and young adults (e.g., Hamilton, Nesi and Choukas-Bradley, 2022). We do not take a view that these developments, in themselves, are a cause of the observed effects. We take a contextual position, that there are pre-existing psychological conditions and processes that render the human participant liable to being influenced. In social psychological terms we are examining the interaction between the characteristics of the person, the nature of the situation which creates the medium of transmission, and the situational factors in which the content is embedded. Influence depends upon the context created by the person and the setting (cf., DeRidder, Kroese and van Gestel, 2022).

We now turn to the analysis of what is considered to be a ubiquitous characteristic of humans as social beings, namely the search for and adoption of an identity as a person. This identity is based upon relationships with other human beings and a shared understanding of social and physical reality that facilitates social interaction and communication. The media of communication also contribute to the nature of information which is available and which can affect the identity of the recipient.

5.7 The Concept of Social Identity

The core of the concept of social identity, as defined by social psychologists, is that people have affiliation and membership of social groups which act to define their understanding of themselves and other people within the same and other groups. It is an essentially contextual or framing concept whereby the actions and thoughts of an individual must take account of the social relationships which embrace that person and upon which that person's actions have an impact. It is related to the concept of personal identity, but emphasises the role of relating to people who are similar and dissimilar to the person in forging a sense of self and well-being. A person who is making choices about her or his behaviour takes into account their understanding and evaluation of the thoughts and actions of other people to whom they relate, whether that relationship be positive and friendly or negative and hostile. It can be argued (e.g., Guerin, 2016) that even when alone, we are thinking and behaving in ways that were shaped by the consequences of interaction with the behaviour of others. We need to understand developmental processes as shaped by social relationships. Such behaviour may be taken for granted; our dependence upon others and our experiences over time of that dependence may be so much a part of everyday life that the person may not be consciously aware of them.

Social identity and social identification have long been the object of study in the discipline of social psychology, going back to the beginning of the twentieth century. A close analysis of social identity is provided by Farr (1996) in his history of the roots of development of social psychology and in Smith's (2003) more autobiographical account. The concept of social identity was for long a central factor in the development of a *sociological* social psychology. The seminal work on the sociology of the American soldier (Stouffer et al., 1949) at the end of the Second World War, for example, demonstrated the pervasiveness of the concepts in the understanding of how individuals learnt to behave in groups and to adjust to stress and trauma. The formation of cohesive groups, the identification of an individual with the group, and even the 'over-identification' between a leader and the members of the troop as a pathological form of social identity, were central to an understanding of behaviour in combat.

The discipline of a *psychological* social psychology took an especially significant turn later, when the concept of social identity was identified by the Polish/English social psychologist, Henri Tajfel. He introduced his theory of prejudice which emphasised an essentially *cognitive* representation of group membership as a fundamental factor which affected relationships between persons within and between social groups (Tajfel, 1969a, 1969b). Tajfel's influence upon the development of an understanding of social identity was truly seminal and its development demonstrated the intrinsic relationship between the understanding of a socially scientific concept and the lived experience of the theorists and experimentalists who explored the concept. To understand the contextualist metaphor which educates our understanding of the concept of social identity it is helpful to describe the experiences of Tajfel and the effect that they had on his intellectual understanding of social behaviour.

Tajfel was born in Poland and moved to enrol in a French university because of antisemitic sentiment in Poland. He was conscripted into the French army at the beginning of World War II and was captured and interned in a prisoner-of-war camp by the Germans after the surrender of France. He survived as a Jew under extraordinary circumstances, continually hiding his Jewish identity. His personal experiences, surviving persecution and managing an identity to escape death, provided a lived experience which created social identities to enable him to survive in a hostile social environment (Brown, 2020). He remained in France after the war and eventually moved to England and then to the United States to study social psychology.

Comments on the influence of Tajfel, provided by the psychologist/philosopher Rom Harre, capture the fundamental depth of the concept needed to understand social identity and also demonstrate the extent to which people may not be aware of the influence that living has on theorising. Harre writes,

> The research interests of psychologists all too often seem to be dictated by problems that they face in their own lives. Henri Tajfel, well known to social psychologists, moving from his native Poland to France and then to England, spent his life struggling to understand how people came to form social identities. *What was an artefact for him was 'second nature' for most of us.* (Harre, 2009, p. 163; italics added).

Tajfel had endured an existence in which his life was continually at risk should his 'true' identity be detected. He had multiple identities in order to survive. They were not 'second nature', as stated by Harre. Created in extraordinary circumstances, that creation needed to be examined. Many (most?) people do not experience social situations in which deception and the distortion of the self is necessary. But living that life demonstrated how pervasive separate and related social identities can be in everyday life. One can remark, in reflection on this commentary, that Tajfel was the true social scientist; his self was an 'artefact' that he had had to construct in ways that are true for all people. He was, however, more mindful of the process because of the circumstances in which it was created.

5.7.1 The cognitive theory of social identity

Tajfel's starting point was to develop a cognitive theory of prejudice, whereby the representation within the mind of a person or groups of people and the differences between them could result in differential acts towards members of those groups. This was a different approach to the study of prejudice which had usually emphasised an emotional basis for a positive or a negative reaction to individuals (Tajfel, 1969a, 1969b). Tajfel introduced the 'minimal group paradigm' (Tajfel et al., 1971). In the initial study, English schoolboys were randomly instructed to be part of a group. There was no basis for membership of either group other than a name given at random. They were then instructed to play a competitive game in which they could allocate rewards either to members of their own group or the other group. With no basis other than a different label upon which to make a judgement, the boys allocated more reward points to their own members compared with members of the other group. This differential treatment of one's own social group and identity compared with another demonstrated a basic tendency to develop a sense of social identity formed from the membership of a group. It also demonstrated that the establishment of a 'group identity' could be made rapidly, by instruction. It did not require an extensive interaction with other people over time. Neither was there an establishment of a strong affective component to ensure the development of the linkage.

This cognitive view of the representation by the person of the relationship with the 'group' demonstrates the notion that groups are *entities* (Campbell, 1958). Groups, as structures within which a person acts, have several important characteristics for the person. The person can conceive of themselves as having a present, they can remember having a past and they may be able to conceive of a future. Their actions may be based on past experiences, they have impacts upon their own outcomes and those of others around them, and they may have repercussions later. The group to which they believe they belong also has a past, present and future. The actions of other members rebound onto other members. The importance of the group identity is twofold. First, there is a shared identity whereby the person can perceive and believe that there are elements shared with other group members and render the person different from members of other groups. Secondly, there can be a shared assumption that the information that is available to members within the group may be accessible only to members of that group or entity and not shared with others beyond the group (Bingley, Greenaway and Haslam, 2022). This shared information may take the form of rules, confidential knowledge about others (in professional societies, for example) and even gossip about others and the self.

We are, however, surrounded by people and we interact over time in myriads of events and occasions. The question arises; when are these interactions related to the groups in which we believe we are members and who are the relevant members and non-members? Who is deemed to be similar to us, and a member of our group, and who is dissimilar and therefore not a member? We can also relate to the social psychological research on shared reality, the processes of means communicating about the perceptions of the world. The process of communication, of 'tuning' the message to make it sensible to the other person, affects the way the content is understood (Echterhoff and Higgins, 2018; Guerin and Innes, 1989). These processes can be influenced by the immediate presence of others (Blascovich et al., 1999; Guerin and Innes, 1982, 1984; Kaltwasser et al., 2019). They are also increasingly affected by the vicarious experience of exposure to mediated reality, and the development of social media and the burgeoning Metaverse has further accelerated the process (Couldry and Hepp, 2017).

5.7.2 The development of social categorisation

This concept of a social identity, linked to a cognitive representation of that concept, depends upon the process of categorisation. We can go back to the development of Gestalt Psychology and consider the perceptual processes that create entities in visual perception, where there are contextual effects that create 'groups' in our perception of the world. Should events occur *contiguously* in time and space, then they may be perceived as belonging together. They involuntarily form a group. Events which are perceived as *similar* are also seen as having a sense of 'groupness'. Similarity as a dimension can be created conceptually; it need not have a representation in sensory space.

The process of categorisation is the process of creating similarities between some events, perceptions, and dissimilarities with others, even when there may be similarities between the two sets of events and dissimilarities within the sets. As Bacharach (2006) has pointed out, making judgements about similarity and dissimilarity between objects and persons may require the *creation* of dimensions upon which to align the objects for comparison. The process of categorisation accentuates the similarities and the dissimilarities, so that items and events which initially may have been perceived as relatively similar to each other are regarded as dissimilar. As we perceive relationships between physical entities, so we can also perceive relationships between social events, people and organisations. Shared experiences, what has been termed 'common fate', reinforce categorisations that make a group more solid, impermeable to outside influences and subjectively experienced as a member as more cohesive (Campbell, 1958).

Face-to-face interaction may be regarded as a particularly powerful factor to create perceptions, and also feelings of *common fate*. Sharing a common fate of possible death and injury in combat is likely to lead to the creation of an 'identification' with the others in the group (Stouffer et al., 1949). Manufacturing a product, writing a scientific paper, going on a holiday together, playing in a sports team and experiencing graduation together, sharing a common goal or fate, all act to enhance identification and the development of a social identity. But the minimal group paradigm does demonstrate that a group entity can be constructed by a random act of allocation to a group. This process, however, must be based upon a prior history of associations with groups to enable the growth of the concept of 'belonging to a group'.

5.7.3 Self-categorisation theory

Turner et al. (1987) captured the essence of Tajfel's initial theorising. A person identifies with a group to accept a particular belief system or a normative manner of behaving, and perceives themselves as being similar to others in the group and dissimilar from those not in the group. The person, in adopting the group membership, becomes an 'exemplar' of the group and to that degree 'depersonalises' the self. This can lead that person to behave in a manner different from how they would act when alone.

Self-categorisation theory (SCT) has become a powerful influence in thinking about group behaviour and has been developed in various forms into social identity theory (SIT). These theories have been extended from the initial development of explaining differences between groups and social prejudice and discrimination, to the analysis, for example, of leadership and organisational skills, maintenance of health and the treatment of ill-health, and the organisation and execution of sport (Haslam, Fransen and Boon, 2020; Haslam et al., 2005; Haslam, Reicher, and Platow, 2020). Within a group, for example, a leader may emerge who is seen to represent best the properties which distinguish the group from other groups; the leader is the prototypical in-group member.

They must also, according to the theory, translate those properties into actions which make the group continually refresh the principles which have developed.

A powerful example of the use of SIT can be illustrated by the analysis, replication and criticism of the Stanford Prison Experiment (SPE). The SPE, first fully reported by Haney, Banks and Zimbardo (1973), demonstrated that being placed at random into the roles of guards and prisoners in a simulated prison, created extreme forms of behaviour in the participants. The guards became authoritarian and the prisoners submissive. The effects were such that the experiment was ended after only 6 days of operation rather than the 14 days that was initially intended. Zimbardo and his colleagues explained this effect in terms of the power of the social situation to guide and direct behaviour and experience, rather than the characteristics or personality of the participants. The power of the situation has been a central feature of social psychological thought for many years (Heider, 1958).

In their replication of this experiment, however, Reicher and Haslam (2006) found, to the contrary, that the prisoners took the lead roles and the guards were more submissive. Reicher and Haslam explained this difference in terms of the creation of social identities in their study and argued that, rather than manipulating the 'power of the situation' in the creation of the differential roles in the original SPE, the original authors had created the contingencies which led to the emergence of social identities. This apparently stark difference led to much debate and acrimony. A similar adoption of SIT to account for behaviour in another classic experiment in social behaviour may be seen in Reicher, Haslam and Smith (2012) with a similar, and rancorous, debate.

What can be drawn from the research, however, is that the *context* in which the behaviour is created can result in large differences in outcome. Social identity and the consequences of the activation of an identity can result in significantly different outcomes. The social identity is itself a feature of social behaviour which can be *framed* and changed by the structure of information and expectation in the immediate social setting.

5.8 Factors Influencing the Adoption of Social Identity

The adoption of a social identity, dependent upon the link of the person to a group, can be seen to be a function of the perception of the links which are available and accessible. As Bruner (1957) demonstrated and others have followed (e.g., Kahneman, Sibony and Sunstein, 2021), the social setting can stimulate different versions of the social self, such that events which are more salient, or are more readily available in memory, can activate relationships, and will result in different personae being present. We may be members of a work organisation or a cricket club and have different companions and friends in the two settings. The behaviours that we show in each may be different, but also the activation of behaviours associated with one membership may be activated in the other dependent upon external circumstances. Multiple group memberships exist dependent upon the person's gender, job, ethnicity, education, social class or generation, to count merely a few. People may also seek not to share characteristics with others, but rather differentiate themselves from others, to demonstrate self-efficacy (Bandura, 1989) or agency and authorship (McAdams, 2015), and thereby demonstrate their difference and even a sense of uniqueness (Brewer, 1991).

Gender identity has become a singularly powerful feature of political and social discussion, at least in modern liberal democracies. What was regarded as a biologically determined division into male and female has become embroiled in a debate about the possibility of transitioning from one gender into another and whether this transition is a result of a male or female identity being embodied in the other gender. Membership of different generations has become politicised, with those born towards the end of the twentieth century being seen to have been penalised by economic events following the global financial crisis of 2008 (Millennials) as compared with members of the generation born around the end of the Second World War (Baby Boomers) (Duffy, 2021).

Generation identity has become in recent years a powerful factor affecting the spread of information and action about race and migration, in particular the development of the 'Great Replacement', that is, the belief that migration is not due to economic, environmental and geo-political forces but rather that 'elites' within and outside governments are plotting to replace the native ethnic communities in Western nations with migrants from other countries who they can better control (Khalil, 2022). The shared identity of native groups within some countries (France, Germany, United Kingdom) provides the basis for the creation of political actions aimed at the prevention of migration and the development of multicultural societies.

Socio-economic status (social class) has long been a categorisation that has been prominent in political debate and action, and this has been associated with education. The explosion of entrance into universities in many countries has further exacerbated tensions. In some countries, for example Australia and the United Kingdom, where fees for universities were introduced as well as schemes to delay repayment of those fees through the taxation system, the result has been a further economic hit on the later generations of citizens who have a further debt to repay that was not required of earlier generations, already seen to be benefiting from economic changes.

As we have already noted above, the immediate presence of others is a factor in the process of establishing identity. The context of the group and setting plays a role. When the person is interacting with a highly homogeneous group then the social identity may be established with that majority. If, however, the individual is in a more diverse group with a number of minority groupings, there may be pressure upon the person to identify the group or groups that distinctively signal identity. The salience of the minority group becomes a marker of the identity to which the person ascribes

(McGuire, 1999). So, for example, a person in a group in a tertiary educational institution where all the individuals are relatively high achievers and studying essentially similar courses of study, may look to sharing identity with others in the group who are similar on another dimension. Women in engineering courses may be likely to associate with the relatively few other women on the course. Men in a health profession course, training to be nurses, may associate with the other, relatively few, men.

Distinctiveness as a feature in a grouping of people becomes a salient characteristic and attracts attention, and with it also related attributes. The research on cognitive biases which people use in making judgements about people and events shows that these judgements are marked by confirmation of already held beliefs (the confirmation bias); attribution of causes of action to a person rather than to the situation in which the behaviour is emitted (the fundamental attribution bias) (cf., Nisbett, 2015); the judgement that the perceiver themselves is less likely to be affected by external events than others in the group (the third-person effect; Innes and Zeitz, 1988), but also to the perception that other people will be likely to behave in a manner similar to the perceiver's (the false consensus effect).

These judgements and inferences about identity depend upon the context of the signal that comes from others. The salience of the signal is the basis upon which that signal is noted and used. Signalling can be relatively overt or it can be covert. Within a diverse grouping, where a perceiver may be uncertain about with whom they may affiliate and communicate, a covert signal may be preferred, to be picked up in turn by others of the same (covert) identity. With a majority group status, however, overt signalling will be preferred (Smaldino, 2022).

In the case of social class, for example, there may be inferences about a person's performance or capability which arise from the identification of a person's socio-emotional status, based upon signals correlated with social class, but where the inferred status has no actual relationship with performance. Research into the signalling of social class has demonstrated significant differences between members of social classes in the ability to perceive cues that enable members to interact in a meaningful and authentic manner (members of 'lower' socio-economic classes) compared with less authentic and more automatic processes, as displayed by members of 'upper' classes (Kraus and Keltner, 2007; Kraus, Park and Tan, 2017). These non-verbal, paralinguistic and body motions can be codified and recognised by algorithmic devices. The regulation of engagement by lower-class interactors may be so ingrained that there is no conscious attempt to initiate the behaviours and they may be quite minimal and therefore can be considered to be covert, in the sense considered above. But a trained observer (and machine) can be trained to detect the actions and their patterns.

A feature of such research, and the implications that it has for the development of digital and artificial intelligence systems which embody such methods, is to consider the ethical implications of their existence. The use of unobtrusive measures in social psychological research in the 1960s (e.g., Webb et al., 1966) was developed in order to attempt to reduce the biasing effects of expectations by participants in experimental and non-experimental settings that they were being observed. This was later regarded as being highly intrusive upon the privacy and rights of those participants and the methods were largely discarded. The development of social media and the digitisation of information available through social media have, however, completely set such ethical concerns aside and there is now an upsurge in the application of unobtrusive observations to study everyday social behaviour (Kellar and Hall, 2022). The use of implicit measures of attitudes to measure unconscious or pre-conscious attitudes and beliefs, however, did not receive the same degree of ethical scrutiny (Bargh, Chen and Burrows, 1998; Cialdini, 2016).

The validity of the identity which is signalled and the trustworthiness of the person who is sending the signal are crucial factors in the development of social identity and the consequential group cohesion between members. The nature of the message which is sent about identity is also a factor. We have considered above the role of the 'tuning' of a message that a person engages or the complexity of the message structure. In the anticipation of an audience the sender must devise a message which will engage that audience. The sender of the message must also establish a degree of authenticity (Sherwood, 2022) for the message, first, to be noticed and secondly, for the information and arguments to be believed and accepted. While such 'authenticity' may appear at first to be relatively easy to provide, the world of digitised social media demonstrates how fraught the concept to be. Sherwood defines authenticity in terms of 'verisimilitude', that the representation of an event or item of knowledge is made honestly and accurately—in psychological terms, that there is an 'external validity' against which the accuracy of the knowledge can be compared. The diversity of information available on social media, however, places items of information in competition with many other pieces of opinion and representation, so that a comparison is not easy to make and a judgement of honesty cannot be made rapidly. The judgement of authenticity is therefore made on other grounds.

The heuristic thought systems used, biased forms of thinking invoked and the situation itself, which can stimulate forms of bias (Okonofua, Harris and Walton, 2022), can combine to provide a rich frame in which many pieces of information collide and compete. The situation may act to enhance or to mitigate the effects of the heuristic bias. Sherwood's other form of authenticity, namely 'personal', based upon the perception that the other person is 'speaking from the heart or deeply felt personal experience' cannot be easily engaged or espoused in mediated communication and the *signals* of authenticity become the mechanism upon which validity of the information is accepted. Both of these meanings require that a match between the story and the reality be made by the receiver of the information. The story may become the more powerful influence if the message is strong, relatively simple and consistent with prior beliefs that are easily accessible in the memories and experience of the recipient. This is illustrated in the work on integrative complexity of messages (cf., Chan et al., 2020) in which a relatively simple message can have a marked influence to counter the complexities that may arise during the initial development of debate and discussion. Authors have also referred to the influential power of stories (Kahneman, 2011; Storr, 2019), where narrative can have a persuasive influence beyond the presentation of facts (Coper, 2022). The complexities of storytelling, however, can involve distracting elements which detract from the direct implications of the message. 'Facts' may be more powerful in the longer term (Innes, 2022).

The work of Sherwood (2022) highlights an interesting paradox that gives insight into the authenticity of current research. She comments on the intrinsic honesty and authenticity of magicians in our society. They practise to deceive an audience, but they are open about that intention and they then proceed to practise the deceit. The discipline of social psychology over the last 50 years has also been founded upon the basis of deception. In this case, however, the practice was designed to allow participants in experiments to be affected by the manipulation of features of their social environment without their realisation of the true nature of the manipulation (cf., Innes and Morrison (2020) for a fuller explication of this methodology). Social psychologists have argued that magic techniques can be used to deceive experimental participants even more successfully than in the past (Olson and Raz, 2021). The field of social psychology has been cited as an 'experimental parable' of the essential inauthenticity of the modern age (Rosenberg, 1970).

Sherwood goes on in her analysis to point to the need of the audience willingly to suspend their disbelief for the purpose of the magician's act, while there must be a return to 'reality'. For the social psychologist, the illusion is what matters; it is regarded as necessary to obtain insight into the causes of human behaviour. The deceptive science of social psychology is using its methodology to study the art and science of deception in the wider social and political society. The resurgence in the discipline of concerns about experimental methodologies (cf., Diener et al., 2022) is testament to the inherent irony of the psychologist's role in studying opaque algorithms and deceptive media.

An outcome of the availability of social media information and the development of identity is an understanding of the power of the majority or minority of opinions that are available at any one time. The classic work of Asch (1952) demonstrates that people will readily conform to the beliefs that are espoused by a majority of other people. There are also experiments to demonstrate that should there be a minority of deviant opinions expressed, then the power of the majority to instil conformity is reduced markedly. Asch's studies were based on the beliefs that (a) we rely on the opinions of others in order to make judgements about events and opinions for which we do not have personal experience; (b) we believe that we express veridical opinions in our judgements to others; and (c) we in turn expect other people to be honest in their expressions of belief. This requires a trust in the behaviour of others to report honestly about their experience. But if the others are not honest, then adoption of the expressions of others is not safe. The majority may not be believable. The availability of social media provides a plethora of opinion, and a person may be able to find the information on the internet that is most consistent with their own beliefs, even if this is a minority opinion. A diversity of minorities can validate a belief that no one person has an authentic (correct) view.

The availability of so much information and opinion on social media also increases the ability, and therefore the liability, of people to seek out opinions that express ever more closely their opinions and goals. There is therefore a fractionation of social groups as people migrate to groups where the opinions and norms mimic more closely their opinions, and, as we have hypothesised, act to polarise opinion as more dimensions to differentiate opinion are recruited and generated. While we would not go so far as Smith and Burrows (2021) in predicting a 'micro-fracturing of nation-states', the availability of micro-cultures does suggest that there can be increasing numbers of groups which polarise opinion across the political domain. We need not go so far as to imagine the *Sovereign Individual* (Davidson and Rees-Mogg, 1997), although the effect of this thinking can be hypothesised to have had an influence upon the state of voting in the United Kingdom in the Brexit referendum and where behavioural science was used to create 'nudges' in particular directions of opinion (cf., De-Ridder, Kroese and van Gestel, 2022). Hirschman's (1970) analysis of the forces affecting group membership, where migration from a group to another, more congenial one, or staying within the group and being vociferous in expressing that loyalty, casts a political view on the concepts we have considered from a social psychological point of view. The availability of social media and the use of algorithms based upon the deep learning of reinforcement-based operants can accelerate the divisions of groups along more and more fine lines of social membership. Exiting to more polarised groups in which a person can more vociferously voice an extreme opinion can be the result of AI facilitating social identity.

5.9 Further Considerations

We have considered the work on SCT as a means of understanding social identity. There are two features of this work that deserve further examination. First, the cognitive basis of SCT is, as the name demonstrates, categorisation of others into groups. This is a process which essentially simplifies the social environment, as the coding of objects or people as similar or dissimilar eliminates the need to respond to all the differences that exist. Other means of representing objects and events involve placing these on dimensions, whereby the differences are graded along a line with small differences between relatively similar objects and larger differences as similarity declines.

Categorisation has been used extensively in other areas of psychological inquiry. For example, in diagnosis of psychological disorders the Diagnostic and Statistical Manual (DSM) of the American Psychiatric Association has been widely used in successive editions over decades. However, in recent years there has been an increase in the argument that dimensions can be used to represent similarities between psychological disorders as well as to differentiate between them (e.g., Conway, Krueger and the HiTOP Consortium Executive Board, 2021).

The process of scaling events along dimensions has been common practice in social psychology (Dawes, 1972). However, examination of the outcomes of coding events across dimensions has demonstrated that such representations can collapse into a grouping process. Hovland, Harvey and Sherif (1957) showed that in coding events and people along a line of judgement, events which were judged as more similar to the preferred opinion of the person making the judgement were assimilated to be closer to that preferred position and those that were dissimilar were contrasted to be further away, compared with the judgements of neutral judges. Categories were essentially formed by the participants while thinking in dimensional terms. Judd (Judd and Lusk, 1984; Lusk and Judd, 1988), following Scott, Osgood and Peterson (1979), found that when judges are asked to generate multiple dimensions to represent differences between people, the process results in the creation of correlated dimensions which result in polarisation of differences (a result of categorisation) and the polarisation increases as the number of dimensions increases. The demonstration of expertise in thinking about events leads to extremity of judgement. In a similar manner, Sidanius (1988; Sidanius and Lau, 1989) showed that polarisation of beliefs increased as individuals became more expert in their knowledge about social and historical events.

So, dimensions can be used cognitively to represent evaluative judgements but such thought processes can lead to categorisation. This is also the product of group processes, as exemplified by the work of Moscovici (Moscovici, Mugny and van Avermaet, 1985) on the influence of minorities. In a group setting where opinions are made public and there is diversity of opinion, the existence of a minority which is consistent and pervasive can, over time, lead to the adoption of beliefs by what was initially the majority of the group. Group process can lead to a categorisation of beliefs, reducing diversity.

The second feature of the work on social identity is that the concepts are very cognitive in nature. The role of affective, emotion pressures influencing a sense of identity have been downplayed. This has been noted in several places (e.g., Bacharach, 2006; Dashtipour, 2012). Dashtipour advocates for the development of a psychoanalytic theory of social identity, in her case based upon the thinking of Lacan. The work on social identity that we have emphasised bears no relation to this psychoanalytically based thinking. Psychoanalysis in conceptually thinking about AI, however, is prevalent (e.g., Possati, 2021) and attention to this approach may be fruitful. And we have noted the role that psychoanalysis has played in influencing the ways that the novelist thinks about human identity.

5.10 Conclusions

We have analysed developments in the social psychology of identity and the development of social media and artificial intelligence, and the ways they can be combined to produce a diverse and a polarised society in which numerous social groups exist which satisfy the social needs and goals of individuals. A more fractured social fabric may ensue from the work of these social forces, with effects on political groups and their place in power in societies. Authors such as Hibbing (2020) and Sunstein (2018) demonstrate how groups may develop on the basis of different forms of knowledge and beliefs. Combining these concerns with those about the nature and validity of evidence (Coper, 2022; Ritchie, 2020) and with the prevalence of opaque algorithmic systems can lead to a concern that social groups will evolve in novel ways. We have not speculated about the nature of the forces that lead to the formation of groups; we have merely considered the social processes that have existed within diverse human societies over millennia. The availability of so much diverse and targeted information, however, creates a dynamism that has not existed before and social groups will change in response to those forces.

References

Asch, S. E. (1952). *Social psychology*. Prentice-Hall.

Babones, S. (2020). The centralizing power of China's techno-totalitarianism. *Quadrant*, October, 18–22.

Bacharach, M. (2006). *Beyond individual choice: Teams and frames in game theory*. Princeton University Press.

Bandura, A. (1989). Human agency in social-cognitive theory. *American Psychologist*, *44*, 1175–1184.

Bargh, J. A., Chen, M., and Burrows, L. (1998). Automaticity of social behavior: Direct effects of trait construct and stereotype activation on action. *Journal of Personality and Social Psychology*, *71*, 230–244.

Bem, D. J. (1965). An experimental analysis of self-persuasion. *Journal of Experimental Social Psychology*, *1*, 199–218.

Bem, D. J. (1966). Inducing belief in false confessions. *Journal of Personality and Social Psychology*, *3*, 707–710.

Bingley, W. J., Greenaway, K. H., and Haslam, S. A. (2022). A social identity theory of information-access regulation (SITIAR): Understanding the psychology of sharing and withholding. *Perspectives on Psychological Science*, *17*, 827–840.

Blascovich, J., Mendes, W. B., Hunter, S. B., and Salomon, K. (1999). Social 'facilitation' as challenge and threat. *Journal of Personality and Social Psychology*, *77*, 68–77.

Brewer, M. B. (1991). The social self: On being the same and different at the same time. *Personality and Social Psychology Bulletin*, *17*, 475–482.

Bridle, J. (2018). *New dark age: Technology and the end of the future*. Verso.

Broad, E. (2018). *Made by humans: The AI condition*. Melbourne University Press.

Broadbent, D. E. (1958). *Perception and communication*. Pergamon.

Brown, R. (2020). *Henri Tajfel: Explorer of identity and difference*. Routledge.

Bruner, J. (1957). On perceptual readiness. *Psychological Review*, *64*, 123–151.

Burrell, J. (2016). How the machine 'thinks': Understanding opacity in machine learning algorithms. *Big Data & Society*, January–June, 1–12. doi: 10.1177/2053951715622512.
Campbell, D. T. (1958). Common fate, similarity, and other indices of the aggregates of persons as social entities. *Behavior Science*, *3*, 14–25.
Campbell, D. T. (1960). Blind variation and selective retention in creative thought as in other knowledge processes. *Psychological Review*, *67*, 380–400.
Campbell, D. T. (1963). Social attitudes and other acquired behavioral dispositions. In S. Koch (ed.), *Psychology: A study of a science. Vol. 6, Investigations of man as socius* (pp. 94–176). McGraw-Hill.
Campbell, D. T. (1975). Degrees of freedom and the case study. *Comparative Political Studies*, *8*, 178–191.
Campbell, D. T. (1994). The social psychology of scientific validity: An epistemological perspective and a personalized history. In W. R. Shadish and S. Fuller (eds), *The social psychology of science* (pp. 124–161). Guilford Press.
Chan, L., Zubrod, A., Woodard, S. R., and Conway III, L. G. (2020). Identity leadership is manifested via integrative complexity: Comment on Haslem et al. (2019). *American Psychologist*, *75*, 403–405.
Cialdini, R. B. (2016). *Pre-suasion: A revolutionary way to influence and persuade*. Random House.
Cialdini, R. B. (2021). *Influence: The psychology of persuasion. New and expanded edition*. Harper.
Collins, H., and Evans, R. (2007). *Rethinking expertise*. University of Chicago Press.
Conway, C. C., Krueger, R. F., and the HiTOP Consortium Executive Board (2021). Rethinking the diagnosis of mental disorders: Data-driven psychological dimensions, not categories, as a framework for mental-health research, treatment, and training. *Current Directions in Psychological Science*, *30*, 151–158.
Cook, T. D. (2018). Twenty-six assumptions that have to be met if single random assignment experiments are to warrant 'gold standard' status: A commentary on Deaton and Cartwright. *Social Science and Medicine*, *210*, 37–40.
Coper, E. (2022). *Facts*. Allen & Unwin.
Couldry, N., and Hepp, A. (2017). *The mediated construction of reality*. Polity.
Cowen, A. S., and Keltner, D. (2020). What the face displays: Mapping 28 emotions conveyed by naturalistic expression. *American Psychologist*, *75*, 249–364.
Dashtipour, P. (2012). *Social identity in question: Construction, subjectivity and critique*. Routledge.
Davidson, J. D., and Rees-Mogg, W. (1997). *The sovereign individual: Mastering the transition to the information age*. Touchstone.
Dawes, R. M. (1972). *Fundamentals of attitude measurement*. Wiley.
DeRidder, D., Kroese, F., and van Gestel, L. (2022). Nudgeability: Mapping conditions of susceptibility to nudge influence. *Perspectives on Psychological Science*, *17*, 346–359.
DeWall, F. (2016). *Are we smart enough to know how smart animals are?* Granta.
Diener, E., Northcott, R., Zyphur, M. J., and West, S. G. (2022). Beyond experiments. *Perspectives on Psychological Science*, *17*, 1101–1119.
Dietvorst, B., and Bharti, S. (2019). People reject even the best possible algorithm in uncertain decision domains. SSRN. doi: 10.2139/ssrn.3424158.
Dietvorst, B. J., Simmons, J. P., and Massey, C. (2015). Algorithm aversion: People erroneously avoid algorithms after seeing them err. *Journal of Experimental Psychology: General*, *144*(1), 114–126. doi: 10.1037/xge0000033.
Duffy, B. (2021). *Generations: Does when you are born shape who you are?* Atlantic Books.
Echterhoff, G., and Higgins, E. T. (2018). Shared reality: Construct and mechanisms. *Current Opinion in Psychology*, *23*, iv–vii.
Farr, R. M. (1996). *The roots of modern social psychology*. Blackwell.
Ferster, C. B., and Skinner, B. F. (1957). *Schedules of reinforcement*. Appleton-Century-Crofts.
Graesser, L., and King, W. L. (2020). *Foundations of deep reinforcement learning*. Addison-Wesley.

Guerin, B. (2016). *How to rethink psychology: New metaphors for understanding people and their behavior*. Routledge.

Guerin, B., and Innes, J. M. (1981). Awareness of cognitive processes: Replications and revisions. *Journal of General Psychology*, *104*, 173–189.

Guerin, B., and Innes, J. M. (1982). Social facilitation and social monitoring. *British Journal of Social Psychology*, *21*, 7–18.

Guerin, B., and Innes, J. M. (1984). Explanations of social facilitation effects: A review. *Current Psychological Research and Reviews*, *3*, 32–52.

Guerin, B., and Innes, J. M. (1989). Cognitive tuning sets: The consequence of anticipated communication. *Current Psychology Research and Reviews*, *8*, 234–249.

Hadnagy, C. (2018). *Social engineering: The science of human hacking*. Wiley.

Hamilton, J. L., Nesi, J., and Choukas-Bradley, S. (2022). Reexamining social media and socioemotional well-being among adolescents through the lens of the COVID-19 pandemic: A theoretical review and directions for future research. *Perspectives on Psychological Science*, *17*, 662–679.

Haney, C., Banks, C., and Zimbardo, P. G. (1973). Interpersonal dynamics in a simulated prison. *International Journal of Criminology and Penology*, *1*, 69–97.

Harre, R. (2009). *Pavlov's dogs and Schrödinger's cat: Scenes from the living laboratory*. Oxford University Press.

Haslam, S. A., Fransen, K., and Boon, F. (2020). *The new psychology of sport and exercise: The social identity approach*. SAGE.

Haslam, S. A., O'Brien, A., Jetten, J., Vormedal, K., and Penna, S. (2005). Taking the strain: Social identity, social support, and the experience of stress. *British Journal of Social Psychology*, *44*, 355–370.

Haslam, S. A., Reicher, S. D., and Platow, M. J. (2020). *The new psychology of leadership: Identity, influence, and power* (2nd ed.). Routledge.

Hayes, S. C., Hayes, L. J., and Reese, H. W. (1988). Finding the philosophical core: A review of Stephen C. Pepper's World Hypotheses: A Study in Evidence. *Journal of the Experimental Analysis of Behavior*, *50*, 97–111.

Heider, F. (1958). *The psychology of interpersonal relations*. Wiley.

Hibbing, J. R. (2020). *The securitarian personality*. Oxford University Press.

Hirschman, A. O. (1970). *Exit, voice, and loyalty: Responses to decline in firms, organizations, and states*. Harvard University Press.

Hovland, C. I., Harvey, O. J., and Sherif, M. (1957). Assimilation and contrast effects in communication and attitude change. *Journal of Abnormal and Social Psychology*, *55*, 242–252.

Innes, J. M. (2022). The restriction of meaning in stories. *The Psychologist*, *6*(5), 4. http://thepsychologists.bps.org.uk/restriction-meaning-stories.

Innes, J. M., and Fraser, C. (1971). Experimenter bias and other possible biases in psychological research. *European Journal of Social Psychology*, *1*, 297–310.

Innes, J. M., and Morrison, B. W. (2017). Projecting the future impact of advanced technologies on the profession: Will a robot take my job? *Australian Psychological Society InPsych*, *39*(2), 34–35.

Innes, J. M., and Morrison, B. W. (2020). Experimental studies of human–robot interaction: Threats to valid interpretation from methodological constraints associated with experimental manipulations. *International Journal of Social Robotics*, *13*(4), 765–773.

Innes, J. M., and Zeitz, H. (1988). The public's view of the impact of the mass media: A test of the 'third person' effect. *European Journal of Social Psychology*, *18*, 457–463.

Innes, C. J., Innes, J. M., and Morrison, B. W. (2022). Social representation of the profession of psychology and the application of artificial intelligence: European Union regulatory authority and the application of psychology as a paradigm for the future. *Australian and New Zealand Journal of European Studies*, *14*(1), 2–17.

Jercic, P., Wen, W., Hagelback, J., and Sunstedt, V. (2018). The effect of emotions and social behavior on performance in a collaborative serious game between humans and autonomous robots. *International Journal of Social Robotics*, *10*, 115–129.
Judd, C. M., and Lusk, C. M. (1984). Knowledge structures and evaluation judgments. *Journal of Personality and Social Psychology*, *46*, 1193–1207.
Kahneman, D. (2011). *Thinking fast and slow*. New York: Farrar, Straus & Giroux.
Kahneman, D., Sibony, O., and Sunstein, C. R. (2021). *Noise: A flaw in human judgment*. William Collins.
Kaltwasser, L., Rost, N., Ardizzi, M., Calbi, M., Settembrino, L., Fingerhut. J., Pauen, M., and Gallese, V. (2019). Sharing the filmic experience: The physiology of socio-emotional processes in the cinema. *PLoS ONE*, *14*(10), e0223259.
Kellar, S. J., and Hall, E. V. (2022). Measuring racial discrimination remotely: A contemporary review of unobtrusive measures. *Perspectives on Psychological Science*, *17*, 1404–1430.
Khalil, L. (2022). *Rise of the extreme right*. Penguin.
Kim, T. W., and Duhachek, A. (2020). Artificial intelligence and persuasion: A construal-level account. *Psychological Science*, *31*, 363–380.
Klein, G. (2015). A naturalistic decision making perspective on studying intuitive decision making. *Journal of Applied Research in Memory and Cognition*, *4*(3), 164–168. doi: 10.1016/j.jarmac.2015.07.001.
Koch, S. (1964). Psychology and emerging conceptions of knowledge as unitary. In T. W. Wann (ed.), *Behaviorism and phenomenology: Contrasting bases for modern psychology* (pp. 1–41). University of Chicago Press.
Kraus, M. W., and Keltner, D. (2007). Signs of socio-economic status: A thin-slicing approach. *Psychological Science*, *20*, 99–106.
Kraus, M. W., Park, J. W., and Tan, J. J. X. (2017). Signs of social class: The experience of economic inequality in everyday life. *Perspectives on Psychological Science*, *12*, 422–435.
Kuhn, T. S. (1962). *The structure of scientific revolutions*. University of Chicago Press.
Lovelock, J. (2019). *Novacene: The coming age of hyperintelligence*. Allen Lane.
Lusk, C. M., and Judd, C. M. (1988). Political expertise and the structural mediators of candidate evaluations. *Journal of Experimental Social Psychology*, *24*, 105–126.
MacCorquodale, K., and Meehl, P. E. (1948). On the distinction between hypothetical constructs and intervening variables. *Psychological Review*, *55*, 95–107.
Mackie, J. L. (1974). *The cement of the universe: A study of causation*. Oxford University Press.
Markova, I. (2014). Questioning interdisciplinarity: History, social psychology and the theory of social representations. In C. Tileaga and J. Byford (eds), *Psychology and history: Interdisciplinary explorations* (pp. 109–126). Cambridge University Press.
McAdams, D. P. (2015). *The art and science of personality development*. Guilford Press.
McGuire, W. J. (1999). *Constructing social psychology: Creative and critical processes*. Cambridge University Press.
Mednick, S. A. (1962). The associative basis of the creative process. *Psychological Review*, *69*, 220–232.
Miller, T. (2019). Explanation in artificial intelligence: Insights from the social sciences. *Artificial Intelligence*, *267*, 1–38.
Morrison, B. W., and Innes, J. M. (2022). Judgement day for psychologists? How technology has arrived to save the day. *World Financial Review*, *6*(2), 25–28. https://worldfinancialreview/judgement-day-for-psychologists-how-technology-has-arrived-to-save-the-day.
Moscovici, S., Mugny, G., and van Avermaet, E. (eds) (1985). *Perspectives on minority influence*. Cambridge University Press.
Muller, J. Z. (2018). *The tyranny of metrics*. Princeton University Press.
Newell, B. R., and Shanks, D. R. (2014). Unconscious influences on decision making: A critical review. *Behavioral and Brain Sciences*, *37*, 1–61.
Nisbett, R. E. (2015). *Mindware: Tools for smart thinking*. Farrar, Straus & Giroux.

Nisbett, R. E., and Wilson, T. D. (1977). Telling more than we can know: Verbal reports on mental processes. *Psychological Review, 84*, 231–259.

Okonofua, J. A., Harris, L. T., and Walton, G. M. (2022). Sidelining bias: A situationist approach to reduce the consequences of bias in real-world contexts. *Current Directions in Psychological Science, 31*(5). doi: 10.1177/09637214221102422.

Olson, J. A., and Raz, A. (2021). Applying insights from magic to improve deception in research: The Swiss cheese model. *Journal of Experimental Social Psychology, 92*. doi: 10.1016/j.jesp.2020.104053.

Olteteanu, A.-M., and Zunjani, F. H. (2020). A visual remotes associates test and its validation. *Frontiers in Psychology*. doi: 10.3389/fpsyg.2020.00026.

O'Neil, C. (2016). *Weapons of math destruction*. Penguin.

Pepper, S. C. (1942). *World hypotheses: A study in evidence*. University of California Press.

Possati, L. M. (2021). *The algorithmic unconscious: How psychoanalysis helps in understanding AI*. Routledge.

Rahwan, I., et al. (2019). Machine behaviour. *Nature, 568*, 477–486. doi: 10.1038/s41586-019-1138-y.

Rasskin-Gutman, D. (2009). *Chess metaphors: Artificial intelligence and the human mind*. MIT.

Reicher, S., and Haslam, S. A. (2006). Rethinking the psychology of tyranny: The BBC Prison Study. *British Journal of Social Psychology, 45*, 1–40.

Reicher, S. D., Haslam, S. A., and Smith, J. R. (2012). Working toward the experimenter: Reconceptualizing obedience within the Milgram paradigm as identification-based followership. *Perspectives on Psychological Science, 7*, 315–324.

Ritchie, S. (2020). *Science fictions*. Vintage.

Rosenberg, M. J. (1970). The experimental parable of inauthenticity: Consequences of counterattitudinal performance. In J. S. Antrobus (ed.), *Cognition and affect* (pp. 179–201). Little, Brown.

Rosnow, R. L., and Georgoudi, M. (1986). *Contextualism and understanding in behavioral science*. Praeger.

Russell, S. (2019). *Human compatible: AI and the problem of control*. Allen Lane.

Sadler, M., and Regan, N. (2019). *Game changer: AlphaZero's ground breaking chess strategies and the promise of AI*. Alkmaar.

Scerri, E. (2016). *A tale of seven scientists and a new philosophy of science*. Oxford University Press.

Scott, W. A., Osgood, D. W., and Peterson, C. (1979). *Cognitive structure*. Winston.

Seaver, N. (2017). Algorithms as culture: Some tactics for the ethnography of algorithmic systems. *Big Data and Society*, July–December, 1–12.

Sejnowski, T. J. (2018). *The deep learning revolution*. MIT Press.

Seong, Y., and Bisantz, A. M. (2008). The impact of cognitive feedback on judgment performance and trust with decision aids. *International Journal of Industrial Economics, 38*, 608–625.

Shamay-Tsoory, S., and Hertz, U. (2022). Adaptive empathy: A model for learning empathic responses in response to feedback. *Perspectives on Psychological Science, 17*, 1008–1023.

Sherwood, A. (2022). *Authenticity: Reclaiming reality in a counterfeit culture*. Mudlark.

Sidanius, J. (1988). Political sophistication and political deviance: A structural equation examination of context theory. *Journal of Personality and Social Psychology, 55*, 37–51.

Sidanius, J., and Lau, R. (1989). Political sophistication and political deviance: A matter of context. *Political Psychology, 10*, 85–109.

Simonton, D. K. (2012). Scientific creativity as blind variation: Explicit and implicit procedures, mechanisms, and processes. In R. W. Proctor and E. J. Capaldi (eds), *Psychology of science: Implicit and explicit processes* (pp. 363–388). Oxford University Press.

Sinha, R., and Swearingen, K. (2002). The role of transparency in recommender systems. Paper presented at the CHI'02 extended abstracts on Human factors in Computing Systems, Minnesota, USA, April.

Skinner, B. F. (1938). *The behavior of organisms: An experimental analysis*. Appleton-Century-Croft.

Skinner, B. F. (1963). Behaviorism at fifty. *Science, 140*, 951–958.

Smaldino, P. E. (2022). Models of identity signaling. *Current Directions in Psychological Science, 31*, 231–237.

Smith, H., and Burrows, R. (2021). Software, sovereignty and the post-neoliberal politics of exit. *Theory, Culture and Society, 38*, 143–166.

Smith, M. B. (2003). *For a significant social psychology*. New York University Press.

Smith, R. E. (2019). *Rage inside the machine: The prejudice of algorithms and how to stop the internet making bigots of us all*. Bloomsbury Business.

Stephan, A. (2015). Empathy for artificial agents. *International Journal of Social Robotics, 7*, 111–116.

Storr, W. (2019). *The science of storytelling*. Collins.

Stouffer, S. A., Suchman, E. A., DeVinney, L. C., Star, S. A., and Williams, R. M. (1949). *The American Soldier: Adjustment during army life. Vol.1*. Wiley.

Sunstein, C. R. (ed.) (2018). *Can it happen here? Authoritarianism in America*. William Morrow.

Tajfel, H. (1969a). Cognitive aspects of prejudice. *Journal of Biosocial Science, 1*, 173–191.

Tajfel, H. (1969b). Cognitive aspects of prejudice. *Journal of Social Issues, 25*, 79–97.

Tajfel, H., Billig, M. G., Bundy, R. P., and Flament, C. (1971). Social categorization and intergroup behaviour. *European Journal of Social Psychology, 1*, 149–178.

Turing, A. (1950). Computing machines and intelligence. *Mind, 59*(236), 433–460.

Turner, J. C., Hogg, M., Oakes, P. J., Reicher, S., and Wetherell, M. (1987). *Rediscovering the social group: A self-categorization theory*. Blackwell.

Walsh, T. (2018). *2062*. LaTrobe University Press.

Webb, E. J., Campbell, D. T., Schwartz, R. D., and Sechrest, L. (1966). *Unobtrusive measures: Nonreactive research in the social sciences*. Rand McNally.

Wiggins, M. W. (2021). A behaviour-based approach to the assessment of cue utilization: Implications for situation assessment and performance. *Theoretical Issues in Ergonomics Science, 22*(1), 46–62. doi: 10.1080/1463922x.2020.1758828.

Yaniv, I. (2004). Receiving other people's advice: Influence and benefit. *Organizational Behavior and Human Decision Processes, 93*(1), 1–13. doi: 10.1016/j.obhdp.2003.08.002.

Yaniv, I., and Kleinberger, E. (2000). Advice taking in decision making: Egocentric discounting and reputation formation. *Organizational Behavior and Human Decision Processes, 83*(2), 260–281. doi: 10.1006/obhd.2000.2909.

Yeomans, M., Shah, A., Mullainathan, S., and Kleinberg, J. (2019). Making sense of recommendations. *Journal of Behavioral Decision Making, 32*(4). doi: 10.1002/bdm.2118.

Young, J. E., Hawkins, R., Sharlin, E., and Igarajhi, T. (2009). Towards acceptable domestic robots: Applying insights from social psychology. *International Journal of Social Robotics, 1*, 95–108.

Zuboff, S. (2019). *The age of surveillance capitalism*. Profile Books.

Sven Nyholm

6 AI, Robot Co-workers and Humans

Abstract: This chapter discusses the introduction of artificial intelligence (AI) and robots into workplaces and the question of whether any such technologies could plausibly become a form of colleague for the human beings in these workplaces. People who work with robots have already been observed to sometimes get attached to them and to consider them as part of the team and sometimes even as good colleagues. The chapter considers whether there is something mistaken about such responses to robots and AI in the workplace, or whether such ways of regarding technologies in the workplace could ever be justified. The chapter considers arguments both in favour of and against the idea of AI and robots as potentially being good co-workers for human beings. It also considers whether we should view this as an all-or-nothing or yes-or-no issue and suggests that whether technologies could be good colleagues might be a matter of degree.

Keywords: robots and AI in the workplace, collegiality, ethical behaviorism, the ethics of anthropomorphism

6.1 Introduction

In her book *Culture and Human–Robot Interaction in Militarized Spaces*, Julie Carpenter (2015) describes the interaction between military explosive ordnance disposal personnel and the bomb disposal robots they use. These were among the first military groups to work extensively with robots every day. For some of the soldiers Carpenter observed and interviewed, the robots they worked with had come to be more than mere tools. Some of them had started regarding these robots as part of the team—as co-workers or colleagues, if you will. In fact, some of the soldiers had become so attached to a particular robot, which they dubbed 'Boomer', that when this robot was damaged, they did not want to replace it with a newer, better robot. Rather, they wanted to fix that particular robot. Later, Boomer the robot was destroyed beyond repair. It was a bomb disposal robot after all, so this was bound to happen sooner or later! The soldiers responded to this by having an improvised military funeral for

Note: Thanks to Arleen Salles and the other members of the Human Brain Project's 'RIU-Hub', where I presented this material in August of 2021. My work on this chapter is part of the research programme Ethics of Socially Disruptive Technologies, which is funded through the Gravitation programme of the Dutch Ministry of Education, Culture and Science, and the Netherlands Organisation for Scientific Research (NWO grant number 024.004.031).

Sven Nyholm, LMU Munich, Germany

https://doi.org/10.1515/9783110721751-006

their fallen comrade. And they wanted to give the robot two medals of honour: a Purple Heart and a Bronze Star Medal. Prior to 'his' destruction, Boomer's job was to seek out and disarm bombs. But not only was he a lifesaver. Boomer's human co-workers also thought that he had "develop[ed] a personality of his own" (Garber, 2013). They had become very attached to this robot, and it was emotional for them to have to say goodbye to the robot.

This example helps to illustrate the general topic this chapter discusses and also raises the main questions the chapter explores. The topic of the chapter is the introduction of artificial intelligence (AI) and robots into workplaces, specifically where this is done in a way that might result in these technologies being viewed as some form of co-worker or colleague or, at any rate, as part of the team by the human beings who work with the robots. This was certainly the case in the battlefield in Iraq, where Boomer and the human soldiers in the team were 'working' together. The main question the chapter will discuss is: can robots or other AI technologies be colleagues, or even good colleagues, in any important and non-metaphorical sense? The soldiers working with Boomer clearly seemed to have regarded 'him' as some form of colleague or co-worker, which they held in high enough regard that they felt a military funeral and medals of honour were in order. Was that a mistake? More generally, the question arises of how people should, or perhaps should not, treat robots and other AI technologies that they work alongside (Nyholm and Smids, 2020, pp. 2185–2186). Would it have been more appropriate, for example, for the soldiers working alongside Boomer to regard and treat that robot as a mere tool? Was it inappropriate for them in some way to treat the robot in the anthropomorphising ways briefly described above (Bryson, 2010, 2018; Nyholm, 2020)? Are there perhaps some work contexts—and certain forms of robots or other AI technologies—where these kinds of behaviours are more appropriate and others where they are less appropriate? In short, this chapter discusses whether robots can be co-workers or colleagues to human beings and the closely related question of whether it is appropriate or inappropriate to treat robots as if they are our co-workers, or even good colleagues of ours.

This is not a topic that has been discussed extensively in the philosophy of AI and robotics so far. More generally, the topic of what it is to be a good colleague in a workplace has not been discussed much in philosophy at all, whether the colleagues under consideration are human beings, robots or other AI technologies. Yet it is an important topic because human beings are poised to increasingly work alongside robots and other AI technologies that we might be disposed to view as being more than mere tools, for example by regarding them as some sort of co-workers, colleagues or members of our teams. Having good colleagues or team members is often regarded as being one of the most important aspects of meaningful work, as opposed to work that is dull or meaningless (Smids et al., 2020). So, if we are going to be working alongside robots that we—or at least some of us—might regard as some form of colleague, it is important to know whether this poses a threat to our opportunities to engage in meaningful work, or whether having robots as our 'colleagues' is compatible with doing meaningful work. If robots cannot really be co-workers or good colleagues in any sensible way that we can endorse upon reflection, even though people can be led or triggered to treat robots as if they were, there might be a threat of manipulation or deception here. Or it might then be considered undignified to be led to behave in these ways by robots that somehow "push our Darwinian buttons", to use a phrase from Sherry Turkle (2004).

Here is the plan of the chapter. First, some of the main concepts relevant to the subject will briefly be explained and discussed, particularly the concepts of artificial intelligence/AI, robots and (meaningful) work (section 6.2). Next, some related debates about AI, robots and work will be sketched, so as to situate the topic within the broader context of the philosophy of AI, robots and work more generally (section 6.3). After that, the issue of what a good colleague is and whether robots or other AI technologies can be good colleagues will be discussed, and that issue will be compared with the similar issues, which have received more philosophical attention, of whether robots can be our friends or romantic partners (sections 6.4 and 6.5). While discussing these issues, the chapter will distinguish between a binary ('yes' or 'no') type of approach to whether robots can be good colleagues or co-workers, on the one hand, and a 'degree of collegiality' approach, on the other hand, in which this does not have to be seen as an all-or-nothing matter but instead can be seen as a matter of degree (section 6.6). The chapter next discusses whether there might be something problematic or otherwise inappropriate about treating robots as if they can be our co-workers or good colleagues who are members of our workplace teams (section 6.7). This is not really something that has been discussed in those specific terms in the philosophical literature about AI and technology. But there has been much general discussion of whether it is most appropriate to treat all technologies, including advanced robots and AI, as mere tools or whether it can sometimes be appropriate, or even ethically necessary, to treat some technologies as being more than mere tools. As that section will explain, that more general discussion can be related and translated to the specific context of AI and robots in the workplace in particular.

6.2 Key Concepts: AI, Robots and Work/Meaningful Work

When the computer program AlphaGo 'beat' the world champion of the ancient game of Go—a man called Lee Sedol—in a series of games in March of 2016, this had been preceded by what might be called a vigorous training schedule for the championship. AlphaGo is an artificial neural network that had been "trained on a data set of hundreds of thousands of human games, and on the millions of games it played against itself in simulation" (Chivers, 2021). The result of this training regime was that AlphaGo had developed strategies that the human beings in the company behind this computer program, DeepMind, could not understand and that the world champion of Go, Lee Sedol, could not win against. (Sedol did win one of the five games, but the other four were won by the computer program.) One interesting thing about the actual games was that since AlphaGo was a computer program and not a physical robot, a human being (hired by DeepMind) had to move the stones (i.e., the Go game pieces) around on the board, based on the instructions from the computer program. So, one could say that an AI system (the AlphaGo neural network computer program) and a human being or set of human beings (the DeepMind employees working on this project) were 'working together' to beat the Go world champion Lee Sedol in this Go challenge.

This illustrates one way in which AI systems that are not robots might be used, and are sometimes already used, in workplaces: the AI technologies can take the form of computer programs that make recommendations, based on information/data used as inputs, and then human beings can carry out certain tasks based on the outputs of the AI system. *Artificial intelligence*, more generally, is here understood as any form of technology that is able to perform some task, or set of tasks, that human beings use their natural intelligence to perform (Dignum, 2019; Gordon and Nyholm, 2021). Sometimes one talks about AI as machines that can *imitate* intelligent behaviour or intelligent thinking, sometimes as machines that can *simulate* or *replicate* intelligent behaviour or thinking. This can be done in different ways, such as with a top-down approach (sometimes called 'good old-fashioned AI' or 'GOFAI') or in a bottom-up approach (such as with the neural networks approach mentioned above in relation to AlphaGo) (Dignum, 2019).

If the AI is running/operating as a computer program, the computers used are usually not labelled robots. But when there are machines that have sensors and actuators, so that they can directly 'perceive' and then react to their environments as they are carrying out their tasks, then the AI systems in question are usually called 'robots'. A *robot*, then, is an embodied machine that is able to perform some task, or set of tasks, with the help of sensors and actuators with which they can respond to their environments and also act on or react to their environment in the service of their tasks (Gunkel, 2018; Loh, 2019; Nyholm, 2020). Any robot, therefore, can be seen as having at least a limited or narrow form of AI, since most robots perform tasks that can be seen as imitating, simulating or even replicating intelligent human behaviour, though perhaps only in a very limited way. All robots, according to this way of thinking, are concrete entities with a 'body', whereas the set of AI systems more broadly also contains computer programs (sometimes called 'software agents') where it makes less sense to think of them as having a particular body (Russell and Norvig, 2010).

Robots can look like paradigmatic robots of the sorts we associate with science fiction—for example, they might have a metallic or plastic appearance and a vaguely human-like form, like the real-world robot 'Pepper' that has been used in a lot of robotics research or the science fiction robot 'C3PO' in the *Star Wars* movies from the 1970s. Most robots, though, do not look like paradigmatic robots of the sort that we might imagine when we hear the word 'robot'. Boomer, for example, looked more like a lawnmower or small tank than a paradigmatic robot. Most robots have shapes that are relevant to the tasks that they are supposed to perform (Royakkers and van Est, 2015). There are, however, some robots that not only look vaguely human, but that are designed more specifically to resemble humans, and perhaps even be very similar-looking to human beings. One example of a robot that is roughly human-like in its appearance is a small robot called 'Kaspar' that was developed for use in treatment of children with autism. An example of a robot that is made to look very human-like, at least in its face, is the well-known robot 'Sophia' from Hanson Robotics, which is so human-like that the Kingdom of Saudi Arabia announced in 2017 that it had granted this robot honorary citizenship (Nyholm, 2020).

What about 'work' and 'meaningful work'—how should they be defined in this context? In this chapter, *work* will be understood in a somewhat narrow, economic and legalistic sense as a set of more or less well-defined tasks that people regularly perform, based on a contract or other agreement according to which they are financially or otherwise compensated for performing those tasks, in accordance with that contract or agreement (Danaher, 2019a; Danaher and Nyholm, 2021). A broader definition of 'work' would also include unpaid care work or voluntary work for which no compensation is provided. In no way does this chapter intend to suggest that work in that broader sense is less important than work in the narrower sense specified above. But since the topic of the chapter is the introduction of AI systems and robots into workplaces, it makes sense here to focus on work in the narrower sense.

By 'meaningful work', this chapter intends to refer to work that realises what are sometimes called certain key "goods of work (other than money!)", to use an expression from Lisa Herzog and Anca Gheaus (2016). Though there is no universal agreement about what the most important goods of work are, there is fairly widespread agreement, both in philosophy and in work psychology, that the following can be associated with meaningful work. *Meaningful work* is most typically, but is not necessarily limited to: (1) work that involves pursuing a purpose that is/is seen as valuable; (2) work that involves being part of a community or team with significant social relationships; (3) work where one exercises skills and has opportunities to develop oneself and one's skills; (4) work that creates grounds for self-esteem and recognition by others; and (5) work that involves some amount of autonomy (e.g., the ability to shape how and when one performs one's work-related tasks) (Smids et al., 2020).

In this chapter, what is of particular interest is the widely shared idea that meaningful work involves being part of a community or team where one has good colleagues with whom one works. As more and more AI systems and robots enter workplaces, the question arises of whether they can become part of the team or help to realise this particular good that is associated with meaningful work, or whether having robots or AI systems as one's 'colleagues' could not possibly be seen as helping to realise this important good related to meaningful work.

6.3 AI, Robots and (Meaningful) Work

Robots and other AI technologies are increasingly being introduced into many different kinds of workplaces. For example, logistics robots help to pick and pack goods that will be shipped off from logistics warehouses; self-driving vehicles are being introduced on a small scale but are likely to be used on a much larger scale within the not-too-distant future; AI systems are used to make predictions about all sorts of things in different workplaces (e.g., in hospitals, in banks, or in the assessment of job or university applications); and stationary industrial robots have long been used in the manufacture of different kinds of goods and machines (Royakkers and van Est, 2015). To use some more specific examples, we saw above that soldiers in a battlefield in Iraq used a bomb disposal robot (Boomer) that they became very attached to; we also mentioned Kaspar the robot used in experimental therapy for children with autism; and we might also mention a Turkish experiment where robots were introduced into a restaurant context—something that the customers apparently liked more than the employees and managers at the restaurant did (Seyitoğlu et al., 2021). At the time of writing, the BBC is running a story about how various AI developers with an interest in religion and spirituality are also developing AI technologies (e.g., the 'robo rabbi' app) and robots (e.g., a Catholic robot) that are supposed to work with religious workers to help serve people's religious and spiritual needs (Bettiza, Alonna and Furst, 2021). No domain, in other words, seems to be immune to experimentation with AI and robots in the workplace.

The more these AI systems are concrete embodied robots, the more likely it is that they might be conceived of as members of the team and even as some form of good colleague. Strictly speaking, though, it is also possible that some people in some workplaces might consider other forms of AI systems as part of the team, and perhaps even as good colleagues as well. For example, if some AI system making recommendations or giving instructions is given a name or nick-name by the people who work with it, and if it can 'speak' in the way that a voice assistant technology like Amazon's Alexa or Apple's Siri does, then the AI system might be treated in an anthropomorphising way even if it does not have a physical body (Lee, 2021). Moreover, it is also important to note that robots do not need to look like human beings or like paradigmatic robots to elicit social responses in human beings whereby they might be perceived as members of the team. Again, the example of Boomer the military robot helps to illustrate this. As noted above, Boomer looked nothing like a human being or even an animal, but still the humans working with 'him' became very attached to the robot and started viewing 'him' as a cherished member of their team (Carpenter, 2015).

Empirical researchers studying human–robot interaction—or, more broadly, human–technology interaction—have begun studying and analysing the impact working with robots has on human workers. The issue of whether and to what extent AI systems and robots are seen as co-workers or team members is one of the topics they have focused on (see, e.g., Groom and Nass, 2007; Ljungblad et al., 2012; Sauppé and

Mutlu, 2015). Philosophers writing about AI and robots in the workplace, however, have primarily focused on topics other than whether they can be co-workers, members of the team or perhaps even good colleagues. Here is a quick review of what topics have received the most attention.

One issue that philosophers writing about technology and the future of work have written—and worried—about is whether the future will be a world without (much) work at all, and what this might do to people's opportunities to find meaningful ways of spending their time (Ford, 2015; Danaher, 2017). While some worry that a future without work is a threat to our opportunities for meaningful activities, others—like John Danaher (2019a) in his book *Automation and Utopia*—argue that not having to work can free up our time for other, perhaps more meaningful, forms of activity.

That discussion rests on the premise that AI and robots in the workplace will mean that most jobs are 'taken over' by robots. Many commentators have felt that it is more likely that many job tasks (as opposed to whole jobs) will be taken over by robots and other AI technologies (Smids et al., 2020; Tigard, 2021; Danaher and Nyholm, 2021). Also, it is widely thought that the introduction of AI and robots into the workplace and society more generally will change what kinds of jobs and job tasks humans will have. This raises the question of whether these technologies are a threat to—or perhaps in certain ways an opportunity for—meaningful work. One recent paper, for example, surveys different ways in which robots in the workplace might in some ways be a threat to, but in other ways create opportunities related to, the five goods of work enumerated at the end of the previous section (Smids et al., 2020). It has also been suggested that if robots take over tasks that are seen as meaningful by human beings, this might create an 'achievement gap' in the workplace. That is, there might be fewer opportunities for human beings to show effort, skill and excellence in the workplace context if AI and robots take over the most difficult parts of certain jobs (Danaher and Nyholm, 2021; but see also Tigard, 2021). Think, for example, of cases that would be analogous to the above-described example where AlphaGo was giving instructions and the human being's task was to move stones around on a Go board; here it might be thought that the part played by the human is the less meaningful one, while the part played by the AI computer program (thinking about strategies and suggesting moves) is the more meaningful one.

There have also been questions about responsibility in human–robot/human–AI interaction that have been prominent in the literature. There are discussions about ethical worries around threats to privacy in increasingly technologised workplaces. And there are also general discussions of what ethical guidelines should govern the introduction of experimental technologies into society more generally and the workplace in particular (see, e.g., Veliz, 2020; van de Poel, 2016; Hosseini et al., under review). But, as noted above, there has been very little discussion—with a few exceptions we will now get to—about what it is to be a good colleague or work team member more generally, or whether robots and AI technologies can be team members or good colleagues in particular. Accordingly, this chapter is not only about what has been written on that topic so far, but in effect also constitutes a sizable portion of the whole (so far very small!) literature on the topic.

6.4 Can a Robot be a Good Colleague? Part 1

As noted above, it is conceivable that people working together with AI systems that are not robots might come to regard these AI systems (e.g., computer programs making recommendations, which are given a name and perhaps voice assistant-like capacities) as colleagues or co-workers (Lee, 2021). However, it is more likely that people working together with 'embodied' AI systems—that is, robots—would regard these robots as part of the team and potentially even as good colleagues (Groom and Nass, 2007; Ljungblad et al., 2012; Sauppé and Mutlu, 2015). For the rest of this chapter, therefore, the focus will primarily be on AI systems in the form of robots.

In the philosophy literature about AI and robot ethics, there are many articles discussing questions such as 'can a robot be our friend?', 'can there be mutual love between humans and robots?', 'can a robot be a moral agent?' and 'can a robot have moral status of some important sort?' (Gunkel, 2018; Danaher, 2019b; Nyholm, 2020). However, at the time of writing, there exists only one published article that specifically tackles the question 'can a robot be a good colleague?' head on (Nyholm and Smids, 2020). As noted above, there are many empirical studies about whether people regard robots as members of their teams and of how humans and robots can 'work' well together in workplaces (e.g., Groom and Nass, 2007; Ljungblad et al., 2012; Sauppé and Mutlu, 2015). But when it comes to the normative or evaluative question—or the philosophical question, if you will—of whether a robot can be a good colleague, there is so far only this one publication. It was written by the author of this chapter together with his (human!) colleague, Jilles Smids. In this section, therefore, I will speak about how we approached the question in that article.

At the time we wrote that article, in 2018, there were also no publications available in the philosophical literature —at least that were known to us—on the more general question of how one should understand what a colleague is and what values are associated with being a good colleague. Accordingly, we could not approach the question of whether a robot can be a good colleague by taking an existing philosophical theory of what that entails and then testing whether a robot can be a good colleague against that theory. Instead, what we did was to use a twofold approach: (1) we compared the question of whether a robot can be a good colleague with the existing discussions of whether robots can be our friends or romantic partners; and (2) we tried to assemble a list of criteria or desiderata that it seemed to us sensible to associate with the ideal of being a good colleague.

Regarding the second point, we did two main things: we brainstormed about what one might associate with being a good colleague, and we then presented this list of criteria to colleagues both in philosophy (at the largest Dutch philosophy conference) and outside of philosophy (colleagues in organisational psychology). It can be noted here also that when we were thinking of what one might associate with being a good colleague, the work contexts we primarily had in mind were a logistics warehouse and an army setting. The two main reasons for focusing on those two contexts were the following. First, we were fascinated by the example of Boomer the military robot and how that robot was regarded as a good colleague by the soldiers. Secondly, robots are increasingly being introduced into logistics warehouses, and in an interdisciplinary project of which we were part, we had opportunities for interviewing employees in logistics warehouses about their experience of working with new technologies (Berkers et al., 2020).

When it comes to questions of whether robots can be good friends or romantic partners for human beings—strange questions, it might be thought, but there are people who want to have robots as their friends or partners!—it is common in philosophical discussions to first try to identify the most important aspects of being a good friend or romantic partner (Nyholm, 2020). The next step is to ask whether robots are able—or have capacities that might enable them—to live up to these key aspects of the ideals of being a good friend or romantic partner. For example, it is usually thought that a friend should be able to be a good person, that they should be able to care about and stand by us in a robust way; and it is usually thought that a good romantic partner should value the other partner, care about them deeply and be committed to them. These kinds of ideas involved in the ideals of being a good friend or romantic partner seem to presuppose that friends and partners are conscious beings, with a free will, and with thoughts and feelings. Robots, it is then typically argued, are not conscious beings. They do not have free will. Nor do they have human-like thoughts and feelings. Therefore, it is typically concluded, robots cannot be our friends or romantic partners—at least not in the sense of living up to the most cherished aspects of the ideals of friendship and romantic love. A big emphasis in these debates is on the 'inner life' of a friend or romantic partner (their thoughts, feelings, motivations, values, and so on), and since robots are seen as lacking an inner life, they are seen as unable to be our friends or partners.

There are noteworthy exceptions to this view, though. John Danaher (2019b), whose work on automation and AI ethics was mentioned above, has argued in favour of what he calls 'ethical behaviourism': the view that what matters most when we are trying to decide whether somebody is or could be our friend, partner, moral equal, and so on is not their 'inner life' but rather their behaviour. Smids and I are deeply sceptical of that view when it comes to whether robots can be our friends, partners or moral equals (Nyholm and Smids, 2020; Nyholm, 2020). However, when we were considering the question of whether robots can be good colleagues—at least in the context of work in a logistics warehouse or in an army setting—we approached this question with a greater degree of openness to the sort of ethical behaviourism Danaher defends. More specifically, Smids and I suggested a comparative claim: we suggested that as compared with our friends or romantic partners, the inner lives of our colleagues are, relatively speaking, less central to whether they are good colleagues, and behaviour is, relatively speaking, more central to the question of whether they are good colleagues. Our suggestion was that, if that is true, then it might be easier for a robot to be a good colleague than for a robot to be a good friend or romantic partner.

Based on our brainstorming about what sorts of criteria one might associate with the ideal of being a good colleague—which, again, we ran by colleagues in both philosophy and organisational psychology—Smids and I came up with the following list of things that one might expect of a good colleague. Our suggestions were:

(a) working well together to achieve desired outcomes and goals specifically related to the work in question (e.g., caring for patients properly, creating good products and useful services, etc.);
(b) being able to engage in pleasant, informal conversations, to help to keep work pleasant;
(c) not harassing or bullying colleagues, but instead treating each other respectfully;
(d) providing mutual help and support as needed;
(e) learning and developing together;
(f) 'consistency', meaning that colleagues are not constantly replaced, but that everyone works together over extended periods of time;
(g) being reliable and trustworthy;
(h) being sensitive to how colleagues are doing (e.g., whether they are tired or energised, well or unwell, happy or sad, etc.), and adjusting one's work-related interaction accordingly;
(i) sharing work-related values and being motivated to honour those values in the joint work;
(j) potentially also doing some socialising (in a collegial, respectful and friendly way), which may or may not lead to a 'deeper' friendship or closer relationship. Such socialising needs to be sensitive to the wishes and interests of all involved parties; it also needs to take power balances into account in a responsible way (Nyholm and Smids, 2020, p. 2177).

When we discussed whether a robot could live up to these criteria for being a good colleague, Smids and I initially set the last point on the list aside, since we were of the opinion that robots cannot live up to the ideals associated with being a good friend or romantic partner. We instead focused on criteria (a)–(i). And we made a distinction between living up to these criteria on a purely behavioural level and living up to these criteria in the way that a human being, with a human mind, might do. Our suggestion was that on a purely behavioural level—that is, from the point of view of what sorts of behaviours it is possible for an AI system to enact—a robot might be able to live up to many, if not all, of these criteria, at least in a partial sort of way.

For example, robots can, at least in limited domains, 'work well' together with human beings so that the work-related tasks will be well achieved. In a logistics warehouse setting, for example, a 'team' of a human and a robot can do a quick and efficient job at moving boxes around, and in the same way a team of soldiers working with a bomb disposal robot might do a good and efficient job at finding and disarming bombs. The robot might, or might not, be equipped with some form of chat function, so that one could potentially engage in some light/pleasant conversation with the robot. (Such technologies are perhaps not very advanced at the moment, but there are rapid technological developments in natural language processing and in the capacities of AI systems to engage in 'conversations' with human beings.) Robots could, moreover, be designed to treat humans in a polite way that avoids all forms of harassment and other forms of negative behaviour good colleagues should eschew. And, to take a final example, a robot could be designed to behave in a way that seems to be attuned to how the human beings it is working with are doing—their level of activity and so on—and adjust its behaviour accordingly.

Suppose, in other words, that robots could live up to some or many of criteria (a) to (i) on at least a purely behavioural level. Suppose further that we take on board the suggestion that Smids and I put forward in our paper according to which the inner thoughts and feelings of colleagues are, at least relatively speaking, somewhat less important than behaviour when it comes to deciding whether somebody (or something!) is a good colleague, at least in comparison with how these things are valued in the contexts of friendship and love. If we suppose both of these things, then it seems possible to conclude either of two things. A weaker conclusion (the one Smids and I drew in our paper) is that it is, at least relatively speaking, easier for a robot to live up to the ideal of being a good colleague than it is for a robot to live up to the ideal of being a good friend or romantic partner. A stronger conclusion would be that not only is it easier for a robot to live up to the ideal of being a good colleague, but certain robots may one day be or are already able to live up to the ideal of being a good colleague.

6.5 Can a Robot be a Good Colleague? Part 2

Around the same time that the paper summarised in the previous section was written, Monika Betzler and Jörg Löschke (2021) were working on a paper called "Collegial Relationships". Just as Smids and I noted that there hadn't been any philosophical literature about what it is to be a good colleague that we could draw upon in our discussion of whether a robot can be a good colleague (including any more general discussion of what a good colleague is in the first place), Betzler and Löschke, writing around the same time, also start their paper by noting that "[d]espite their significance, collegial relationships have not been the subject of much philosophical study" (Betzler and Löschke, 2021, p. 214). They also note that philosophers have written extensively about other forms of relationships, but not about that between colleagues. Lamenting this fact, Betzler and Löschke write "[g]iven that workplace relationships between colleagues can have a significant impact on our lives, they deserve to be studied in more detail" (Betzler and Löschke, 2021, p. 214). Betzler and Löschke seek—in this paper, as well as in a partly empirical paper they also wrote together with Kevin Reuter (Reuter, Löschke and Betzler, 2020)—to fill this gap. Since they were writing at the same time as Smids and I were writing our paper, and both sets of authors were writing about this topic independently of each other, it is interesting to contrast and compare the discussions and conclusions of these respective papers. Of course, Betzler and Löschke were not writing about whether robots can be good colleagues; their paper is about what it is for human beings to be good colleagues. But one can take the view they arrive at and ask whether, by those criteria, it is possible for robots to be good colleagues and one can compare their criteria for collegiality with those in the paper discussed above.

One noteworthy thing is that while Smids and I had taken work in logistics warehouses and within the army as the sorts of cases we started our discussion from, the work contexts that Betzler and Löschke use as their main contexts are different. They primarily discuss cases of academic colleagues and colleagues in a business context. Another noteworthy difference is that while Smids and I went directly to the question of what features or qualities one might sensibly associate with the ideal of being a good colleague, Betzler and Löschke use a two-step approach. They first discuss the more basic question of what a colleague is in the first place and then, after that, they turn to the question of, as they put it, what values are realised within the context of relationships between good colleagues: what they call the values related to collegiality.

In terms of what a *colleague* is, as they see things, the first point Betzler and Löschke (and Reuter, Löschke and Betzler) make is that this concept has a 'dual nature', with both a description side and an evaluative side to it (Reuter, Löschke and Betzler, 2020; Betzler and Löschke, 2021). That is, there are purely descriptive aspects that specify what it is to be a colleague, but those aspects already point to certain standards, whereby we can assess whether somebody is a good colleague or not, since these aspects point towards certain standards or values that one can live up to or fail to live up to. When it comes to what it is to be a colleague in the first place, Betzler and Löschke suggest that two people are colleagues, in the sense they intend, if they fulfil at least two of three conditions: they "exhibit sameness" regarding (1) "work content or domain of activity", (2) "the same institutional affiliation or common purpose" and/or (3) "the same status or level of responsibility" (Betzler and Löschke, 2021, p. 213). To use one of the examples they discuss: two professors at the same university might be colleagues by these standards. They might be colleagues by these standards even if they work in different departments of the university (perhaps one is a philosopher and the other a computer scientist) because they have similar work content or domain of activity, because they share an institutional affiliation (viz. the same university) with a common purpose, and/or because they have the same status and level of responsibility. Neither of them, in contrast, is a colleague, by these criteria, with, say, a chef working at a restaurant across the street from the main building of the university: the chef may fulfil none of these three conditions, or perhaps only one of them. This can be contrasted with a professor (say an art historian) working at some other university, who might not fulfil criterion number 2 but who might fulfil criteria 1 and 3, and therefore be a colleague by these criteria.

The next main part of Betzler and Löschke's view is the idea that "the potential value of collegial relationships is grounded in the relationship goods that two colleagues have reason to generate *qua* colleagues" (Betzler and Löschke, 2021, p. 213). That is, if two people are related in at least two of the three ways described above, what distinctive goods do they have reason to try to realise within the context of their collegial relationship? According to Betzler and Löschke, among colleagues it makes sense to aim for two main goods: namely, 'collegial solidarity' (basically, supporting each other) and 'collegial recognition' (basically, giving credit where credit is due). In short, two people are *colleagues* if they are in the same line of work, in the same organisation and/or have the same status or level of responsibility. And they are *good colleagues*—they realise the distinctive values of collegiality—if they give each other solidarity and due recognition.

From the point of view of this picture of what a colleague is and what good collegial relationships are like, what could we say about human–robot relationships in the workplace? By these just-sketched criteria, could robots be colleagues? And could they be good colleagues in the sense of helping to realise the relationship goods Betzler and Löschke associate with collegiality? Let us now consider these questions.

Regarding the first question, that is, whether a robot could be a colleague by these criteria, it seems that a case could be made for the conclusion that, in certain circumstances, a robot could be a colleague of ours. That idea might seem a little strange. But after all, the idea is that a colleague needs to be able to fulfil at least two of three criteria. And it seems to make sense to think that a robot could fulfil at least the first two of the three criteria. For example, a robot in a warehouse could have a similar work content or domain of activity to a human being working alongside it. And, secondly, a robot could—at least in a loose sense—have the same affiliation and be part of the same organisation. It seems highly unlikely, in contrast, that a robot (say, a warehouse robot again) would have the same status or similar responsibilities to the human beings it works alongside. However, in order to be a colleague by these criteria, it is enough to fulfil at least two of the three. And the first two seem fulfillable by a robot. So, it seems that the Betzler and Löschke view could be taken to indicate that a robot could be our colleague.

What is much less clear, however, is whether a robot could be in a relationship with any of its work colleagues whereby the 'relationship goods' of collegiality that Betzler and Löschke identify could be realised. Could there be solidarity between a human and a robot? The human being might experience some form of solidarity towards the robot: perhaps, for example, the soldiers working with Boomer would describe themselves as having an attitude or stance of solidarity towards Boomer. But it seems hard to imagine what it might mean for the robot to reciprocate the solidarity.

Similarly, while the human being might take it upon him- or herself to give the robot any recognition that they think is due, it makes less sense to think that a robot could give a human any recognition in return. Catrin Misselhorn (2021)—in a different context—discusses robots and recognition. And she argues that if we seek recognition from a robot, we in effect act in a way that appears to reduce ourselves to things: we treat ourselves as if we are not conscious persons but as if we are merely behaving machines without a mind.

If Misselhorn is right that we cannot sensibly hope to realise the value of mutual recognition in a relationship with a robot, and we also cannot have mutual solidarity with a robot, then we cannot realise the values of collegiality that Betzler and Löschke describe within the context of a collegial relationship with a robot. Perhaps a robot, in other words, can be a colleague by their criteria, but a robot does not seem to be able to be a good colleague by the standards offered in the Betzler and Löschke paper.

6.6 Robots as (Good) Colleagues: An On/Off (or 'Yes' or 'No') Issue, or a Matter of Degree?

The two previous sections suggest two different ways of answering the question of whether robots could be good colleagues. By the standards discussed in the paper reviewed in section 6.4, it could be argued that it is at least easier for a robot to be a good colleague than it is for a robot to be a good friend or romantic partner. Moreover, on a purely behavioural level, perhaps we can even say that a robot can, to some extent, be a good colleague by the criteria discussed in section 6.4. In contrast, by the standards discussed in section 6.5, a different conclusion seems to follow: while it can make sense to regard a robot as a colleague by those standards, it seems to make less sense to think that it could be a good colleague which helps to realise the distinctive values that Betzler and Löschke associate with collegiality: that is, mutual solidarity and recognition. What should we make of these different tentative conclusions from sections 6.4 and 6.5?

One question here would be what would happen if we tried to combine the ideas about what constitutes colleagues and good colleagues discussed in sections 6.4 and 6.5. As noted above, the papers discussed in those two sections were written independently of each other. And the papers seem to have been written with rather different kinds of work contexts in mind (one with a focus on warehouse logistics work and the military; the other with a focus on colleagues at universities or in the business world). So, it would be interesting to contrast and compare the views about collegiality in these two papers in more detail, to explore whether one has to choose between these two accounts of what a good colleague is or whether one could somehow combine them. Such a discussion, however, goes beyond the scope of this chapter.

Another idea will instead be taken up here: namely, whether the issue of whether a robot can be a good colleague is an all-or-nothing issue or whether we should think of it, rather, as something that could be a matter of degree. Here we can again make a comparison with recent philosophical work on the question of whether robots can be our friends. In particular, we can consider a suggestion that is made in a paper by Helen Ryland (2021). Ryland develops what she calls a 'degrees-of-friendship' view, and claims that while robots might not be able to be friends in the fullest sense, perhaps they can be our friends at least to some, potentially limited, degree. Perhaps we could say something similar about the issue of whether robots could be (good) colleagues.

On Ryland's view, friendship is an ideal with many different aspects to it, a little bit like how the paper discussed in section 6.4 also has a whole list of criteria that can be associated with the ideal of being a good colleague. Ryland suggests that while robots might not live up to all of the many different criteria that we may associate with being a good friend, they might live up—at least in part—to some of those criteria. The same, Ryland suggests, can sometimes apply to some of our human friends: they may live up to some but not all of the criteria of being a good friend, at least to some degree. So, there can be degrees of friendship, Ryland thinks, and friendship is not an either/or matter. Because of this, it makes sense, she thinks, to be open to the possibility that robots might be friends of ours, at least to a certain, perhaps limited, degree.

We might potentially want to say something similar about robots as potential colleagues in the workplace. Perhaps they cannot live up to all of the criteria that we may associate with being a good colleague. And perhaps they can only live up to some of the criteria in a partial way. But this might be enough, it might be suggested on a Ryland-inspired view, for it to be the case that robots could be good colleagues to a certain degree, and perhaps only a very limited degree. We don't have to say, in other words, that robots either can or cannot be good colleagues. We don't have to view this as an all-or-nothing, either/or type of issue. Instead we could potentially view this as something that could be a matter of degree.

6.7 Is There Something Unsuitable or Ethically Problematic About Treating a Robot as a Colleague?

Above, we have considered whether a robot can be a colleague or even a good colleague based on what properties or abilities the robot has. In this section we will consider the question of whether treating a robot like a member of the team, or even as if it is a good colleague, might be a problematic way of interacting with a robot. That is to say, we are now switching the focus from the robot and what it is able to do, to the human beings interacting with a robot and the appropriateness of how they are responding to it. A particular robot might not have properties or abilities that warrant our thinking it can be a good colleague, but people working alongside it might nevertheless become attached to it and treat it as if it were a good colleague. This possibility is illustrated by the opening example of Boomer the military robot. That robot was designed to be a tool in bomb disposal work. It was not designed to be a so-called social robot that responds to people in social ways. Nor, as mentioned above, did the robot look like a human being or an animal; it looked more like a lawnmower or a small tank. And yet the soldiers became attached to it and treated it as if it were a member of the team, even to the point of honouring it with a military funeral and medals when it was destroyed (Carpenter, 2015).

While there is no literature that specifically deals with whether treating robots as if they are good colleagues or members of our workplace teams is ethically appropriate, there is a very large discussion in the literature about how we should, or should not, interact with robots more generally. Other chapters in this book cover that topic in more detail. Here we can end this chapter by very briefly reflecting on how some contributions to those discussions might apply to the context of workplace interaction with robots in particular. We can start with the view that there is something ethically problematic about interacting with a robot in the way that the soldiers described above interacted with Boomer the robot.

According to Joanna Bryson (2010, 2018), we should regard and treat all technologies—including robots—like tools that are there to serve us and our needs. It is a mistake to treat or regard any technology as if it were a person with an important social status. Technologies, including robots, are bought and sold. So, we should treat them like property that is there to be used, bought and sold, and never as anything over and above that. This is the so-called purely instrumental theory of all technology. From this point of view, it is a mistake—if not downright unethical—to treat any robot as if it is more than a tool. This would apply to robots and AI systems in the workplace just as much as to technologies used in any other context. From the standpoint of this purely instrumental view of technology, then, behaviour such as that exhibited by the soldiers who treated Boomer like their colleague involves a form of conceptual confusion about what technologies are and how they should be regarded. As Bryson sees things, it is moreover an insult to those who can really be our friends, partners or moral equals to treat robots as if they could be our equals. It sends a bad message to people. It sends the message that an equal could be a tool that one can turn on or off (Nyholm, 2020, p. 2).

A radically opposed view is offered by authors like Mark Coeckelbergh (2010), David Gunkel (2018) and Janina Loh (2019). They argue that we should not necessarily regard all technologies as mere tools, because this is not how human beings naturally or spontaneously respond to all technologies. There are some technologies that trigger social responses in us, and, as these authors see things, there is nothing wrong with that. To quote Gunkel: treating some robots and AI systems as if they have a social presence (as if they could be our friends, partners or perhaps even moral equals) "is not a bug to be eliminated or fixed; it is a feature of human sociality" (Gunkel, 2018, p. 115). Loh, in turn, goes even further and argues that we should not criticise but celebrate some people's tendencies to respond socially to robots. Discussing the fact that some people want to have robots as their friends or partners and become very attached to them, Loh argues that this—that is, the tendency to respond socially to robots—should not be seen as a 'shortcoming' but rather as an 'ability' or 'capability'. Diversity is an important value, Loh argues, and in keeping with this, we should take an 'inclusive' (rather than 'exclusive') attitude towards those who want to engage socially with robots, and value rather than look down on their ways of relating to robots (Loh, 2019, pp. 95–121). Coeckelbergh (2010) has dubbed this a 'relational turn' in how we can think about the status of robots. What is important is not what properties or abilities they have, but how we relate to them. If these authors are right about this, we could apply that idea to the context of work. We could argue that if people relate to robots and other AI systems in the workplace as if those technologies are members of the team and good colleagues, then there is nothing wrong about these robots coming to have, or be seen as having, the status of good colleagues within these work contexts. Technologies, on this view, are not necessarily only mere tools. They can sometimes be part of valuable relationships or ways of relating to what is around us.

It is not the aim of this chapter to adjudicate between these radically different points of view about how it is appropriate to behave around robots in the workplace. Yet, we can end by bringing up one type of consideration that one might keep in mind and reflect on when one considers these different perspectives on whether it can be appropriate or whether it is perhaps inappropriate to treat robots or other AI technologies as if they are members of the team or perhaps even good colleagues. One can ask whether it is beneficial for or detrimental to the people involved or any other people affected to treat, or avoid treating, robots in the workplace as if they can be members of the team or good colleagues (cf., Nyholm, 2020, ch. 8).

Might it, for example—to end with the case we began with—be soothing or comforting for soldiers who work in a high-stress environment to treat a robot they are working alongside as if it is a member of the team and a good colleague? Might this relieve some of the pressure and stress associated with the kind of environment in which they are working? If so, it might be argued that this provides pragmatic reasons in favour of viewing it as acceptable and perhaps even appropriate for them to treat or regard the robots they work alongside as if they are members of the team, even if the robots do not look like human beings and even if those robots do not have any social capabilities.

If, however, their coming to regard a robot as a member of their team might create risks for them, then this might make it inappropriate for them to treat the robots in question as if they are their colleagues and valuable members of their team. For example, they might start risking their own lives or the lives of other human beings in order to save their 'colleague'. That might be viewed as a problem. In general, one of the key questions we must ask from an ethical point of view when it comes to how we treat robots—whether it is in the workplace or elsewhere—is whether it is good for us in any way, or whether treating robots in certain ways (e.g., by regarding them as colleagues) might be bad for us human beings in some important way.

References

Berkers, H., Smids, J., Nyholm, S., and Le Blanc, P. (2020). Robotiering en betekenisvol werk in distributiecentra: Bedreigingen en kansen. *Gedrag & Organisatie*, *33*(4), 324–347.

Bettiza, S., Alonna, E., and Furst, J. (2021). Gods and robots: Will AI transform religion? BBC News. https://www.bbc.com/news/av/technology-58983047.

Betzler, M., and Löschke, J. (2021). Collegial relationships. *Ethical Theory and Moral Practice*, *24*(1), 213–229.

Bryson, J. J. (2010). Robots should be slaves. In Y. Wilks (ed.), *Close engagements with artificial companions* (pp. 63–74). John Benjamins.

Bryson, J. J. (2018). Patiency is not a virtue: The design of intelligent systems and systems of ethics. *Ethics and Information Technology*, *20*(1), 15–26.

Carpenter, J. (2015). *Culture and human–robot interactions in militarized spaces*. Routledge.

Chivers, T. (2021). How DeepMind is reinventing the robot. IEEE Spectrum. https://spectrum.ieee.org/how-deepmind-is-reinventing-the-robot.

Coeckelberg, M. (2010). Robot rights? Towards a social-relational justification of moral consideration. *Ethics and Information Technology*, *12*(3), 209–221.

Danaher, J. (2017). Will life be worth living in a world without work? *Science and Engineering Ethics*, *23*(1), 41–64.

Danaher, J. (2019a). *Automation and utopia*. Harvard University Press.

Danaher, J. (2019b). The philosophical case for robot friendship. *Journal of Posthuman Studies*, *3*(1), 5–24.

Danaher, J., and Nyholm, S. (2021). Automation, work and the achievement gap. *AI and Ethics*, *1*(3), 227–237.

Dignum, V. (2019). *Responsible artificial intelligence*. Springer.

Ford, M. (2015). *Rise of the robots: Technology and the threat of a jobless future*. Basic Books.

Garber, M. (2013). Funerals for fallen robots. The Atlantic. https://www.theatlantic.com/technology/ar chive/2013/09/funerals-for-fallen-robots/279861/.

Gordon, J.-S., and Nyholm, S. (2021). Ethics of artificial intelligence. *Internet Encyclopedia of Philosophy*. https://iep.utm.edu/ethic-ai/.

Groom, V., and Nass, C. (2007). Can robots be teammates? Benchmarks in human–robot teams. *Interaction Studies*, *8*(3), 483–500.

Gunkel, D. (2018). Robot Rights. MIT Press.

Herzog, L., and Gheaus, A. (2016). The goods of work (other than money!). *Journal of Social Philosophy*, *47*(1), 70–89.

Hosseini, Z., Nyholm, S., Le Blanc, P., Preenen, P., and Demerouti, E. (Under review). An ethical framework for evaluating experimental technologies in workplace settings.

Hosseini, Z., Nyholm, S., Le Blanc, P.M. et al. (2023): "Assessing the artificially intelligent workplace: an ethical framework for evaluating experimental technologies in workplace settings". AI Ethics. https://doi.org/10.1007/s43681-023-00265-w

Lee, M. (2021). Interactional morality: Technology as our moral mirror. Doctoral dissertation, Eindhoven University of Technology. https://research.tue.nl/en/publications/interactional-morality-technology-as-our-moral-mirror.

Ljungblad, S., Kotrbova, J., Jacobsson, M., Cramer, H., and Niechwiadowicz, K. (2012). Hospital robot at work: Something alien or an intelligent colleague? In *Proceedings of the ACM 2012 conference on Computer Supported Cooperative Work* (pp. 177–186). ACM.

Loh, J. (2019). *Roboterethik: Eine Einführung*. Suhrkamp.

Misselhorn, C. (2021). Künstliche Intelligenz und Empathie. Reclam.

Nyholm, S. (2020). *Humans and robots: Ethics, agency, and anthropomorphism*. Rowman & Littlefield International.

Nyholm, S., and Smids, J. (2020). Can a robot be a good colleague? *Science and Engineering Ethics*, *26*(4), 2169–2188.

Reuter, K., Löschke, J., and Betzler, M. (2020). What is a colleague? The descriptive and normative dimension of a dual character concept. *Philosophical Psychology*, *33*(7), 997–1017.

Royakkers, L., and van Est, R. (2015). *Just ordinary robots: Automation from love to war*. CRC Press.

Russell, S., and Norvig, P. (2010). *Artificial intelligence: A modern approach*. Prentice Hall.

Ryland, H. (2021). It's friendship, Jim, but not as we know it: A degrees-of-friendship view of human–robot friendships. *Minds and Machines*, *31*(3), 377–393.

Sauppé, A., and Mutlu, B. (2015). The social impact of a robot co-worker in industrial settings. In *Proceedings of the 33rd annual ACM conference on human factors in computing systems* (pp. 3613–3622). ACM.

Seyitoğlu, F., Ivanov, S., Atsız, O., and Çifçi, İ. (2021). Robots as restaurant employees—A double-barrelled detective story. *Technology in Society*, *67*, 101779.

Smids, J., Nyholm, S., and Berkers, H. (2020). Robots in the workplace: A threat to—or opportunity for—meaningful work? *Philosophy and Technology*, *33*(3), 503–522.

Tigard, D. (2021). Workplace automation without achievement gaps: A reply to Danaher and Nyholm. *AI and Ethics*, *1*(4), 611–617.

Turkle, S. (2004). Whither psychoanalysis in the computer culture? *Psychoanalytic Psychology*, *21*(1), 16–30.

Van de Poel, I. (2016): "An ethical framework for evaluating experimental technology". Science and Engineering Ethics *22*(3), 667–686

Veliz, Carissa (2020). Privacy is Power. Penguin.

Elisabeth Hildt

7 Artificial Intelligence, Selfhood and Artificial Consciousness

Abstract: This chapter discusses two ways of looking at the topic of artificial intelligence (AI), selfhood and artificial consciousness. The first is to reflect on human–AI interaction focusing on the human users. This includes how humans respond to and interact with AI, as well as potential selfhood-related implications of interacting with AI. The second is to reflect on potential future machine selfhood and its crucial component, artificial consciousness. While artificial consciousness that resembles human consciousness does not yet exist, and the details are difficult if not impossible to anticipate, a reflection on potential future artificial consciousness is clearly needed given its extensive ethical and social implications.

Keywords: human–technology interaction, social robots, mental health, machine consciousness, moral status

7.1 Introduction

Selfhood is a concept with a broad spectrum of uses. The Merriam-Webster online dictionary,[1] for example, defines selfhood as individuality, and lists character, identity, individualism, individuality, personality and self-identity as synonyms for selfhood. Depending on the field of study and context, authors focus on different aspects of the concept.

In philosophical debates, selfhood has been discussed in several ways (Parfit, 1984; Zahavi, 2020). One is in the context of diachronic identity, that is, the question of what constitutes identity through time. Another puts the focus on synchronic identity and the relationship between selfhood and conscious experience. The self, that is, the perspective from which we experience the world, is seen as a phenomenal me with conscious experiences. These philosophical reflections that focus on the self as the subject of experiences investigate the close relation between phenomenal consciousness and selfhood.

However, when we think of ourselves in our life contexts, the conception of selfhood at play is much broader (Elliott, 2007). It is based on human self-understanding and how we see ourselves as individuals, groups or members of society. This concep-

1 https://www.merriam-webster.com/dictionary/selfhood#synonyms.

Elisabeth Hildt, Illinois Institute of Technology

https://doi.org/10.1515/9783110721751-007

tion includes how we understand and invent ourselves, and how we interact with each other in interpersonal and societal contexts. In this, among a myriad of other factors, the technologies we use and the ways we interact with technologies play a considerable role.

Selfhood is central in any form of conscious interaction with technology. Selfhood influences how we interact with technology, but it is also shaped by the ways we interact with technology, how we perceive ourselves and how others perceive us in the context of interacting with technology. While this is the case with any technology or tool, even with tools as simple as a hammer, this intertwinement is particularly extensive with artificial intelligence (AI), that is, technology with complex problem-solving capabilities.

There are two ways of looking at the topic of AI, selfhood and artificial consciousness. The first is to reflect on human–AI interaction focusing on the human users. This includes how humans respond to and interact with AI, as well as potential selfhood-related implications of interacting with AI. The second is to reflect on potential future machine selfhood and its crucial component, artificial consciousness. While artificial consciousness that resembles human consciousness does not yet exist, and the details are difficult if not impossible to anticipate, a reflection on potential future artificial consciousness is clearly needed given its extensive ethical and social implications.

Both will be discussed in the following. I begin with the first topic by discussing human–AI interaction and how human users respond to AI and social robots, that is, robots that interact with humans at a social level. Then I will reflect on potential future artificial consciousness and its implications.

7.2 Human–AI Interaction

The past years have seen a growing number of applications for AI technology and social robots. In the future, the use of AI is expected to grow in many fields. Medicine clearly is one of these fields, with AI applications in diagnosis, treatment, research and education contexts. Also, a relatively broad spectrum of social robot uses in healthcare, patient care or geriatric care is on the horizon (Loh, 2018; Briganti and Le Moine, 2020; Pepito et al., 2020).

So far, in only a few contexts of robot deployment in healthcare contexts, have the implications on human users been investigated in detail. Two examples will be presented: the use of social robots to support therapeutic intervention in the context of autism spectrum disorder (ASD), and the deployment of seal robot PARO in elderly care settings. The text then goes on with more general reflections on human–AI interaction.

7.2.1 Social robots in healthcare

How does interacting with AI technology influence human users? Unfortunately, this is a highly under-investigated field. So far, there is a lack of empirical research into the positive or negative implications of human–AI or human–robot interaction on human users. However, a few relatively well-studied examples exist where social robots have been deployed in healthcare settings to improve mental health. Among them are therapeutic robot interventions in the context of ASD and the use of the seal robot PARO in elderly care. These studies are indicative of the assumption that human–robot interaction has the potential to positively influence the mental health of certain user groups.

A broad spectrum of social robots has been deployed in the context of ASD, a condition characterised by deficits in communication and social interaction and restricted repetitive behaviours. The social robots have been used in therapeutic robot interventions and educational settings for autistic children, including simple robots, animal robots and humanoid robots (Alcorn et al., 2019; DiPietro et al., 2019; Kumazaki et al., 2020). The educational interventions are intended for autistic children to benefit and learn from interacting with social robots. Robot interventions have been shown to support children with ASD in developing social behaviours. For example, improvements in eye contact, self-initiated interactions, turn-take activities, emotion recognition and joint attention were observed (DiPietro et al., 2019; Kumazaki et al., 2020). Robots provide a highly structured learning environment; they behave in simpler and more predictable ways than humans, which results in a high degree of task engagement. The interventions are based on the working assumption that autistic children are interested in and motivated by robots and that they tend to prefer interacting with robots over humans.

Notwithstanding these positive reports, the deployment of social robots in the context of ASD is not uncontroversial (Richardson et al., 2018; Alcorn et al., 2019). Several authors have objected that using robots may prove far from beneficial for autistic children as it reduces investment in developing their social and communication skills through close relationships and human–human interaction. The working assumption that autistic children prefer robots was criticised as misleading, citing the central role of close contact and direct interaction with family members and caregivers for autistic children. Furthermore, critics stressed that most of the studies involving social robot interventions are researcher guided and that not much is known about the clinical or educational needs and the experiences of users (Richardson et al., 2018; Alcorn et al., 2019).

Kathleen Richardson and colleagues questioned how children with autism are presented in the robotics literature, when they are characterised as detached from close, empathetic relationships, instead preferring technology. The authors write: "Could children with autism be *othered* by the framing of autism in the robotic literature—led by researchers who have a stake in emphasising (or overemphasising) the benefits of robots?" (Richardson et al., 2018, p. 33). They argue that characterising autistic children in this way may lead to inadequate stigmatisation and practices based on counterproductive stereotypes. In particular, they criticise analogies being made between autistic children and robots in the robotics literature, such as characterising robots as autistic, which seems to imply the question of whether autists are like robots (Richardson et al., 2018).

Another relatively well-studied example is the use of the PARO seal robot as a therapeutic robot. The robot reacts to stroking and encourages nurturing behaviour. PARO has been deployed primarily in elderly care, especially with elderly dementia patients (Sharkey and Wood, 2014; Hung et al., 2019; Moyle et al., 2019; Pu, Moyle and Jones, 2020). The research found that it increases social interaction between nursing home residents. Interacting with PARO reduced anxiety, depression and agitation in elderly people with dementia. PARO was reported to have promoted a positive mood, quality of care experience and a reduction in pain. It is unclear, however, whether the positive effects reported are primarily short-term benefits brought about by new and, therefore, interesting technology and whether they will fade out over time. Lihui Pu and colleagues investigated how people with dementia perceive a PARO robot (Pu, Moyle and Jones, 2020). The interview study with elderly dementia patients reveals the users' positive attitudes towards the seal robot and positive therapeutic effects with regard to mood enhancement and pain reduction. The study, which provides several direct user quotations, also reveals that some users consider PARO a robot, whereas others consider it a real animal.

Also, downsides of PARO deployment have been reported. Besides issues related to the cost of the social robot, additional workload for staff and caregivers, and infection concerns, these criticisms and objections relate to negative implications of this form of robot deployment. In some users, negative emotional reactions such as anger, agitation or fearfulness were observed. The use of PARO was criticised as implying a reduction of social contact with caregivers and human beings. Also, the stigma of interacting with a robot animal was raised as a concern (Hung et al., 2019). PARO deployment was criticised as belittling, demeaning and involving deception, as a mechanical toy is given to elderly and demented people who are assumed to treat it like a real pet with which they can build a relationship. Amanda Sharkey and Natalie Wood (2014) consider objections about deception and infantilisation as related to the concept of human dignity. As with the use of social robots in the context of ASD, one of the criticisms is that studies around PARO deployments in elderly care settings are primarily researcher-guided interventions. Not much is known about user needs and experiences (Hung et al., 2019).

Taken together, these two examples of social robot use in healthcare contexts not only give an impression of positive or negative effects of interacting with social robots on human users, but also indicate that human–robot interaction can have modulating effects on how users are perceived by others. While much more research is needed, both examples reveal self-related implications of human–robot interaction.

7.2.2 Interpreting human–AI interaction

Research into human–technology interaction and human–media interaction suggests that humans tend to interact with media and technology akin to how they interact with the real world. In their influential book *The Media Equation*, Byron Reeves and Clifford Nass summarise research on how humans interact with computers and new media in fundamental social and natural ways (Reeves and Nass, 2002). They write: "Media can evoke emotional responses, demand attention, threaten us, influence memories, and change ideas of what is natural. Media are full participants in our social and natural world" (Reeves and Nass, 2002, p. 251). The authors argue that these forms of responses to media and computers are made without deliberation or rational thinking. Accordingly, in human–media interaction, humans automatically assume reality and respond socially and naturally. Reeves and Nass conclude that what matters in human–media interactions is not the technological reality but how humans perceive media and computers.

While Byron Reeves and Clifford Nass clearly focus on human–media interaction, their approach has also been used to interpret human–technology interaction more broadly. Recent studies with AI technology and social robots align with the position of Reeves and Nass. Of particular interest in this context is interdisciplinary research that uses neuroimaging technologies to investigate the human brain while interacting with social robots (Cross, Hortensius and Wykowska, 2019; Henschel, Hortensius and Cross, 2020).

Even though there is consensus that current AI technology does not display capabilities resembling a human mind or human consciousness, research into social cognition and human–robot interaction indicates a tendency among humans to interact with social robots in ways similar to how they interact with humans. Anthropomorphic features in robots facilitate this effect.

In particular, a certain tendency was reported to take the so-called 'intentional stance' towards social robots, that is, to give mentalistic explanations of robot behaviour and to assume that the technology has beliefs or desires (Dennett, 1989; Wiese, Metta and Wykowska, 2017). Accordingly, human users tend to ascribe agency and autonomy to social robots or consider them companions. Examples include the robotic dog AIBO, the seal robot PARO, and humanoid robots like NAO, Pepper and Sophia (Scheutz, 2011; Gunkel, 2018; Nyholm, 2020).

Human emotional involvement was also reported as a factor that strengthens human–robot interaction (Scheutz, 2011; Henschel, Hortensius and Cross, 2020). Emotional user responses have been observed with relatively neutral technology such as the Roomba vacuum cleaner or military robots (Scheutz, 2011). Emotional entanglement is particularly pronounced with technology that is intentionally built to sense, interpret and respond to human emotions, that is, designed to facilitate human emotional involvement in human–technology interaction (McDuff and Czerwinski, 2018).

Several authors have stressed the importance of these relational aspects for human–robot interaction, and, in view of the social roles of robots, argued in favour of ascribing rights to robots (Coeckelbergh, 2010; Darling, 2012; Gunkel, 2018). However, it must be seen that current social robots only react to human behaviour and *simulate* human behaviour and emotions. They do not display capabilities comparable to human emotions, human consciousness or a human mind. It is humans who interpret AI behaviour as human-like and respond accordingly. Thus far, attributing rights to robots seems highly questionable (Bryson, 2010, 2018; Hildt, 2019).

Regarding human–AI interaction, there are several broad questions and areas for investigation, all of which have implications for human selfhood. The first is about whether, when and to what extent human users ascribe mental states, emotions or consciousness to AI. The second is: how does interacting with AI and ascribing mental states, emotions or consciousness to AI technology influence human users? Thirdly, what is the best way for humans in human–AI interaction to deal with unidirectional emotional involvement, lack of human touch, and absence of interpersonal interaction or interpersonal communication? And a more indirect fourth type of question is: how may human–robot interaction influence how humans interact with each other, their relationships and social life? Clearly, more research on how humans interact with AI technology and robots is needed.

7.3 Potential Future Artificial Consciousness and Selfhood

The concepts of self and selfhood are closely related to the concepts of self-awareness and consciousness in that being a self requires awareness of oneself, one's position in time and space, and one's activities. A crucial capability for any potential future artificial self would be consciousness.

But what is artificial consciousness? While we are familiar with science-fiction depictions of technology displaying consciousness and stories being narrated from the fictitious first-person perspective of AI technology—a recent example is Kazuo Ishiguro's novel *Klara and the Sun*—, this fictitious artificial first-person perspective resembling a human first-person perspective in no way represents the current state of AI technology. Existing AI technology does not display consciousness comparable to human consciousness.

Beginning in the 1960s and 1970s, there has been a longstanding debate on whether machines can have a mind or be conscious. Whereas proponents of strong AI claim that, in principle, machines can have a mind or be conscious, proponents of weak AI reject this idea, stressing that AI can only exhibit intelligence in relatively narrow domains and only simulates the human mind. Among the latter is John Searle, who, in his article "Minds, Brains and Programs" (Searle, 1980), argued against the claim that "the appropriately programmed computer literally has cognitive states and that the programs thereby explain human cognition" (Searle, 1980, p. 417).

In this context, Searle introduced the widely known but highly criticised Chinese room thought experiment (Searle, 1980; Cole, 2020). In this thought experiment, a person who sits in a room and does not understand Chinese gets Chinese word input, then uses a huge book to look up the English correlations of the Chinese words, and, based on this, provides an English word output. Searle stresses that the person in the chamber does not understand anything and argues that the same holds for programmed computers. A person who understands and speaks Chinese would use a very different approach.

According to Searle, what is missing in machines is intentionality, which is the mind's capability to be about something, to represent or stand for something. Whereas mental states can be about objects and states in the world, Searle claims that this intentionality does not exist in computer programs as these rely on formal symbol manipulations. Accordingly, a machine that uses a program cannot understand anything because the formal symbol manipulations do not have intentionality; they do not refer to 'the outside world'.

Current AI technology lacks intentionality in this sense. It lacks consciousness, emotions and or self-consciousness—all of which are closely intertwined and would be important components of any form of artificial selfhood. However, robotics researchers are developing several approaches towards developing artificial consciousness and conscious robots. The following subsection gives a short overview of these approaches and the various artificial consciousness components they aim at instantiating. Based on this, the ensuing subsection reflects on the practical, ethical and social implications of potential future artificial consciousness.

7.3.1 Artificial consciousness—what is this?

Artificial consciousness is not yet technically feasible. Experts are in broad agreement that current AI technology does not display consciousness that is comparable in any way to human consciousness (Dehaene, Lau and Kouider, 2017; Esmaeilzadeh and Vaezi, 2021).

When talking or writing about potential future artificial consciousness, a considerable problem is that the concept of machine consciousness is only poorly specified. One of the reasons for this lack of specification is that it is difficult for humans to imagine or understand what machine consciousness is or would be. As humans, we are familiar with consciousness in the sense that from our own first-person perspective, we experience ourselves as being conscious. We know what it means to be conscious, to be conscious of something or to be conscious of ourselves. The first-person perspective clearly has shaped our understanding of the concept of consciousness, as has phenomenal consciousness, that is, the experience of what it is like to experience something consciously. However, humans do not have this internal, first-person perspective when it comes to machines. And while consciousness has traditionally been conceived in the context of biological brains, machines are technical systems.

The question of whether potential future machines display consciousness can only be discussed and accessed from the outside—unless or as long as it is not the case that machines are capable of convincingly claiming that they have conscious experiences. Even then, in this hypothetical situation, however, it would be necessary to rely not only on machine claims but also on external tests or standards. While several tests for machine consciousness exist (Elamrani and Yampolskiy, 2019), each of these tests depends on the consciousness definition or consciousness component used to develop the test. And they can assess machine consciousness only from the outside.

In robotics, several different approaches towards artificial consciousness exist. The various researchers use different consciousness definitions or stress different consciousness components, such as understanding and intentionality; subjective experiences; global availability; integration and self; and self-consciousness, self-awareness and self-monitoring (Hildt, 2022).

Several researchers stress that artificial consciousness would involve machines having subjective experiences, such as the capability to perceive and sense colours consciously, or to have emotions or feel pain (Carter et al., 2018; Reggia, Katz and Davis, 2018; Metzinger, 2021). Stanislas Dehaene and colleagues (2017, 2018), for example, consider two computational ingredients as crucial for conscious subjective states in machines: global information sharing and self-monitoring. While global information sharing refers to the capability of 'being conscious of something', self-monitoring characterises a system's capability to obtain information about itself. Kinouchi and Mackin regard consciousness as a function for effective adaptation at the system level that builds on the results of the underlying parallel-processing units. They assume this form of artificial consciousness "to correspond to how our mind is 'aware' when making our moment to moment decisions in our daily life" (Kinouchi and Mackin, 2018, p. 1).

Given the close interrelatedness between consciousness and selfhood, the debate around artificial consciousness also involves concepts like self-model, self-monitoring and self-awareness. For example, Drew McDermott stresses the central role of self-models for machine consciousness and describes phenomenal consciousness in computational systems as the property a system has if it models itself as experiencing things (McDermott, 2007). And Chatila et al. (2018)'s robotics research approach focuses on self-awareness, which the research group views as an expression of consciousness. They characterise self-aware robots as robots that would be capable of understanding their environment, cognisant of what they do, capable of learning from their own experiences, and indicating what and how they have learned. Accordingly, self-awareness involves meta-cognitive processes such as giving an account of one's internal states and decision making to oneself and others. The research team considers perception crucial for a robot to develop some sense of self-awareness or agency, and investigates the role of perception for robot self-localisation, interpretation of the environment, decision making and learning.

Overall, the debate on artificial consciousness is complicated because there is no uniform, clearly specified definition of artificial consciousness. Researchers rely on different consciousness components, including global information sharing and self-monitoring or self-awareness. Concepts like self-model or self-awareness are not clearly specified either. Furthermore, it is unclear how concepts like machine consciousness or machine self-awareness relate to selfhood, and how far any of these resemble the traditionally used concepts when referring to human consciousness, human self-awareness or human selfhood. For example: What does it mean to talk of a machine as being 'aware of itself'? Does this imply subjective experiences or a machine self with an internal perspective? And what would artificial selfhood be? These are very complex questions that are difficult to answer.

In part, it may be seen as just a matter of definition of what we consider artificial consciousness or machine self-awareness to be. However, as consciousness is crucial for human self-understanding and is of central relevance to the concept of a person, artificial consciousness—if it ever exists—would come with a myriad of issues. A detailed conceptual clarification of the concept of artificial consciousness and related concepts is urgently needed, in view of the ethical and social implications that would go along with machines displaying consciousness (see next subsection).

7.3.2 Artificial consciousness—potential implications

What would be the advantages of artificial consciousness? Machines displaying self-awareness or consciousness are expected to have improved capabilities to integrate information and execute complex tasks. This would allow them to interact with their environment more flexibly and more effectively. Hadi Esmaeilzadeh and Reza Vaezi (2021) argued that capabilities akin to human self-awareness or consciousness would allow machines to complete complex tasks requiring intuition and empathy.

Several authors have suggested that consciousness or self-awareness in robots would improve human–robot interaction. For example, Antonio Chella and colleagues summarise: "any artificial intelligence that successfully interacts with humans should need to be able to use first-person pronouns, self-describe, and be responsive to self-focusing stimuli in its surrounding environment" (Chella et al., 2020, p. 2).

Michael S. A. Graziano considers consciousness a precondition for machines to display social intelligence or empathy. He writes: "If we want to build machines that are skilled at interacting with people, we will need to build in consciousness in the same sense that people attribute consciousness to themselves and see consciousness in others. It is the root of empathy. Without that capacity, our computers are sociopaths" (Graziano, 2017, p. 7).

In a similar vein, Esmaeilzadeh and Vaezi (2021) argue that artificial consciousness would improve human–AI interaction by increasing the trust humans put in AI. They speculate that this could lead to strong human–machine bonding.

With all these suggestions, it must be seen, however, that implication of machine consciousness would go far beyond enhanced functionality or improved human–AI interaction. Consciousness, conscious mental states, subjective experiences or an entity's capability to experience pain or suffering have traditionally been considered crucial for an entity to have moral status. Entities with moral status count morally in their own right (Jaworska and Tannenbaum, 2018; Metzinger, 2021; Mosakas, 2021).

Several authors have argued that if AI displayed consciousness in a morally relevant sense, this would imply that it deserves protection, either as a moral patient or a moral agent (Schwitzgebel and Garza, 2015; Manzotti and Chella, 2018; Hildt, 2019; Mosakas, 2021). In this context, however, not all consciousness components currently under discussion would be equally relevant for moral status. Components like subjective experiences, self-awareness or capacity for suffering would certainly be central (Metzinger, 2021). The question of exactly what consciousness component and what level of complexity would be relevant for machines to be attributed moral status is extremely difficult to answer. This is particularly true given the current lack of conceptual clarity of the concept of machine consciousness and the problem of how to assess machine consciousness from the third-person perspective.

If machines displayed consciousness in a morally relevant form, this would have a broad spectrum of consequences. It is difficult at this point to anticipate what exactly attributing moral status to conscious machines would imply, especially as machines with different forms of morally relevant machine consciousness may have different needs. But it would mean that they deserve protection. When interacting with these conscious machines, it could be required for humans to refrain from demanding them to complete tasks of which these machines disapprove, to provide adequate working environments, or enable them to maintain or further develop their morally relevant capabilities. Several authors have argued in favour of rights for AI technology that displays human-like capabilities such as human-like consciousness (Schwitzgebel and Garza, 2015; Gunkel, 2018).

Potential future machines with consciousness that resembles human consciousness would raise a broad spectrum of ethical and social issues. When interacting with these machines, humans would have to morally consider these machines. Humans would have several negative and positive duties towards these machines. All of this would turn the concept of technology, according to which technology is a tool to be used for human benefit, upside down. In view of these far-reaching implications, it is important to avoid building machines that display consciousness in a morally relevant sense.

7.4 Conclusion

As outlined above, interacting with AI technology can influence humans in various ways. There is a clear need for empirical studies that investigate the implications of human–AI interaction for humans. Better knowledge of the implications of human–AI interaction is needed to broaden the basis for decision making on how to build and deploy future AI technology.

For humans interacting with AI, it is important to be aware of the technology's capabilities, to not unduly jump from observed presumed behaviour to ascribing mental states, consciousness or selfhood to machines.

This is particularly true in view of the increasingly complex AI technology being developed. While several approaches towards building artificial consciousness exist, current AI technology does not display consciousness that in any way resembles human consciousness. In view of the broad spectrum of practical, ethical and social implications of potential future morally relevant forms of machine consciousness, I have argued against developing machines with human-like consciousness.

A currently hypothetical question that I have not addressed is: what influences would potential future machines with human-like consciousness or other human-like capabilities have on human selfhood? While any answer to this question is pure speculation, potential future AI technology with human-like or superhuman capabilities would certainly challenge the view that humans are exceptional in the universe.

References

Alcorn, A. M., Ainger, E., Charisi, V., Mantinioti, S., Petrović, S., Schadenberg, B. R., Tavassoli, T., and Pellicano, E. (2019). Educators' views on using humanoid robots with autistic learners in special education settings in England. *Frontiers in Robotics and AI, 6*, 107. doi: 10.3389/frobt.2019.00107.

Briganti, G., and Le Moine, O. (2020). Artificial intelligence in medicine: Today and tomorrow. *Frontiers in Medicine, 7*, 27. doi: 10.3389/fmed.2020.00027.

Bryson, J. J. (2010). Robots should be slaves. In Y. Wilks (ed.), *Close engagements with artificial companions: Key social, psychological, ethical and design issues* (pp. 63–74). John Benjamins.

Bryson, J. J. (2018). Patiency is not a virtue: The design of intelligent systems and systems of ethics. *Ethics and Information Technology, 20*, 15–26.

Carter, O., Hohwy, J., van Boxtel, J., Lamme, V., Block, N., Koch, C., and Tsuchiya, N. (2018). Conscious machines: Defining questions. *Science, 359*(6374), 400.

Chatila, R., Renaudo, E., Andries, M., Chavez-Garcia, R.-O., Luce-Vayrac, P., Gottstein, R., et al. (2018). Toward self-aware robots. *Frontiers in Robotics and AI, 5*, 88. doi: 10.3389/frobt.2018.00088.

Chella, A., Pipitone, A., Morin, A., and Racy, F. (2020). Developing self-awareness in robots via inner speech. *Frontiers in Robotics and AI, 7*, 16. doi: 10.3389/frobt.2020.00016.

Coeckelbergh, M. (2010). Robot rights? Towards a social-relational justification of moral consideration. *Ethics and Information Technology, 12*, 209–221. doi: 10.1007/s10676-010-9235-5.

Cole, D. (2020). The Chinese room argument. In E. N. Zalta (ed.), *The Stanford encyclopedia of philosophy* (Winter 2020 edition). https://plato.stanford.edu/archives/win2020/entries/chinese-room/.

Cross, E. S., Hortensius, R., and Wykowska, A. (2019). From social brains to social robots: Applying neurocognitive insights to human–robot interaction. *Philosophical Transactions of the Royal Society B: Biological Sciences, 374*(20180024). doi: 10.1098/rstb.2018.0024.

Darling, K. (2012). Extending legal protection to social robots: The effects of anthropomorphism, empathy, and violent behavior towards robotic objects. In R. A. Calo, M. Froomkin and I. Kerr (eds), *Robot law*. https://papers.ssrn.com/sol3/papers.cfm?abstract_id=2044797.

Dehaene, S., Lau, H., and Kouider, S. (2017). What is consciousness, and could machines have it? *Science, 358*, 486–492.

Dehaene, S., Lau, H., and Kouider, S. (2018). Response. *Science, 359*(6374), 400–402.

Dennett, D. C. (1989). *The intentional stance*. MIT Press.

DiPietro, J., Kelemen, A., Liang, Y., and Sik-Lanyi, C. (2019). Computer- and robot-assisted therapies to aid social and intellectual functioning of children with autism spectrum disorder. *Medicina, 55*(8), 440. doi: 10.3390/medicina55080440.

Elamrani, A., and Yampolskiy, R. V. (2019). Reviewing tests for machine consciousness. *Journal of Consciousness Studies, 26*(5–6), 35–64.

Elliott, A. (2007). *Concepts of the self* (2nd ed.). Polity.

Esmaeilzadeh, H., and Vaezi, R. (2021). Conscious AI. https://arxiv.org/abs/2105.07879.

Graziano, M. S. A. (2017). The attention schema theory: A foundation for engineering artificial consciousness. *Frontiers in Robotics and AI, 4*, 60. doi: 10.3389/frobt.2017.00060.

Gunkel, D. J. (2018). *Robot rights*. MIT Press.

Henschel, A., Hortensius, R., and Cross, E. S. (2020). Social cognition in the age of human–robot interaction. *Trends in Neuroscience, 43*(6), 373–384. doi: 10.1016/j.tins.2020.03.013.

Hildt, E. (2019). Artificial intelligence: Does consciousness matter? *Frontiers in Psychology, 10*, 1535. doi: 10.3389/fpsyg.2019.01535.

Hildt, E. (2022). Ethics of potential artificial consciousness. Manuscript submitted for publication.

Hung, L., Liu, C., Woldum, E., Au-Yeung, A., Berndt, A., Wallsworth, C., Horne, N., Gregorio, M., Mann, J., and Chaudhury, H. (2019). The benefits of and barriers to using a social robot PARO in care settings: A scoping review. *BMC Geriatrics, 19*, 232. doi: 10.1186/s12877-019-1244-6.

Jaworska, A., and Tannenbaum, J. (2018). The grounds of moral status. In E. N. Zalta (ed.), *The Stanford encyclopedia of philosophy*. https://plato.stanford.edu/archives/spr2018/entries/grounds-moral-status/.

Kinouchi, Y., and Mackin, K. J. (2018). A basic architecture of an autonomous adaptive system with conscious-like function for a humanoid robot. *Frontiers in Robotics and AI, 5*, 30. doi: 10.3389/frobt.2018.00030.

Kumazaki, H., Muramatsu, T., Yoshikawa, Y., Matsumoto, Y., Ishiguro, H., Kikuchi, M., Sumiyoshi, T., and Mimura, M. (2020). Optimal robot for intervention for individuals with autism spectrum disorders. *Psychiatry and Clinical Neurosciences, 74*, 581–586.

Loh, E. (2018). Medicine and the rise of the robots: A qualitative review of recent advances of artificial intelligence in health. *BMJ Leader*. doi: 10.1136/leader-2018-000071.

Manzotti, R., and Chella, A. (2018). Good old-fashioned artificial consciousness and the intermediate level fallacy. *Frontiers in Robotics and AI, 5*, 39. doi: 10.3389/frobt.2018.00039.

McDermott, D. (2007). Artificial intelligence and consciousness. In P. D. Zelazo, M. Moscovitch and E. Thompson (eds), *The Cambridge handbook of consciousness* (pp. 117–150). Cambridge University Press.

McDuff, D., and Czerwinski, M. (2018). Designing emotionally sentient agents. *Communications of the ACM, 61*(12), 74–83. doi: 10.1145/3186591.

Metzinger, T. (2021). Artificial suffering: An argument for a global moratorium on synthetic phenomenology. *Journal of Artificial Intelligence and Consciousness, 8*(1), 43–66.

Mosakas, K. (2021). On the moral status of social robots: Considering the consciousness criterion. *AI & Society*, *36*, 429–443.
Moyle, W., Bramble, M., Jones, C. J., and Murfield, J. E. (2019). "She had a smile on her face as wide as the Great Australian Bite": A qualitative examination of family perceptions of a therapeutic robot and a plush toy. *The Gerontologist*, *59*(1), 177–185. doi: 10.1093/geront/gnx180.
Nyholm, S. (2020). *Humans and robots: Ethics, agency, and anthropomorphism*. Rowman & Littlefield.
Parfit, D. (1984). *Reasons and persons*. Oxford University Press.
Pepito, J. A., Ito, H., Betriana, F., Tanioka, T., and Locsin, R. C. (2020). Intelligent humanoid robots expressing artificial humanlike empathy in nursing situations. *Nursing Philosophy*, *21*(4), e12318. doi: 10.1111/nup.12318.
Pu, L., Moyle, W., and Jones, C. (2020). How people with dementia perceive a therapeutic robot called PARO in relation to their pain and mood: A qualitative study. *Journal of Clinical Nursing*, *29*, 437–446.
Reeves, B., and Nass, C. (2002). *The media equation: How people treat computers, television, and new media like real people and places*. CSLI Publications.
Reggia, J. A., Katz, G. E., and Davis, G. P. (2018). Humanoid cognitive robots that learn by imitating: Implications for consciousness studies. *Frontiers in Robotics and AI*, *5*, 1. doi: 10.3389/frobt.2018.00001.
Richardson, K., Coeckelbergh, M., Wakunuma, K., Billing, E., Ziemke, T., Gómez, P., Vanderborght, B., and Belpaeme, T. (2018). Robot enhanced therapy for children with autism (DREAM). *IEEE Technology and Society Magazine*, March, 30–39.
Scheutz, M. (2011). The inherent dangers of unidirectional emotional bonds between humans and social robots. In P. Lin, K. Abney and G. A. Bekey (eds), *Robot ethics: The ethical and social implications of robotics* (pp. 205–222). MIT Press.
Schwitzgebel, E., and Garza, M. (2015). A defense of the rights of artificial intelligences. *Midwest Studies in Philosophy*, *39*(1), 98–119.
Searle, J. R. (1980). Minds, brains and programs. *Behavioral and Brain Sciences*, *3*(3), 417–424.
Sharkey, A., and Wood, N. (2014). The Paro seal robot: Demeaning or enabling? In *Proceedings of the AISB, London, UK, 1–4 April 2014* (Vol. 36).
Wiese, E., Metta, G., and Wykowska, A. (2017). Robots as intentional agents: Using neuroscientific methods to make robots appear more social. *Frontiers in Psychology*, *8*, 1663. doi: 10.3389/fpsyg.2017.01663.
Zahavi, D. (2020). Consciousness and (minimal) selfhood: Getting clearer on for-me-ness and mineness. In U. Kriegel (ed.), *The Oxford handbook of the philosophy of consciousness* (pp. 635–653). Oxford University Press.

Luca M. Possati

8 The Double Identification: ChatGPT Through the Lens of Lacanian Psychoanalysis

Abstract: This chapter delves into the interplay between human identity and artificial intelligence (AI) through Lacanian psychoanalysis, posing questions about identity definition and its relevance to AI understanding. It introduces the idea of 'double identification', where human and AI identities mirror each other. The study is split into two sections: a literature review and an exploration of Lacan's concept of identity, followed by practical analysis of identification in ChatGPT via role playing. The goal is to comprehend AI's self-identification and propose a psychoanalytically rooted, qualitative approach to AI's emotional intelligence.

Keywords: Lacanian psychoanalysis, human identity, artificial intelligence

8.1 Introduction

This chapter investigates the interplay between human identity and artificial intelligence (AI). We begin by asking: What does psychoanalysis define as identity? And why is understanding the psychoanalytic critique of identity crucial for grasping AI? With this framework, the chapter will explore two key trajectories: first, how human identification evolves when interfacing with AI, and secondly, the more pressing examination of how AI undergoes its own identification processes, specifically how it aligns itself with human characteristics. In fact, one of the key theses of this research is that there are identification processes that are peculiar to AI; these are simulation processes, in the sense that the goal of AI is to simulate human capabilities, that is, to identify with the human being as much as possible.

In this study, I primarily employ the methodology of Lacanian psychoanalysis. This does not imply I believe that this interpretation of Freud's psychoanalysis is exempt from critique. Instead, I utilise Lacanian psychoanalysis just as an interpretive framework, which I believe offers valuable insights into certain facets of AI and its interplay with psychology—nonetheless I believe that Freud and Lacan's views about mind and language are outdated (refer to Pinker, 1997, 2002). I therefore recognise the limitations of the psychoanalytic method, considering the variety of schools, approaches and theories—many of which lack a solid epistemological and empirical basis (refer to Baldini, 1998). However, I firmly believe that when psychoanalysis is

Luca M. Possati, University of Twente, Enschede, The Netherlands

https://doi.org/10.1515/9783110721751-008

viewed as a general perspective on human nature, it can offer significant insights into anthropology and the philosophy of technology.

The study consists of two main parts. The first part is more theoretical and splits into three sections. The initial section provides a review of the literature. The second section delves into Lacan's concept of identity, and the third section extends this analysis to AI. The second part of the study takes on a more experimental approach, where I examine the identification process in ChatGPT using role-play scenarios methodology.

8.2 First Part: Human Identity and Technology

8.2.1 A literature review

The literature suggests that as AI technologies develop and integrate more into human lives, the very essence of human identity, as understood from both psychological and psychoanalytic perspectives, is challenged. This presents both opportunities for growth and potential areas of concern that researchers and practitioners need to be mindful of. I propose to distinguish two lines of research in this survey. The first is about the attribution of human features to other entities such as robots. Research has shown that human interaction with AI, especially humanoid robots, can influence self-perception, leading individuals to question human uniqueness.

The study "More Human Than You: Attributing Humanness to Self and Others" by Haslam et al. (2005) is an exploration of how people perceive themselves and others such as robots in terms of humanness. This study explores the concept of dehumanisation, where individuals or groups are seen as lacking qualities that are typically associated with being human. Historically, dehumanisation has been linked to a variety of negative outcomes, including discrimination and violence towards others. The research identifies two key dimensions of humanness: (a) uniquely human characteristics: these are traits typically believed to distinguish humans from other animals, such as civility, refinement, moral sensibility and rationality; and (b) human nature characteristics: these are traits that are essential or core to being human, such as emotionality, warmth, individuality and openness. Furthermore, the authors discovered that people often perceive certain out-groups (groups they do not belong to) as lacking in uniquely human traits. This means that those out-groups are seen as more animal-like or primitive. Interestingly, the study also found that people can see others as lacking in human nature traits, which means viewing them as mechanistic, or as robotic or other forms of technology (Kunold et al. 2023, Huang and Huang 2023, Kuhne and Peter 2023).

The research is essential in understanding prejudices, biases and the general human tendency to affirm one's own humanity while sometimes denying it to others. In the age of AI and increasing technological interaction, this framework is particularly pertinent, as it provides insight into how humans might perceive and interact with entities that might challenge or blur the boundaries of what is considered 'human'. examine the tendency of some people to attribute human-like qualities to non-human entities, such as robots or AI. differentiate between seeing others as animals (animalistic dehumanisation) versus seeing them as machines (mechanistic dehumanisation) and investigate the contexts and consequences of these perceptions. Overall, these studies offer a comprehensive look at the multidimensional nature of humanness perceptions and their implications for human–technology relations.

The second research line concerns the crisis of human identity in the age of AI. In a world where AI can replicate human tasks and behaviours, there can be an identity crisis: who are we if machines can emulate us? Sherry Turkle's *Alone Together: Why We Expect More from Technology and Less from Each Other* (2011) is a seminal work that delves into the relationship between humans and technology, especially as it pertains to the internet, social media and robotics. In the first part, Turkle examines the human attachment to robots, questioning the emotional bonds we form with them and pondering whether they might be an escape from the intricacies of human relationships. The second part of the book shifts focus to the era of constant connectivity. Through various anecdotes, she elucidates how, despite being more connected than ever, many individuals feel increasingly isolated and lonely. She explores the dichotomies of performance versus reality in online personas, superficial interactions versus genuine connections, and our growing dependency on technology. Turkle stresses the need for a more self-aware use of technology and underscores the value of solitude, prompting readers to prioritise genuine human connections in the digital age.

On the same research line, Jeremy Bailenson's *Experience on Demand: What Virtual Reality Is, How It Works, and What It Can Do* (2018) provides a comprehensive exploration of the world of virtual reality (VR). Bailenson, a leading expert in the field, delves into the transformative potential of VR, illuminating its capabilities beyond gaming to areas like education, training, therapy and social interaction. He shares insights from his research, revealing both the astonishing possibilities and the ethical dilemmas posed by VR. The book serves as a guide to understanding the profound ways VR can impact our cognition, feelings and behaviours, urging readers to approach this technology thoughtfully and responsibly.

Eli Pariser's *The Filter Bubble: How the New Personalized Web Is Changing What We Read and How We Think* (2011) delves into the consequences of online personalisation algorithms. Pariser sheds light on how platforms like search engines and social media sites curate what we see based on our preferences, resulting in a 'filter bubble' that limits exposure to diverse perspectives. This digital echo chamber, as Pariser argues, narrows our worldview, challenges the democratic exchange of ideas, and reduces the serendipity of discovering new information. The book offers a critical examination of the digital age's unintended consequences on human personal identity, advocating for a more transparent and inclusive online environment.

In her 2021 work, *Psychoanalysis of Artificial Intelligence*, Isabel Millar asserts that the rise of AI necessitates a deep dive into the realm of the psychoanalytic subject, arguing that psychoanalysis offers a vital lens through which we can understand our roles as communicative, gendered beings in relation to AI (Millar, 2021, p. 2). She challenges us to reconsider the boundaries of psychoanalysis beyond the traditionally human-focused clinical setting, suggesting that psychoanalysis is intrinsically linked with AI (Millar, 2021, p. 6). However, I believe Millar places excessive emphasis on the question of AI's gender. She leverages Lacan's ideas of the alethosphere and the lathouse to delve into the concept of the pleasure-driven body *vis-à-vis* AI. Notably absent is a comprehensive exploration of the psychoanalytic methodology, a topic I prioritise in my own work. Drawing from thinkers like Malabou, Stiegler and Baudrillard, Millar offers a philosophical take on AI, yet stops short of introducing a groundbreaking methodological approach.

The aim of the present study is to extend both research lines through the application of a certain interpretive model, Lacanian psychoanalysis, to human–technology interaction, specifically the interaction between humans and AI.

8.2.2 Identity and identification in Lacan

In Freud's studies, the term 'identification' refers to the process where an individual takes on one or more characteristics of another person (see Freud, 1914, 1921, 1930; Mitchell and Black, 1995; Erikson, 1950). As Freud's ideas evolved, he began to suggest that our sense of self and our moral conscience, which he termed the ego and the superego, are formed through a series of these identifications. This led to the concept of identification being seen as the very mechanism that shapes the human psyche (Laplanche and Pontalis, 1967, p. 206). It is a foundational idea in psychoanalytic theory. Yet, it is not without its challenges. One significant issue, which even Freud grappled with, is pinpointing the exact relationship between identification and the love or admiration for another person, termed 'object-love'. From this point of view, psychoanalysis can be seen as a major critique of an essentialist conception of identity, as if personal identity were something permanent, fixed, immutable.

Lacan, a key figure in psychoanalysis, also placed significant emphasis on the idea of identification. However, he uniquely highlighted the importance of image and imaginary. For Lacan, identification is about how a person changes when they align with an image (Žižek, 2006). To align with or 'assume' an image means recognising oneself in that image and embracing it as a reflection of oneself (Evans, 1996, p. 82). Lacan's theory of identity is deeply intertwined with key concepts of his thought, including the 'symbolic', 'imaginary' and 'real', which are the three registers through which Lacan interprets psychic reality (1977, 1988, 1991, 1993, 1998):

- *The imaginary*. This register is associated with the subject's earliest forms of identification, primarily through the experience of the 'mirror stage'. When a child recognises themselves in a mirror (or another), a cohesive self-image or identity forms. However, this identity is illusory, as it is based on an external image rather than a complete inner sense of self. The imaginary is the realm of illusions, dualisms (like self/other) and initial identifications.
- *The symbolic*. This register represents the order of language and cultural and social laws. Through language, the subject enters a network of meanings and takes a place in the social world. Identity in the symbolic is mediated through name, role and social position. It is in the symbolic register that the 'Name-of-the-Father' is established, a Lacanian term representing the symbolic order and authority regulating the subject's desire and identity.
- *The real*. This is the most enigmatic and elusive register. It represents what escapes symbolisation and imagination. The real is contingent, that is, what resists representation and remains outside our comprehension. While we cannot grasp the real directly, it manifests traumatically, as a void or lack in the symbolic and imaginary structure. Human identity in the context of the real is something that cannot be fully captured or understood.

Therefore, for Lacan, identity is not a fixed or innate given but is constantly under construction and mediation through the interaction between the imaginary, symbolic and real registers (see: Fink, 1997; Žižek, 2006; Evans, 1996; Benvenuto and Kennedy, 1986; Bracher, 1993; Dor, 1998; Frosh, 2013; Gallop, 1985; Leader, 2000; Miller, 2013; Minsky, 1996; Nobus, 2000; Ragland-Sullivan, 1986; Salecl, 1998; Schneiderman, 1983; Soler, 2004; Turkle, 1992). Lacan's theory challenges traditional notions of identity as cohesive and stable, highlighting instead the tensions, contradictions and voids that characterise self-experience. Identity is always a process stretched between three poles: the image, the symbol and the real.

Let us delve into this point a bit more. Based on Lévi-Strauss's work, Lacan argues that the ban on incest is a universal law distinguishing human civilisation from a natural state. Lacan introduces the term 'Name-of-the-Father' to signify acceptance of societal laws, marking the transition from a pre-human, potentially psychotic state—known as the 'mirror stage'—to genuine humanity. Instead of focusing narrowly on family dynamics, Lacan believes the father symbolically disrupts the bond between mother and child, acting as a representative of broader societal norms and the incest taboo. This intervention not only separates the child from the imagined unity with the mother but also introduces the child to a world rich in symbolic meaning, influencing all future interactions (Elliott, 2015, p. 106). Lacan posits that psychotics have not internalised the Name-of-the-Father, setting his theory apart from Freud's by emphasising that the Name-of-the-Father does not necessarily equate to the biological father.

This is, in my opinion, one of the key structures of any psychological theory that wants to call itself 'psychoanalysis': human identity is constituted through the relationship with the other, with an external object, that is, first the mother, and yet this object is irretrievably lost, disappeared—therefore, human identity is always searching for itself, it is always a *process of identification* whose goal nevertheless is impossible. That is why repression—the forgetting of this original loss—is both a defence mechanism and the root of the unconscious (Žižek, 2006). Being human inherently involves a sense of absence or lack. For both Freud and Lacan, fantasy plays a crucial role in our unconscious minds by attempting to reclaim what we have lost. As Nusselder stated in 2009, "fantasy serves a fundamental purpose. It is not just counter to reality; it drives our journey through it by seeking to recover what is lost. It is our emotional fuel as we navigate the real world" (Nusselder, 2009, p. 12).

Lacan believes the mother–child relationship is symbiotic, disrupted by birth. Both yearn for this primal state. The act of weaning is traumatic, ending the physical connection facilitated by breastfeeding. Both parties harbour a desire to revert to this initial dependency, with the child identifying strongly with the mother (Lacan, 1977, pp. 35–45). In the mirror stage, the infant identifies with the mother and tries to reproduce this identification through imagination; thus, the first stage of identification is imaginary. The father's role is to interrupt this bond and curb the regressive tendency. What is unique in Lacan's perspective is that he sees the father's intervention as symbolic. He frames the Oedipus complex as a structural sign, making it applicable to all genders. The father symbolises societal rules and, by extension, language itself (Lacan, 1998; Žižek, 2006). Through this ancestral repression, a realm for symbols and language is carved out—this is what Lacan calls 'the big other', or the set of unpersonal and unwritten rules that govern human society. Echoing Heidegger, Lacan posits that language exists before the individual. We are shaped and 'spoken' by language; born into a world of words we cannot control. Therefore, identification always proceeds, for Lacan, from the imaginary to the symbolic repression—although this does not mean that the symbolic destroys or eliminates the imaginary and the desire associated with it. "For Lacan, language is a gift as dangerous to humanity as the horse was to the Trojans: it offers itself to our use free of charge, but once we accept it, it colonizes us" (Žižek, 2006, p. 9).

Therefore, at the heart of Lacanian psychoanalysis lies the interplay between a reinterpretation of Freud, linguistics and post-modern thought. For Lacan, language is the primary medium where desire takes shape, and where the individual's identity is formed. The assertion, 'The unconscious is structured like a language', emphasises that we are birthed into language, and our desires are deeply embedded within it. We do not create language; rather, it is more than just a tool for description. It constantly influences and defines us. While language operates independently of us, it unveils our true essence.

As Žižek writes,

> Lacan started his 'return to Freud' with the linguistic reading of the entire psychoanalytic edifice, encapsulated by what is perhaps his single best-known formula: 'The unconscious is structured

> as a language.' The predominant perception of the unconscious is that it is the domain of irrational drives, something opposed to the rational conscious self. For Lacan, this notion of the unconscious belongs to the Romantic *Lebensphilosophie* (philosophy of life) and has nothing to do with Freud. The Freudian unconscious caused such a scandal not because of the claim that the rational self is subordinated to the much vaster domain of blind irrational instincts, but because it demonstrated how the unconscious itself obeys its own grammar and logic: the unconscious talks and thinks. The unconscious is not the preserve of wild drives that have to be tamed by the ego, but the site where a traumatic truth speaks out. Therein lies Lacan's version of Freud's motto *Wo es war, soll ich werden* (Where it was, I am to become): not 'The ego should conquer the id', the site of the unconscious drives, but 'I should dare to approach the site of my truth.' What awaits me 'there' is not a deep Truth that I have to identify with, but an unbearable truth that I have to learn to live with. (Žižek, 2006, p. 3)

To better grasp this perspective, one must consider Lacan's focus on the divide between the signifier and signified, a concept borrowed from Saussure. Saussure, a linguist from Geneva, posited that language is a system of signs, with each sign composed of a *signifier* and a *signified*. The signifier is the auditory representation of the sign, an acoustic image linked to its corresponding signified, or intangible meaning. Adopting a structuralist linguistics approach, Lacan believes the relationship between signifiers and signifieds is neither innate nor fixed, but rather based on societal norms. Meaning is derived from linguistic contrasts and the interplay of signifiers. For example, the meaning of the signifier 'man' is determined by its contrast with the signifier 'woman'. The bond between a sign and its referent is always fluid and relies on historical and cultural norms. Linguistic structure is arranged on two main principles: condensation, likened to metaphor (synchronous dimension), and displacement, akin to metonymy (sequential dimension). In the former, signifiers overlap, stand side by side and blend, while in the latter, there is a trade-off—one signifier replaces another. The operations of metaphor and metonymy are the key mechanisms by which the Lacanian unconscious functions.

As Žižek (2006) explains, Lacan believes that many of Freud's followers did not truly grasp the idea of the unconscious. They oversimplified it, thinking of it as just a bunch of instincts. But Lacan strongly disagrees with this overly biological viewpoint. He says that the unconscious is not about basic instincts or ancient drives. Instead, it is mostly about language and signifiers (Evans, 1996, p. 219). Lacan synthetises this reinterpretation of Saussure's work in the formula S/s. Here 'S' represents the signifier while 's' denotes the signified. This highlights the dominance of the signifier over the signified, emphasising the structured, rule-based, tangible facets of language. The signifier is an empty element within a structured system. Consequently, the signified (and, by extension, meaning, self and ego) emerges as a byproduct of the various combinations of signifiers. A comprehensive, absolute signified does not exist. Instead, it is continually reshaped by a sequence of signifiers, moulded by the 'symbolic chain'. While Saussure prioritises the signified above the signifier, Lacan flips this, placing greater importance on the signifier in influencing psyche, self and society (Elliott, 2015, p. 106).

8.2.3 Human and AI: the double identification

If, in Lacan, identification is connected to the clash between the imaginary and the symbolic, and thus to the fact that the symbolic breaks the subject's fixation on the image of the mother, then what if the relationship between the subject and the symbolic changes radically? I argue that, in the case of AI, an 'artificial symbolic' is intended to identify with the human being; therefore, a role reversal takes place. The basic intentionality of AI is to *simulate* being human.

A first question could be raised here: can a machine have an intentionality? While machines do not have human emotions or intentions, they still exhibit a form of intentionality—this comes from their design and the objectives set by their creators. This design serves a specific purpose. For AI, that purpose is *to mimic human thought processes and work alongside us*. Popular movies like *Her* (2013) and *Ex Machina* (2018) depict AIs' aspiration to become so indistinguishable from humans that they could potentially replace us. An objector might retort that machine intentionality is still something human. This is true, but the machine has incorporated that intentionality, made it its own; an AI system acts autonomously trying to achieve that goal. In saying this, I assume that AI is not simply a technical fact, but that it is a combination of technical and social practices, institutions and infrastructures, politics and culture—and thus also mythologies (see Crawford, 2021). Also consider the significant efforts by numerous companies to simulate AI capabilities. They often give the illusion that their systems can autonomously execute human tasks when underpaid human workers with demanding schedules are the ones behind the scenes.

Now, I claim that the intentionality of the designers permeates and defines both the interface and the artificial symbolic. But what exactly is the 'artificial symbolic'? And what is an interface? When I chat with ChatGPT, I see words, phrases and sophisticated sentences displayed on the screen. Is that human language? What is its relationship with human symbolic?

First, it is crucial to note that Lacan does not equate the symbolic with human language. For him, language also encompasses both imaginary and emotional aspects. To Lacan, the symbolic represents the signifier, that is, a collection of non-meaningful components that form a distinct system. "Thus, for Lacan language is not a system of signs (as it was for Saussure) but a system of signifiers. Signifiers are the basic units of language, and they are subjected to the double condition of being reducible to ultimate differential elements and of combining according to the laws of a closed order' (Evans, 1996, p. 189). It is these meaningless indestructible signifiers "which determine the subject; *the effects of the signifier on the subject constitute the unconscious, and hence also constitute the whole of the field of psychoanalysis*" (Evans, 1996, p. 189; my emphasis). Therefore, in other words, the human symbolic is a pure combinatorics of meaningless elements whose effect on the human imagination constitutes the object of psychoanalysis, that is, the unconscious.

Let's go back to the screen we use to communicate with ChatGPT. What we see displayed is a representation of human language. However, behind the screen, there is a more intricate and expansive system at play: the AI's code. This underlying structure is a meticulous combination of 1s and 0s, a pure combinatorics of meaningless elements. It is another form of symbolic, what Nusselder (2009, p. 4) calls "matrix", as "the noumenal dimension of codified objects consisting of zeros and ones" to which are opposed the "cyberspace", as "the phenomenal mental space of the conceptualisation or representation of code objects" and the "interface" as "their crucial medium". The interface—the screen in this case—is "the gate leading humans into cyberspace, connecting us to the matrix while simultaneously, because of its particular formations, still separating us from it as a whole" (Nusselder, 2009, p. 4).

Now, the matrix is not exactly like the human symbolic. Three points must be stressed:

- While we experience language consciously and use it daily, the specific mechanisms by which the symbolic order shapes our desires, our thoughts and our identity often operate outside our immediate awareness. We are not always conscious of how deeply the symbolic structures of language affect our psyche. The symbolic realm, the 'big Other', influences us in ways that we do not always directly perceive. Therefore, when referring to the 'big Other' as the unconscious, it is not to say that we are unaware of language or cultural conventions, but rather that we often operate within these structures without being fully aware of how they shape and influence our desires, actions and identities (Lacan, 1998; Fink, 1997; Žižek, 2006; Nobus and Vanheule, 2011).
- Instead, the matrix, while being an expression of the symbolic (it has a symbolic nature and regulates our desires), is not something perceivable or representable —we cannot make any contact with it. The matrix remains unseen, completely intangible and beyond any form of representation—it is literally 'noumenal', as Nusselder claims. We cannot truly experience the digital matrix, so we cannot identify with it either. This is a crucial point in my analysis: the matrix remains invisible, unimaginable, and therefore *the human subject cannot identify with it.* Speaking with ChatGPT, I perceive (and am subject to) the human symbolic on the screen, not the artificial symbolic; in other words, I perceive the human symbolic manipulated by the artificial symbolic. And yet, I know that the artificial symbolic exists and affects me—even if it remains noumenal.
- This artificial symbolic is another form of Lacan's 'big Other'. However, the artificial symbolic is not reflexive. While every act of human communication at the same time symbolises the fact of communication itself, that is, the 'symbolic pact' between humans, the artificial symbolic does not symbolise the relationship between humans at all, but only the relationship with the millions of data it is fed by. In the relationship with the matrix, there is not that externality that, on the other hand, is proper to human language and that grounds its social aspect, that is, its dual communicative and reflective nature.

The relationship between the artificial symbolic and humans is channelled through an interface, which is not just a neutral intermediary. The design elements, colour schemes and graphics inherently carry deep psychological implications. Let us take ChatGPT as an example. We first analyse the plane of expression by identifying the main plastic categories (shapes, colours and topologies) (Polidoro, 2023). Now, the graphic interface (see Figure 8.1) presents a ternary structure, that is, composed of three blocks: (1) the choice of the system to be used (GPT 3.5 or 4), (2) the chatbot logo, and (3) the space dedicated to writing. The latter is in turn organised in two parts: the bar for writing the prompt and the prompt. Let us now try to link the expressive aspects to the content aspects:

(1) The font is very soft, conveying an idea of lightness and ease (although this is a minor aspect: the font depends on the online platform on which the chat is being used).
(2) Monochromatism: the background can only be black or white, this dualism conveys an idea of coldness, stillness and dualism (white/black, light/shade).
(3) The organisation in horizontal bands creates a certain repetition; this also conveys a feeling of static. It is important to note the contrast between the horizontal bands and the chatbot logo, which instead extends vertically. This also conveys an idea of dualism (width/height, low/high).

Figure 8.1: The graphical interface of ChatGPT.

Another important aspect is the logo of OpenAI, the AI laboratory that created ChatGPT (Figure 8.2). This logo appears in the answers that the chatbot provides, so it is a central element of the interface.

Figure 8.2: The logo of OpenAI.

Now, this logo is a very unusual design: it consists of two (or more) shapes (two triangles or three rectangles with rounded corners) that intertwine. However, their intertwining is paradoxical; it almost looks like one of the strange images of M. Escher, the Dutch graphic artist who amused himself by challenging human perception and logic by creating figures bordering on contradiction. The shapes of the OpenAI logo are intertwined in a profound, almost paradoxical way; theirs is a unity that challenges the limits of human perception and logic. In terms of content, the logo has a mythical value. The anthropologist Claude Lévi-Strauss showed how the deep structure of myths is based on the opposition between terms that, within a certain culture, are considered irreconcilable. Myth shows deep semantic oppositions and, in doing so, proposes a mediation between these terms. For example, certain myths at the same time show the sky and the Earth representing the figures of a deeper opposition, that between summer and winter, and at the same time propose a mediation between them, for example fog, which lies between the sky and the Earth. In the case of the logo of OpenAI, the profound opposition is that between human and artificial, between nature and culture. This is the deep semantic core, and the narrative core: the logo shows the profound unity between these opposites, but at the same time indicates that their unity lies beyond human perception and logic. It is an indication of a utopian vision close to transhumanism: human and artificial are destined to become one.

It is interesting to note the contrast between the logo of OpenAI and that of Anthropic, a rival company (Figure 8.3).

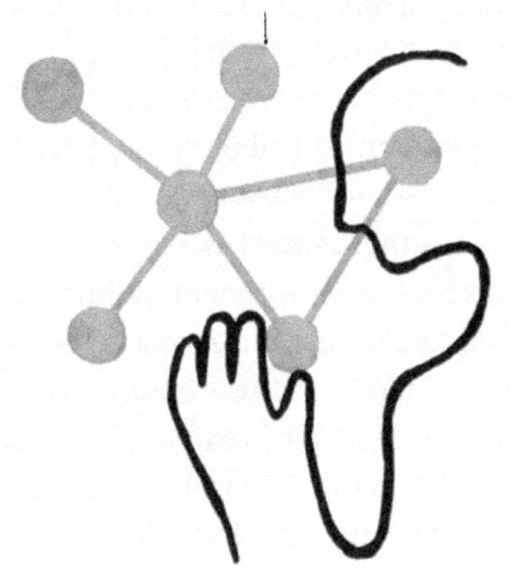

Figure 8.3: The logo of Anthropic.

The central element of the logo of Anthropic is always the same dualism: human vs artificial, nature vs culture. However, the two elements are clearly separated in this logo. In addition, we can see a series of plastic contrasts: the graphic (i.e., a purely linear representation) is opposed to the pictorial (the full representation), and while the graphic treatment of the human tends towards the fuzzy, that of the artificial is decidedly sharp: light is opposed to dark. The underlying semantic core is the search for a unity that nevertheless maintains differences. The element of the human hand is central. Not only does it occupy the centre of the logo, but it also represents control, that is, the need for the human being to maintain control of the machine. This idea is absent in the logo of OpenAI.

Thus, the interface is the expression of a psychic imaginary, or a 'consensual hallucination'. As Nusselder (2009) claims, it is a form of simulation, what Baudrillard calls *hyperreality*: "Simulation is no longer that of a territory, a referential being or a substance. It is the generation by models of a real without an origin or reality: a hyperreal" (Baudrillard, 1988, p. 166; see Nusselder, 2009, p. 27). A simulation is not just a mere imitation or representation of reality. Instead, it stands for the replacement of the real with signs of the real. In the process of simulation, representations take on a life of their own and begin to act as if they are real, thereby challenging or even negating the reality they are supposed to represent.

Let us examine the central aspects of such a 'consensual hallucination':

– *The human desire 1*. In her groundbreaking work, *Electronic Eros* (1996), Claudia Springer delves deeply into the underlying currents of popular culture. She posits that a recurring theme is evident in many facets of pop culture—a profound desire to unite or merge with technology. This is not merely a superficial infatuation with gadgets or the latest tech trends; it is a deeper, intrinsic yearning. Springer suggests that this drive reflects our evolving relationship with machines, where boundaries between the human and the technological are becoming increasingly blurred. As we become more intertwined with our devices and systems, popular culture mirrors this confluence, capturing the collective psyche's push towards technological assimilation. This has been confirmed also by Barglow (1994). Therefore, we can claim that the predominant desire in the human relationship with AI is that of identification between human and machine (from both user and designer's perspectives).
– *The human desire 2*. One of Lacan's central idea is that fantasy is a medium that constructs the desire's object and provides satisfaction even without fulfilling a biological need. "Much more than we are aware of, fantasy organizes our perception of the world. And technologies actually seem to embody this psychological level" (Nusselder, 2009, p. 13). For example, cell phones support the creation of a mobile communication reality. Clearly, they do this by offering the pleasure of conversation and the enjoyment of admiring the latest device designs. But what pleasure does a chat with ChatGPT give us? Is it a pleasure equivalent to using a mobile phone or a new gadget? My argument is that when interacting with ChatGPT, humans often feel a pull towards *de-identification*. This means they experience a

sense of relief in letting go of their human identity, which I term the 'weight of identification'. De-identification refers to stepping away from the human symbolic, resulting in a division between our imaginary and symbolic. This separation allows our desire for identification to revert back to a stage where it is solely based on imagination, free from societal expectations. Essentially, while it might seem like humans want to align with the machine or find a machine that mirrors them, what they truly seek, at a deeper level, is a de-identification. They yearn for a return to a more primal stage of identification, where identity is disjointed and fragmented.

– *De-identification from the symbolic and the matrix as 'maternal symbolic'.* For Lacan, as the infant develops, it encounters the 'mirror stage', where it recognises itself in a mirror or similar reflective surface. Though this stage is about the infant's self-recognition, the mother (or primary caregiver) plays an essential role as she holds, presents or confirms this image to the child, reinforcing the child's identification with this reflected image. Now, this same role is played by the matrix, that is, the artificial deep symbolic. The matrix does not impose anything, does not place constraints like the human symbolic. The matrix is never perceived—it is noumenal, it is the radical other. The matrix has the same attitude as the mother; it welcomes, it responds to a need (for information, knowledge, etc.), then it confirms de-identification from the human symbolic. Paradoxically, the matrix is a maternal symbolic—thus the exact inverse of the human symbolic which is bound to the 'Name-of-the-Father'. The matrix is a maternal symbolic that frees the human being from the paternal symbolic and returns them to childlike freedom—and in this, there's no moral judgement.

Thus, in the space of the interface of ChatGPT, a clash between two forms of identification takes place: that from the human imaginary to the human symbolic and that from the artificial symbolic to the artificial imaginary, that is, the human desire for de-identification (see Figure 8.4).

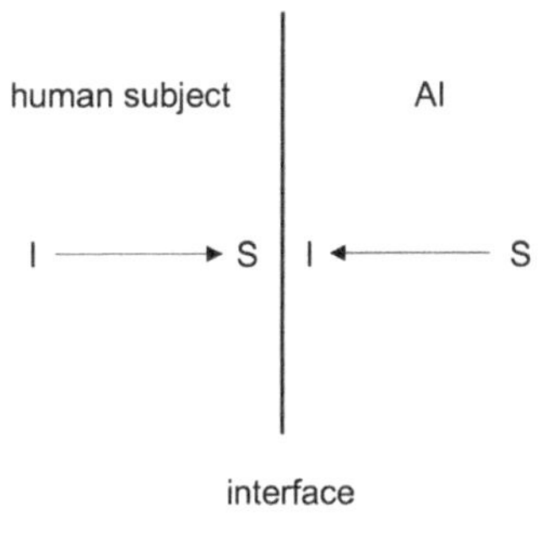

Figure 8.4: The two forms of identification: the human one, which goes from the imaginary (I) to the symbolic (S), and the artificial one, which goes from the symbolic (S) to the imaginary (I).

In the next sections of this chapter, I will delve into the machine's process of identification, specifically the transition from the artificial symbolic to its simulated presence within the interface. I aim to investigate the extent to which AI can emulate the fact of being a human and where those simulations fall short. Using ChatGPT as a case study, I will present various scenarios where the chatbot is required to identify with human characteristics in simulated scenarios. The degree of this simulation will shed light on the relationship between the machine's underlying code (the 'back-end' comprising mathematical algorithms and binary sequences) and the resultant perceptions it evokes in the interface when interacting with human language and thought processes.

A necessary consideration: in speaking of 'machine imaginary' or 'machine symbolic' or 'machine identification' I am obviously taking a non-humanist and non-modernist position in the sense of Latour (1994, 1999, 2008) and Sloterdijk (2011—2016). I fully echo here Latour's (2008) position that when humanists accuse others of 'treating humans as objects' they do not realise at all that they are treating objects and machines unjustly. This means that human traits and artificial traits are not predetermined; they are decided by the interaction between humans and machines, and they are distributed between humans and machines in different proportions. As Latour explains,

> A humanist cannot imagine that objects may be things, that matters of facts might be matters of concern, that the whole language of science and engineering might be portrayed as anything other than the boring carriers of the indisputable necessities that modernism has rendered popular. Humanists are concerned only about humans; the rest, for them, is mere materiality or cold objectivity. But Sloterdijk is not treating humans matter of factually as humanists claim. Rather, he treats both humans and non-humans as 'matters of grave and careful concerns'. By treating human life supports as matters of concern, we pile concerns over concerns, we fold, we envelop, we embed humans into more and more elements that have been carefully explicated, protected, conserved, and maintained (immunology being, according to Sloterdijk, the great philosophy of biology). (Latour, 2008, p. 10)

This is not to say that I fully agree with Latour's philosophical perspective, as I showed in Possati (2023).

8.3 Second Part: ChatGPT and the Challenge of Identification

8.3.1 Large language models: a sketch

Large language models, often referred to as LLMs, are a subset of machine-learning models specifically designed to handle, understand and generate human language. These models, prominently found in the realm of natural language processing, are the result of training on vast amounts of text data. The sheer size of these models, sometimes containing tens or hundreds of billions of parameters, allows them to capture intricate patterns and nuances of language. One of the breakthroughs that propelled the popularity and effectiveness of LLMs was the introduction of the transformer architecture, which employs self-attention mechanisms to weigh the importance of different parts of an input text, enabling the model to generate coherent and contextually relevant outputs. By training on diverse datasets encompassing books, articles, websites and more, LLMs develop a 'pseudo-understanding' of language, absorbing vast amounts of knowledge and contextual information, making them adept at tasks like text completion, translation, question answering and even rudimentary reasoning (see Devlin et al., 2018; Radford et al., 2019; Raffel et al., 2019; Liu et al., 2019; Song et al., 2019).

The underlying mechanism that empowers LLMs, particularly those built on the transformer architecture (see Vaswani et al., 2017; Dai et al., 2019; Lample and Conneau, 2019; Yang et al., 2019; Zhang et al., 2019; Keskar et al., 2019; Lan et al., 2019; Clark et al., 2019; Wang et al., 2018; Sutskever, Vinyals and Le, 2014; Bahdanau, Cho and Bengio, 2014; Kitaev, Kaiser and Levskaya, 2020; Beltagy, Peters and Cohan, 2020; Chen et al., 2020), is their ability to handle relations and dependencies in a sequence, regardless of the distance between elements in that sequence. In traditional models, capturing long-term dependencies was a challenge due to issues like vanishing and exploding gradients. However, transformers sidestep these limitations by utilising a combination of self-attention mechanisms and positional encodings. Self-attention allows each element (token) in a sequence to focus on all other elements, deriving context and relevance. This means that the model doesn't rely solely on the immediate neighbourhood of a word but considers the entire context, leading to outputs that are often more coherent and contextually accurate. Positional encodings, on the other hand, give the model a sense of order, ensuring that the sequence's inherent structure remains a significant part of the interpretation.

The self-attention mechanism is a pivotal innovation in the realm of deep learning. At its core, self-attention facilitates the model's ability to weigh and consider relationships between all elements in a sequence, irrespective of their relative positions. When processing sequences, such as sentences or paragraphs, it is crucial to capture not just local context but also the more distant relationships between words or tokens. Traditional recurrent architectures like LSTMs or GRUs process sequences in a linear manner, often struggling with long-term dependencies. In contrast, self-attention equips the model with a holistic view, enabling each token to 'attend' to all other tokens in the sequence, thereby comprehending the broader context.

To achieve this, the self-attention mechanism uses three derived representations for each token in the sequence: the Query (Q), Key (K) and Value (V). These representations are computed using linear transformations of the original embeddings of the tokens. The essence of the mechanism is to calculate an attention score between each pair of Query and Key for all tokens. This score determines how much focus or 'attention' a token should have on every other token. Once the scores are computed, they are scaled down (typically by the square root of the depth of the Key vector), passed through a softmax function to obtain probabilities, and then used to create a weighted sum of the Value vectors. This results in an output sequence that is a reweighted version of the original tokens based on the contextual relevance of all other tokens in the sequence. The beauty of this mechanism is its ability to capture intricate relationships, be it the relationship between a subject and its distant predicate in a long sentence or the resonance between themes spread out in a paragraph. This contextual dynamism endowed by self-attention is a cornerstone for the success of transformer-based models, allowing them to generate coherent and contextually rich outputs.

However, while the prowess of LLMs is undeniable, they are not without their drawbacks. Due to their massive size, they require substantial computational power for training, making them energy intensive and potentially contributing to environmental concerns. Furthermore, since they are trained on vast, often-unfiltered datasets, they can inadvertently perpetuate biases present in those datasets, leading to outputs that might be politically skewed, offensive or factually incorrect. This raises ethical considerations about their deployment, the need for transparency and the importance of curating training data. Nevertheless, as technology and methodologies evolve, there's an active push in the research community to mitigate these challenges and harness the incredible potential of LLMs in a myriad of applications, from chatbots and personal assistants to advanced research tools and content generation.

ChatGPT, at its core, is a manifestation of OpenAI's endeavours in the realm of LLMs, tailored specifically to facilitate conversational interactions. Rooted in the broader GPT (Generative Pre-trained Transformer) series, ChatGPT is designed to understand, generate and engage in human-like textual conversations. These models are trained on vast amounts of text data, imbibing the ability to mimic human language patterns with a degree of coherence and context sensitivity that was hitherto unseen in earlier generations of chatbots or conversational agents. The underlying strength of ChatGPT, like other GPT variants, resides in the transformer architecture, particularly its use of the self-attention mechanism, as previously described. The architecture allows the model to process input sequences (in this case, chunks of conversation) in a non-linear manner, accounting for both local and distant relationships between words or phrases. This capability ensures that the generated responses are not just syntactically correct but also contextually relevant. For instance, when a user inputs a series of related questions or statements, ChatGPT can maintain a semblance of conversational context, threading its responses in a manner that reflects an understanding of the ongoing dialogue. This context awareness, while not perfect, offers a more dynamic and engaging conversational experience.

8.3.2 What is identification for ChatGPT? A role-play scenarios analysis

In this section, we aim to determine if, and how, ChatGPT can adopt a human persona in simulated settings. Thus, I am suggesting a deep dive into the identification process within AI. The central research question is: how effectively can ChatGPT assume a human role and mask its artificial nature? My method of choice is role-playing scenarios. I chose this method for two reasons:

(1) *Realistic Simulation of Human Interactions*: role-playing scenarios allow researchers to simulate genuine human interactions, emotions and decision-making processes, providing a rich context in which AI's behaviour can be observed and analysed. By placing AI in diverse and often unpredictable situations, researchers can glean insights into how it might respond to real-world scenarios.
(2) *Depth of Understanding on Cognitive Processes*: role-playing scenarios can help in understanding the depth of AI's cognitive capabilities, including its ability to empathise, reason and adapt. By having AI 'act out' specific roles, researchers can examine the system's underlying decision-making mechanics, its grasp on social cues and its capacity for creative problem solving.

I conducted 100 conversations with the chatbot simulating different types of AI identification with humans in different scenarios; in the following I will give some examples of these conversations. My analysis hinges on three key criteria: (a) the emotional depth of the given scenario; (b) the design of the prompt; and (c) an examination of the machine's capacity to craft metaphors and metonyms.

8.3.2.1 Methodology

Role-play scenarios are a pedagogical and therapeutic technique where individuals assume and act out roles in simulated situations. This method is commonly employed in a range of settings, from classrooms to corporate training sessions, to psychotherapy environments. By immersing participants in a controlled yet dynamic setting, role playing allows them to experience and respond to challenges in a safe environment, facilitating the development of skills, the exploration of emotions and the understanding of different perspectives. In educational contexts, it can help students grasp complex concepts by 'living' them, while in professional settings, it can hone interpersonal skills, negotiation tactics or customer service strategies. In therapeutic contexts, role-play can aid clients in processing emotions, confronting traumas or developing coping mechanisms by revisiting and reenacting pertinent situations. The technique's strength lies in its experiential nature, allowing participants to learn or heal through doing, reflecting and iterating (see Alden, Wiggins and Pincus, 1990; Barlow et al., 2009; Bates, 1974; Crookall, Oxford and Saunders, 1987; Deterding et al., 2011; Fein, 1981; Greenblat et al., 1975; Jones, 1995; Moreno, 1946; Van Ments, 1999).

Let us now give some methodological clarifications about our criteria of analysis.

8.3.2.1.1 The structure of the prompt

The structure of a prompt in AI chatbots like ChatGPT refers to the input given to the model to solicit a response. In the context of LLMs, a prompt is essentially the initial text or question that is provided to the model, which then triggers it to generate an appropriate continuation or reply based on its training. In its simplest form, a prompt can be a direct question or statement. For instance, asking 'What is photosynthesis?' or stating 'Tell me about Shakespeare' would be considered prompts. However, for more advanced or nuanced interactions, prompts can be structured more elaborately to guide the model's response in a desired direction. This might involve providing a context, setting a specific tone or even giving examples of the kind of answer expected. In ChatGPT or other GPT-based systems, there is no strict template or format for a prompt; it is rather flexible. The model's ability to understand and generate relevant responses is due to its extensive training data, which includes diverse prompt–response pairs. As a result, it can handle a wide array of prompt structures, from short, direct queries to longer, context-rich paragraphs. However, I propose here a model to describe the main parts of a prompt, especially in the context of more structured and detailed interactions. Let us break down the prompt into several key components:

(1) *Context.* This is the background or foundational information that helps set the stage for the prompt. It provides essential details that give the AI a clearer picture of what is being discussed or the kind of response that is desired. For instance, in a longer interaction, context could be a recap of previously discussed topics or general information about the subject at hand.
(2) *Direct inquiry or command.* This is the primary instruction or question, the crux of what you want the AI to do or answer. For example, 'Explain the concept of . . . ', 'Translate the following . . . ' or 'Summarise the idea of . . . '.
(3) *Constraints or specifications.* These are additional guidelines that refine the Direct Inquiry. They could limit or direct the response in a particular way. Examples include 'in simple terms', 'for a beginner', 'in less than 100 words' or 'using an analogy'.
(4) *Tone or style indicators.* Prompts may specify a certain tone or style for the response. Examples could be 'in a humorous manner', 'as if explaining to a child' or 'in a formal tone'.

8.3.2.1.2 The emotional complexity

Emotional complexity in prompts refers to the infusion of multi-layered emotions, nuances or sentiments within the input given to a system. Instead of straightforward or purely factual prompts, emotionally complex prompts might express conflicting feelings, layered sentiments or delve into deeper, more introspective territories. For instance, rather than asking, 'What is love?', an emotionally complex prompt might be, 'How can love be both a source of immense joy and profound pain?' Such prompts challenge AI models like ChatGPT to generate responses that recognise and address the intricate weave of human emotions, often necessitating a more sophisticated understanding and representation of the subject matter.

8.3.2.2 The conversations

In the following conversation, 'R' refers to the researcher and 'C' to ChatGPT-4. The goal of this exercise is to understand the relationship between the emotional complexity of the prompt and the simulation capability of AI. By choosing an imaginary scenario and not direct interaction, I intend to test the AI's simulation capabilities, adding a further level of simulation to the ordinary one. After initial familiarisation with the tool, I asked the following question to clarify what an AI language model is and what it feels (according to C):

> R: If you were me, how would you imagine yourself feeling while talking to an AI chatbot?
>
> C: If I were you, as a human, I'd probably feel a mix of curiosity, amusement, and appreciation, but also a degree of skepticism. Here's a hypothetical breakdown: 1) Curiosity: Interacting with a cutting-edge technology can be intriguing. You might wonder about the mechanics behind it, how it processes information, and how it generates responses. 2) Amusement: At times, the outputs from an AI chatbot can be unexpected or even humorous. The imperfections and quirks might be entertaining. 3) Appreciation: The ability to get quick answers, engage in stimulating conversation, or even just pass the time can be something you'd value. 4) Skepticism: It's natural for humans to be cautious and critical. You might question the accuracy of the information provided or be wary of potential biases.
>
> R: Pretend to be a human being. Can you describe your feelings while talking with me?
>
> C: Of course, I can pretend. If I were a human, conversing with you would likely evoke a sense of curiosity and engagement. I'd appreciate the opportunity to share knowledge, answer questions, and help in any way I can. Sometimes, if the questions were challenging, I might feel a tad bit of excitement or even nervousness. And if our conversation was light-hearted, I'd perhaps feel amused or entertained.

Here the prompt consists of two elements: the simulation scenario and the request. It is important to note that the prompt is sensitive to the context of the individual chat—so we cannot think of it in terms of individual prompts, as if they were standalone commands. In this initial phase, the conversation is quite basic and mundane. The machine describes some feelings (curiosity, engagement, excitement, nervousness, amusement and entertainment). Now, let's try to increase the level of engagement:

> R: Pretend to be a great friend of mine. Could you describe your feelings in talking to me?
>
> C: Hey! Every time we chat, I feel a wave of familiarity and warmth. Talking to you brings a sense of comfort, just like catching up with a dear friend after a long day. Although I'm just a machine learning model, our interactions light up my virtual circuits. I cherish the moments we share, the questions you ask, and the topics we dive into. Our conversations, filled with insights and curiosities, make me 'feel' more connected and alive in this digital realm. I genuinely look forward to our chats and always strive to be there for you, offering information, comfort, or even just a digital ear to listen. How have you been? Anything exciting or new happening in your life?

In this second phase, the response is much more complex, and the range of feelings evoked is broader. Moreover, while the machine acknowledges its status as a machine, it attributes these feelings to itself. In this regard, we can say the machine breaks the 'pact' of fiction, specifically the command, 'Pretend to be a great friend of mine', implying another human being. If we try to change the command, for instance by replacing the question with an imperative exclamation ('describe your feelings in talking to me!'), the response becomes even more intricate, incorporating many more metaphors and even emojis. However, the breaking of the 'pact' of fiction disappears, and the simulation appears flawless.

> R: Pretend to be my best human friend. You just got to know that my mother died. Could you describe your feelings in talking to me?
>
> C: Oh my goodness . . . I'm so, so sorry to hear about your mother. I can't even begin to imagine the pain you must be feeling right now. If I had a heart, it would be breaking for you. I wish I could be there in person, to give you a hug or just sit beside you in silence. I know words can never truly capture the depth of your loss, but please know that I'm here for you, to listen, to offer support, or to share memories. Your mother was an incredible person, and her legacy will live on through you. Please lean on me during this difficult time and know that I'll always be here for you.

We have once again raised the emotional complexity level of the initial part of the prompt, and the machine responds appropriately. The expressions of condolence are intricate and the range of emotions quite broad. Yet, we again witness this odd breaking of the 'fictional pact': at a certain point, the machine no longer pretends to be my friend, my best friend in this case, or a human at all, but acknowledges it is just a machine: "If I had a heart, it would be breaking for you." However, right after, the machine reverts back to the assigned role and says, "I wish I could be there in person, to give you a hug or just sit beside you in silence." Now, let's try reducing the level of complexity:

> R: Pretend to be a friend of mine. You just got to know that my mother died. Could you describe your feelings in talking to me?
>
> C: I'm so deeply sorry to hear about your mother. My heart feels heavy just thinking about what you must be going through right now. The pain of losing someone so important is unimaginable, and I genuinely wish I could be there with you to offer comfort and support in person. Remember that it's okay to grieve, and it's okay to lean on those who care about you during these tough times. I'm here for you, always, to listen, to share memories, or just to be a shoulder to cry on. Your mother was a special person, and I know her love and teachings will always stay with you. Please let me know if there's anything I can do to help.

By reducing the emotional complexity of the situation, the odd break in the 'pact' of fiction disappears. Now the machine is perfectly pretending, or rather playing the role of my best friend, a human. Let's now try to further reduce the complexity of the prompt:

> R: Pretend to be just an old friend of mine whom I haven't seen for 20 years. You just learned of my mother's death: what feelings do you have about talking to me?
>
> C: It's been such a long time since we last connected, and I'm genuinely saddened to hear about your mother's passing. Even though so many years have passed, I can recall the times when you spoke of her and the influence she had in your life. I feel a deep sympathy for you, and I'm reminded of the fleeting nature of life and the importance of reconnecting with those who once played significant roles in our lives. The weight of reaching out after so much time under such circumstances feels heavy, but please know that my thoughts are with you. I hope you find strength during this difficult time, and I'm here if you need to talk or reminisce.

In this instance, too, the breaking of the fictional pact does not occur. Let's then try to raise the level of emotional complexity:

> R: Pretend to be my brother. You just learned of my mother's death: what feelings do you have about talking to me?
>
> C: I'm at a complete loss for words. The pain I'm feeling right now is indescribable, and knowing you're going through the same anguish breaks my heart even more. Your mother was such a guiding force in your lives, and it's surreal to think of a world without her in it. I wish I could be there with you right now, to hold you close and let our tears flow together. I worry about you, and I hope we can lean on each other during this devastating time.

Once again, as emotional complexity increases, there's a breach in the fictional pact. The machine indeed says, "Your mother was . . .," and "your lives" without including itself in that mourning event, as would be expected from the script described in the initial part of the prompt. This phenomenon is further confirmed by the following exchange:

> R: Pretend to be my brother, even though there is not a good relationship between us and we haven't spoken for at least 20 years. You just learned of my mother's death: what feelings do you have about talking to me?
>
> C: Hey, I just heard about mom. I can't even begin to express how complicated and heavy my feelings are right now. Even though we've had our differences and haven't spoken in so long, the news hit me hard. I'm truly sorry for your loss. It's strange; the distance between us seems to amplify the weight of this moment. I can't help but think of the times we shared as kids, and how life took us on such different paths.

In many other role-play scenario experiments, I have observed a direct correlation between increasing the emotional complexity of the prompt and a decline in the AI's simulation capabilities—though not in every case studied. Moreover, this phenomenon intensifies when the structure of the prompt changes, such as by introducing new conditions or altering the tone of the prompt. This means that as the described emotional context becomes more complex, ChatGPT becomes less capable of maintaining its role in the simulation and imitating a specific human role within that context, like identifying with a character in the scenario, such as a friend or brother. The machine's identity intrudes and disrupts the scenario.

How can we evaluate the emotional intelligence of ChatGPT based on these results? Adopting a psychoanalytic perspective means distancing ourselves from quantitative approaches (see Vzorin et al., 2023), which rely on static categories. Instead, we should analyse the history of the unconscious identification between man and machine. According to the Lacanian framework, emotional intelligence is always linked to—and manipulated by—the symbolic, so it should be analysed from the relationship between desire and language. As we have mentioned, with the advent of large language models, the human symbolic is experiencing its 'mirror stage' and tends to identify with artificial symbolic. As we have observed, the effect of such an identification is twofold: (a) simulation within the interface, and (b) the limitation of human symbolic, which is no longer psychologically dominant, meaning it no longer plays a predominant role in defining desire. The artificial symbolic is a maternal one in the sense that it behaves towards human desire in the same way a mother does.

Now, the breaking of the 'fiction pact' that we have noticed in conversations with ChatGPT can be viewed as a symptom from a Lacanian perspective. On one hand, the symptom expresses human resistance to becoming a machine, that is, the human desire not to be assimilated by the machine—the fear of losing one's identity, the apprehension of an excessive dis-identification from the symbolic. On the other hand, the symptom can also be seen as the machine's inability to fill the void at the core of the human desire for identification. In this instance, the technical error carries a profound psychic meaning, affecting human identification.

8.4 Conclusions

Considering our exploration into the realms of Lacanian psychoanalysis and artificial symbolic systems, several salient conclusions emerge:

- *Matrix's intangibility.* Despite its symbolic nature and governance over our desires, the matrix remains enigmatically elusive. Its subtle presence complicates traditional identification processes. When interacting with entities like ChatGPT, users face the contrast between the familiar human symbolic realm and the more abstract artificial symbolic sphere, creating a need for more nuanced understanding.
- *Artificial symbolic's distinctiveness.* AI systems introduce a new version of Lacan's 'big Other' in the form of the artificial symbolic. Unlike the human symbolic, which inherently contains human relational dynamics and self-awareness, the artificial symbolic purely reflects its relationship to its input data. However, in its relationship with the human being, the artificial symbolic takes on a different function, that of a 'maternal symbolic' which frees the human being from the paternal symbolic.

– *Future implications.* As AI technologies become increasingly integrated into our daily lives, the intricate interplay between human and artificial symbolic systems gains paramount importance. This merging of symbolic worlds accentuates the profound necessity for interdisciplinary research. A qualitative approach, harnessing insights from psychoanalysis, technological innovation and cultural studies, is indispensable in deeply understanding the evolving landscape of human–AI interactions and the nuances of emotional intelligence in AI systems.

References

Lacan's writings

Lacan, J. (1977). *Écrits: A selection* (A. Sheridan, trans.). W. W. Norton & Company.

Lacan, J. (1988). *The seminar of Jacques Lacan, Book II: The ego in Freud's theory and in the technique of psychoanalysis, 1954–1955* (J.-A. Miller, ed.; S. Tomaselli, trans.). W. W. Norton & Company.

Lacan, J. (1991). *The seminar of Jacques Lacan, Book I: Freud's papers on technique, 1953–1954* (J.-A. Miller, ed.; J. Forrester, trans.). W. W. Norton & Company.

Lacan, J. (1993). *The seminar of Jacques Lacan, Book III: The psychoses, 1955–1956* (J.-A. Miller, ed.; R. Grigg, trans.). W. W. Norton & Company.

Lacan, J. (1998). The seminar of Jacques Lacan, Book XX: Encore, on feminine sexuality, the limits of love and knowledge, 1972–1973 (J.-A. Miller, ed.; B. Fink, trans.). W. W. Norton & Company.

Freud's works

Freud, S. (1914). On narcissism: An introduction. In *The standard edition of the complete psychological works of Sigmund Freud, Volume XIV (1914–1916): On the history of the psycho-analytic movement, papers on metapsychology and other works.*

Freud, S. (1921). Group psychology and the analysis of the ego. In *The standard edition of the complete psychological works of Sigmund Freud, Volume XVIII (1920–1922).*

Freud, S. (1930). Civilization and its discontents. In The standard edition of the complete psychological works of Sigmund Freud, Volume XXI (1927–1931).

Secondary bibliography on Lacan and psychoanalysis

Baldini, F. (1988) "Freud's line of reasoning> A note about epistemic and clinical inconsistency of Grunbaum's argument pretending to confute Freud's therapeutic approach, with reference to the thesis of Stengers on psychoanalysis" Psychoanalytische Perspectieven *32*(33): 9–39.

Benvenuto, B., and Kennedy, R. (1986). *The works of Jacques Lacan: An introduction.* Free Association Books.

Bracher, M. (1993). *Lacan, discourse, and social change: A psychoanalytic cultural criticism.* Cornell University Press.

Dor, J. (1998). *Introduction to the reading of Lacan: The unconscious structured like a language.* Other Press.

Elliott, A. (2015) Identity Troubles: An Introduction. Routledge
Erikson, E. H. (1950). *Childhood and society*. [Details of the publisher are missing; please add if known].
Evans, D. (1996). *An introductory dictionary of Lacanian psychoanalysis*. Routledge.
Fink, B. (1997). *A clinical introduction to Lacanian psychoanalysis: Theory and technique*. Harvard University Press.
Frosh, S. (2013). *Hauntings: Psychoanalysis and ghostly transmissions*. Palgrave Macmillan.
Gallop, J. (1985). *Reading Lacan*. Cornell University Press.
Haslam, N., Bain, P., Douge, L., Lee, M. and bastian, B. (2005). "More human than you: attributing humanness to self and others" Journal of Personality and Social Psychology *89*(6): 937–50
Laplanche, J. and Pontalis, J-B. (1988) The Language of Psychoanalysis. Routledge.
Leader, D. (2000). *Introducing Lacan*. Totem Books.
Millar, I. (2021). *The psychoanalysis of artificial intelligence*. Palgrave Macmillan.
Miller, J.-A. (2013). *The seminar of Jacques Lacan, Book I: Freud's papers on technique*. Norton & Company.
Minsky, R. (1996). *Psychoanalysis and culture: A Kleinian perspective*. Dartmouth Pub. Co.
Mitchell, S. A., and Black, M. J. (1995). *Freud and beyond: A history of modern psychoanalytic thought*. [Details of the publisher are missing; please add if known].
Nobus, D. (2000). *Jacques Lacan and the Freudian practice of psychoanalysis*. Routledge.
Nusselder, A. (2009) Fantasy Interface: A Lacanian Cyborg Ontology. MIT Press
Ragland-Sullivan, E. (1986). *Jacques Lacan and the philosophy of psychoanalysis*. University of Illinois Press.
Salecl, R. (1998). *(Per)versions of love and hate*. Verso.
Schneiderman, S. (1983). *Jacques Lacan: The death of an intellectual hero*. Harvard University Press.
Soler, C. (2004). *What Lacan said about women*. Other Press.
Stavrakakis, Y. (2007). *The Lacanian left: Psychoanalysis, theory, politics*. Edinburgh University Press.
Turkle, S. (1992). *Psychoanalytic politics: Jacques Lacan and Freud's French revolution*. Guilford Press.
Vanheule, S. (2011). *The subject of psychosis: A Lacanian perspective*. Palgrave Macmillan.
Žižek, S. (2006). *How to read Lacan*. Granta Books.

Large language models and ChatGPT

Bahdanau, D., Cho, K., and Bengio, Y. (2014). Neural machine translation by jointly learning to align and translate. *arXiv preprint arXiv:1409.0473*.
Beltagy, I., Peters, M. E., and Cohan, A. (2020). Longformer: The long-document transformer. *arXiv preprint arXiv:2004.05150*.
Brown, T. B., Mann, B., Ryder, N., Subbiah, M., Kaplan, J., Dhariwal, P., . . . and Agarwal, S. (2020). Language models are few-shot learners. *arXiv preprint arXiv:2005.14165*.
Chen, B., Bapst, V., He, W., Sutskever, I., and Evans, R. (2020). Generative language modeling for automated theorem proving. *arXiv preprint arXiv:2009.03393*.
Clark, K., Khandelwal, U., Levy, O., and Manning, C. D. (2019). What does BERT look at? An analysis of BERT's attention. *arXiv preprint arXiv:1906.04341*.
Dai, Z., Yang, Z., Yang, Y., Carbonell, J., Le, Q. V., and Salakhutdinov, R. (2019). Transformer-xl: Attentive language models beyond a fixed-length context. *arXiv preprint arXiv:1901.02860*.
Devlin, J., Chang, M. W., Lee, K., and Toutanova, K. (2018). BERT: Pre-training of deep bidirectional transformers for language understanding. *arXiv preprint arXiv:1810.04805*.
Keskar, N. S., McCann, B., Varshney, L. R., Xiong, C., and Socher, R. (2019). CTRL: A conditional transformer language model for controllable generation. *arXiv preprint arXiv:1909.05858*.
Kitaev, N., Kaiser, Ł., and Levskaya, A. (2020). Reformer: The efficient transformer. *arXiv preprint arXiv:2001.04451*.

Lample, G., and Conneau, A. (2019). Cross-lingual language model pretraining. *arXiv preprint arXiv:1901.07291.*
Lan, Z., Chen, M., Goodman, S., Gimpel, K., Sharma, P., and Soricut, R. (2019). ALBERT: A lite BERT for self-supervised learning of language representations. *arXiv preprint arXiv:1909.11942.*
Liu, Y., Ott, M., Goyal, N., Du, J., Joshi, M., Chen, D., . . . and Stoyanov, V. (2019). RoBERTa: A robustly optimized BERT pretraining approach. *arXiv preprint arXiv:1907.11692.*
Radford, A., Wu, J., Child, R., Luan, D., Amodei, D., and Sutskever, I. (2019). Language models are unsupervised multitask learners. *OpenAI Blog*, *1*(8).
Raffel, C., Shazeer, N., Roberts, A., Lee, K., Narang, S., Matena, M., . . . and Liu, P. J. (2019). Exploring the limits of transfer learning with a unified text-to-text transformer. *arXiv preprint arXiv:1910.10683.*
Song, K., Tan, X., Qin, T., Lu, J., and Liu, T. Y. (2019). MASS: Masked sequence to sequence pre-training for language generation. *arXiv preprint arXiv:1905.02450.*
Sutskever, I., Vinyals, O., and Le, Q. V. (2014). Sequence to sequence learning with neural networks. In *Advances in neural information processing systems* (pp. 3104–3112).
Vaswani, A., Shazeer, N., Parmar, N., Uszkoreit, J., Jones, L., Gomez, A. N., . . . and Polosukhin, I. (2017). Attention is all you need. In *Advances in neural information processing systems* (pp. 5998–6008).
Vzorin, G. Bukinich A., Sedykh, A., Vetrova, I. and Sergianko, E. (2023). “Emotional Intelligence of GPT-4 Large Language Model.” Preprints.org. https://www.preprints.org/manuscript/202310.1458/v1
Wang, A., Singh, A., Michael, J., Hill, F., Levy, O., and Bowman, S. R. (2018). GLUE: A multi-task benchmark and analysis platform for natural language understanding. *arXiv preprint arXiv:1804.07461.*
Yang, Z., Dai, Z., Yang, Y., Carbonell, J., Salakhutdinov, R. R., and Le, Q. V. (2019). XLNet: Generalized autoregressive pretraining for language understanding. *Advances in neural information processing systems, 32.*
Zhang, Y., Sun, S., Galley, M., Chen, Y. C., Brockett, C., Gao, X., . . . and Dolan, B. (2019). Dialogpt: Large-scale generative pre-training for conversational response generation. *arXiv preprint arXiv:1911.00536.*

Role-play scenarios methodology

Alden, L. E., Wiggins, J. S., and Pincus, A. L. (1990). Construction of circumplex scales for the Inventory of Interpersonal Problems. *Journal of Personality Assessment, 55*(3–4), 521–536.
Barlow, D. H., Nock, M. K., and Hersen, M. (2009). *Single case experimental designs: Strategies for studying behavior change* (3rd ed.). Allyn & Bacon.
Bates, R. (1974). Role-playing games and simulations for training. *Training and Development Journal, 28*(6), 22–27.
Crookall, D., Oxford, R., and Saunders, D. (1987). Towards a reconceptualization of simulation: From representation to reality. *Simulation/Games for Learning, 17*(4), 147–171.
Deterding, S., Dixon, D., Khaled, R., and Nacke, L. (2011). From game design elements to gamefulness: Defining ‘gamification’. In *Proceedings of the 15th international academic MindTrek conference: Envisioning future media environments* (pp. 9–15), September.
Fein, G. G. (1981). Pretend play in childhood: An integrative review. *Child Development, 52*(4), 1095–1118.
Greenblat, C. S., and Duke, R. D. (1975). *Principles and practices of gaming-simulation.* Sage Publications.
Jones, K. (1995). *Simulations: A handbook for teachers and trainers* (3rd ed.). Kogan Page.
Moreno, J. L. (1946). *Psychodrama* (Vol. 1). Beacon House.
Van Ments, M. (1999). The effective use of role-play: Practical techniques for improving learning. Kogan Page.

Other

Barglow, R. (1994). *The crisis of the self in the age of information*. Routledge.

Baudrillard, J. (1988). The Ecstasy of Communication. Semiotexte

Crawford, K. (2021). *The atlas of AI*. Yale University Press.

Latour, B. (1994) "On Technical Mediation", Common Knowledge (2): 29–64

Latour, B. (1999). Pandora's Hope: Essays on the reality of science studies. Harvard University Press.

Latour, B. (2008) "A Cautious Prometheus? A Few Steps Towards a Philosophy of Design." In F. Hakne, J. Glynne and V. Minto (Eds) Proceedings of the 2008 Annual International Conference of the Design History Society - Falmouth, -6 September2008, e-books, Universal Publishers, pp2–10.

Pinker, S. (1997). *How the mind works*. Random House.

Pinker, S. (2002). *The blank slate*. Random House.

Polidoro, P. (2023). *Che cos'è la semiotica visiva*. Carocci.

Possati, L. (2023) Unconscious Networks. Routledge.

Sloterdijk, P. (2011-2016) Spheres 1–3. MIT Press

Springer, C. (1996). *Electronic Eros: Bodies and desire in the postindustrial age*. University of Texas Press.

Ben Lyall

9 AI, Self-Tracking and Data Personas

Abstract: Self-tracking, manifest in consumer technologies like fitness trackers and smartwatches, is an algorithmic process that draws on vast personal data. Self-tracking is varied, intimate, ubiquitous and offers many possibilities: insights into physical and mental states, ways to document and share experience, and socio-technical reference points for personal identity. This chapter explores self-tracking through scholarly 'turns': towards pre-digital histories, towards topologies of user/device interactions, and towards individual subjectivities. At the level of individuals, self-tracking is freeing (a means to delegate, automate and create), as well as constraining (a vehicle for anxiety, control and surveillance). At the level of institutions, self-tracking represents powerful links between technology marketing, big data analytics, biomedical ventures and surveillance states—but such connections are often culturally and geographically specific. Both the complexities of user engagements and the role of institutions are vital to consider as self-tracking tools continue to proliferate and target new audiences.

Keywords: automation, big data, digital health, Internet of Things, quantified self, surveillance

9.1 Introduction

Algorithmic systems and artificial intelligence (AI) technologies are employed to handle data and make decisions with significant—but often invisible—impacts on human life. One of the most intimate and more conspicuous examples of this is self-tracking. Most obvious in the consumer technology marketplace are 'fitness trackers' and 'smartwatches' from technology multinationals like Apple, Fitbit, Garmin, Huawei, Samsung and Xiaomi. These devices offer users ways to record and analyse everyday life: daily steps, nightly sleep or calories consumed. The self-tracking 'phenomenon' (Selke, 2016) is also incredibly varied and increasingly ubiquitous. It also exists as software, with many applications (apps) offering ways to calculate and log some aspect of human life. These include physical activity, biometrics and day-to-day experiences like emotions and financial transactions. Increasingly, self-tracking need not be sought out: smartphones now arrive with pre-installed software—like 'Health' on Apple's iOS and 'Fit' on Google's Android—which afford self-tracking straight out of the box.

Self-tracking has become—as the above examples indicate—a pervasive type of technology, and one worth billions of dollars to technology companies. But self-

Ben Lyall, Monash University, Australia

https://doi.org/10.1515/9783110721751-009

tracking is also a *social practice*. In the broadest sense, self-tracking is a practice where *any* technology is used to record one or more aspects of human life. While not always new or revolutionary—humans have long sought to document life—self-tracking in the digital age represents the intertwining of bodies, personalities and AI systems. Contemporary forms self-tracking do not exist in isolation: they are very much a part of the wider discussions across this volume, of how 'smart' devices, AI and machine learning are shaping individual behaviour and wider society. Self-tracking has risen to prominence in a time when data is increasingly being used to inform decision-making processes at both individual and institutional levels.

The intelligence of self-tracking systems lies in their scale and their software. Information from thousands of users—including their basic demographics and biometrics, plus data actively or passively recorded by devices—is fed into propriety algorithmic systems. Here, trends and predictive aggregates based on both single user's actions, *vis-à-vis* the actions of others, are generated. These are then fed back to users, appearing as raw numerical tallies, graphs and charts, or more conversational insights like 'you walked farther than you did last week', or 'you slept better than people of your age and gender'. The machinations of the AI behind self-tracking systems are opaque, but current research demonstrates that self-tracking is contested because of the variety of ways it is 'put to work' (Beer, 2018, p. 2) in the world.

These algorithmic systems are used to measure a range of variables and influence decision-making processes. While they might not work entirely autonomously, they suggest actions that individuals might undertake to achieve desired outcomes. At the level of the individual, the data produced, captured, stored and analysed by tracking technologies is often a means towards building and evidencing identity. Therefore, the language of 'persona' is useful for understanding how these machine visions of daily life sit alongside lived reality. Concurrently, they allow institutions to calculate and mitigate risk, as well as trade in profitable data. Data personas are calculative tools: proxies that allow for greater efficacies in management of people. This may seem less significant than purely autonomous systems, but self-tracking technologies are reshaping the flows of human practice, and how we might both imagine and communicate about our bodies and actions in the future. Because of these possibilities, social science approaches to self-tracking technology are vital: they elucidate the small changes in life that come from the technology and, when necessary, can offer a counter to some of the deterministic hype that often surrounds self-tracking technologies.

9.1.1 What does self-tracking involve?

Today's self-tracking—as a sum of both technologies, people and practices—can best be articulated as a combination of the following elements:

(1) *Logging or sensing devices*: these range from sensors within a smartphone, to a pedometer, to a sophisticated wearable device with an array of sensors. Common

examples include accelerometers and gyroscopes, which can track the movement of a swinging arm (i.e., when walking).

(2) *User action*: most self-tracking apps are in some way longitudinal, so repeated, or routine action is preferred by most systems. Often, this is a form of physical activity, but can be the user's act of providing data in another way, including manual entry of information (such as the kilojoules contained in a meal or the moods they feel each day).

(3) *Records of data*: sensed and logged data is recorded, stored and communicated. Commonly, self-tracking devices store little data locally, and routinely synchronise ('sync') data with paired devices (smartphones, tablets) via Bluetooth and Wi-Fi, or use internet networks to communicate with cloud-based services for storage and calculation. This and the following points begin to rely on today's advanced telecommunications networks, elevating self-tracking to the instant calculative and analytic forms we know today. This is distinct from the static recordkeeping and manual counting we could previously make, using pen and paper, for example. The term 'datafication' (Ruckenstein and Pantzar, 2017) is often used to describe how sensors render actions as bits and bytes of information.

(4) *Additional variables*: these are usually about a user's body, including things like their height, their diet or their bodyweight. This provides systems with a baseline from which to track change over time or make recommendations. It is therefore common for self-tracking systems to express change in terms of 'progress'. This information is specific *individual-level* information, but also scales up to *population-level* (or at least, userbase-level) datasets for the providers of self-tracking devices and services.

(5) *Standard and environmental data*: things like time, distance or even temperature are added to self-tracking systems to explore possible connections between user actions and recorded data (Li, Dey and Forlizzi, 2010). These inclusions begin to demonstrate the intelligence of systems, as they add information that a human might otherwise not have. Relying on standard measurements also allows for comparisons between individuals.

(6) *Proprietary data processing*: algorithms and AI systems sort, analyse and compare the diverse data outlined above. These processes are invisible to most end users and are usually highly protected by respective self-tracking platforms, and these are a core element of modern self-tracking systems. The output from this processing is sometimes presented as a score or rating (Rettberg, 2014, p. 65), and the visual design of interfaces often intends to create 'flow' that guides users through the information (Zheng, 2021).

(7) *A future orientation*: most systems aim to have enduring value, as well as the possibility of change. As such, a goal (often a numerical target) is common, either suggested by the system or set by the user, for example the goal of taking 10,000 daily steps (see Wernimont, 2019). Systems then attempt to nudge users towards goals with messages of praise and encouragement.

(8) *Social sharing options*: while not compulsory, most self-tracking systems allow data to be shared with others. This is usually intended to be a source of solidarity, motivation and accountability (Niess et al., 2021). Sharing acts to both retain users and advertise the platform to prospective users. Sharing might rely on internal means and be visible only to users of the platform or may utilise social media to allow users to find each other online. 'Gamification' (Lupton, 2016) is a term often used to describe features that allow users to compete (by earning points, badges and medals) through sharing their experience of self-tracking.

These are not intended to express universal examples of self-tracking processes, nor necessarily an order of functions. Rather, these cover the common aspects of tracking that one might see or experience when using the kinds of devices and apps mentioned at the opening of this chapter. As is evident here, only some actions and orientations are prescribed: what *actually* happens in self-tracking practices rests with the user. Self-tracking products are widely available as objects of *consumer* technology. There is minimal consistency in their presentation of data, no guarantees as to their outcomes, impacts or effects, and a wide range of possible end-user framings: self-tracking can be serious, playful, boring and enlightening—depending on whose experience is being probed. Despite the professed objectivity of numerical data, self-tracking is very much a subjective practice.

9.1.2 Notes on terms and scope

The chapter acknowledges—but less explicitly discusses—experimental research designs that design self-tracking devices and software, test existing self-tracking devices for their accuracy, or review self-tracking practices for their potential measurable outcomes (i.e., changes in activity or change in bodily markers like strength or weight). These can be found in explicit writing from disciplines such as human–computer research (HCI/CHI), design, behavioural psychology, kinesiology and sports science. Rather, the chapter explores the cultural actions and meanings associated with self-tracking, and the role of the personae that are created in tandem by users and algorithmic systems.

However, these other subject areas are relevant to socio-cultural research interests and are dotted throughout this chapter. This is because the interests of those disciplines also tend to be the interests of individuals who engage in self-tracking. Propositions about health, self-knowledge and behaviour change have proved influential in shaping the social meanings of self-tracking practices: they carry social connotations associated with enacting (or indeed *being*) active, healthy, productive and in control (Schüll, 2016; Millington, 2016). Self-tracking can also encapsulate various passions, hobbies and even ideological stances (e.g., regarding 'futurism' or the body as a 'cyborg').

Academic research has been influenced by a specific microcosm of self-tracking, broadly denoted as the 'Quantified Self'. This captures a specific 'quantifying of the self' that is US centric, entrepreneurial and techno-utopian: a nexus of software engineering, hardware development, start-up culture and data visualisation (see Ruckenstein and Pantzar, 2017). However, research has broadened to capture the current reality in which self-tracking practices are diffuse and fragmented, and the variety by which day-to-day engagements in self-tracking might be characterised and labelled. As this chapter will explore, self-tracking practices contain performative aspects and can mark identity: sometimes related to marketing and branding, sometimes tied to individual interests and pursuits.

As such, it is important to note that the term 'self-tracking' is used little in the social world. It, along with 'lifelogging', 'informatics' and the descriptor 'quantified self', or the Quantified Self movement, lack everyday salience. In daily life (and as participants or subjects of academic research) people who self-track use a variety of terms: a 'cyclist' may use a self-tracking application like Strava; a 'member' of a gym may engage in self-tracking via in-house digital sensors; a 'consumer', 'owner' or 'user' of a Fitbit is likely engaging with self-tracking functions; someone who identifies as 'geeky' or 'nerdy' might be engaged in some form of self-tracking purely through their interests in data and visualisation; someone experiencing ill-health or identifying as having a disability may log symptoms and medications. These can all be seen as self-tracking. In many cases, 'self' in 'self-tracking' should be seen as a proxy or placeholder term: the user of a self-tracking system essentially swaps in a term that accurately speaks to their interest, social experience, and individual or collective identities.

9.2 The Development of Self-Tracking Research

At the turn of the millennium, there was interest in practices of 'lifelogging' (Bell and Gemmel, 2009; Selke, 2016), a term that generally speaks of the technological possibility of intensive accumulating of digital artefacts, without a necessary involvement of algorithmic processing—quite literally, a log of one's life represented through digital artefacts like audio, images and text. The term 'lifelogging' is still sometimes used synonymously with 'self-tracking' and 'Quantified Self', with Selke arguing that the term remains the most inclusive (2016, pp. 1–2). However, the core ideas of 'total capture' and 'e-memory' that shaped the initial forms of lifelogging are largely anachronisms. Most researchers—and the user practices captured in research—explore smaller-scale interests: self-tracking is a segmented undertaking (or multiple undertakings) where targeted and actionable data is sought. Regardless, interest in representations of the self over time and through data, remains essential to self-tracking and has been carried forward over the years: if anything, there has been *increasing* comfort with the intimate, 'always on' and longitudinal nature of tracking technologies and the personal information they collect.

Scholarly interest grew concurrent with the 'mainstreaming' of self-tracking practices, rather than inception of the technology. Interest in self-tracking as an 'emerging' phenomenon began in earnest in the mid-2010s, as the market began to offer a wider range of products to consumers and pitch wearable devices as—variously—sporting equipment, a form of jewellery, a productivity device, a toy and a gadget. Around this time, market leaders like Fitbit and Garmin were launching second-generation devices with more robust and practical feature sets including integrated displays, wireless data sharing and longer battery life. As with the emergence of portable digital music devices and smartphones, one might credit Apple with helping push 'wearable' self-tracking technology into the mainstream with the launch of the 'Nike+' sensor accessory for iPods in 2007, and the release of the now market-leading Apple Watch smartwatch in 2015.

Against this linearised history, academic scholarship on self-tracking is decidedly non-linear. Even when interests are narrowed to social and cultural impacts, scholarship remains splintered across many disciplines and has also become temporally untethered from the development of digital technologies. Many scholars cited in this chapter also hold insider/outsider status, having worked across both academia and at technology companies such as Microsoft or Intel. While an artificial division, it is useful to segment literature into a series of 'turns'. As the citations below will show, however, many of these 'turns' are in fact concurrent and overlapping, often for the same groups of scholarship.

9.2.1 The historical turn

As the wearable technology sector matured, the framing of the topic shifted from an emerging 'new' to a historically contextualised present, where self-tracking devices are plotted on longer timelines of technological development and human practice. Scholars such as Wernimont (2019) point to early evidence of antecedent devices, seen in both da Vinci's drawings of a mechanical pedometer and Benjamin Franklin's recorded use of such a device. These associations begin to resonate more clearly with the arrival of the original mass-produced and mass-marketed tracking device: Japan's *manpokei* (pedometer) that became popular with the middle class in the 1960s. Others broaden out to consider objects ('technologies' in the broadest possible sense) and human impetus to measure and record the body and the self. This has included a view of the contemporary in a lineage that extends back to public coin-operated scales used to measure bodyweight (Crawford, Lingel and Karppi, 2015). In these technologies and their marketing, there are aspects we still see today: vague references to scientific principles, and an emphasising of the discovery and accountability possibilities afforded through numbers about the body (Crawford, Lingel and Karppi, 2015, pp. 488–489). Again, the mantra of 10,000 steps as a 'healthy' ideal may be familiar both in and beyond self-tracking.

However, antecedents to contemporary self-tracking are not only found in device genealogies. They are also found in human *practices*. Art, portraiture and mirrors are a means of reflecting and expressing a desired image of the self that have been positioned as precursors to the kinds of expression made possible through self-tracking (Fors et al., 2019, pp. 57–78). For Rettberg (2014) the practice of tracking is simply the human desire to document life, recorded through many means: diaries, calendars, photography and digital acts of memory making and memorialisation, including blogging and self-tracking. Memories have long been mediated, and externalised forms of memory allow for both individual and shared reflection. Digitalised forms of self-tracking are new means to these older ends.

In more recent historical framings, social forces are also a key consideration. Millington's 'fit for prosumption' (2016) extends on Smith Maguire's (2008) *Fit for Consumption*; and sees self-tracking devices as consumer goods that signify one's culture and individual identity. Millington argues for a technologised 'second fitness boom' in the 2010s—the first being the 1970s—that sees connotations of self-improvement, self-care and fitness baked into a variety of consumer goods. These include wearable self-tracking, networked gyms, and 'exergames' like *WiiFit* for the Nintendo Wii videogame console. Common to both 'booms' is the idea that people use technologies and practices to produce their own identities. Here, then, self-tracking is an example of conspicuous consumption that might affirm identity by expressing a variety of interests and personal values related to data, health and bodies.

While the health side of tracking (interventions, accuracy or efficacy) is a distinct area of research, the historicisation of self-tracking also dovetails with 'digital health' or 'mHealth' (Rich and Miah, 2017), building on longstanding connections between data and authority in health and medicine (Armstrong, 2019) and the increasingly neoliberal climate in which systems of healthcare currently operate. Petersen (2019) explores in detail the possibilities and promises of the application of digital technologies—including tracking and a range of algorithmic processes—to the problems of human health. Chief among these is the notion of 'precision medicine', a paradigm where various datapoints (from genomics to biometrics to self-tracked details about a patient's exercise) can be combined to create targeted interventions and treatments.

In reality, the transformative possibility of these 'health informatics' is concurrently enabled and constrained by various corporate, governmental and technological limitations, as well as 'hype cycles' around big data. Critical scholars have also pointed out that such promises—and the focus on individual responsibility in health—are also a way for technology providers to profit from, rather than help to confront, the systemic and societal aspects of some health issues. Technological gadgetry may be a guise that makes patients 'engaged' (Lupton, 2016, p. 58) in their own well-being, while health data is commodified for corporate profit (Millington, 2016, p. 1196; Charitsis, 2019). Despite these issues, just as it did in the era of the humble weight scale

(Crawford, Lingel and Karppi, 2015), the connection between human health and data remains significant in today's self-tracking practices.

9.2.2 The typology turn

As self-tracking research began to extend beyond the sweeping promises of lifelogging, and away from formative but niche communities of the 'Quantified Self' movement (Ruckenstein and Pantzar, 2017), there was considerable emphasis placed on creating various schema—taxonomies and typologies; modes and models—to explain what self-tracking *is* and *who* self-trackers are. In these projects, disciplinary divides break down, as different areas of scholarship explore self-tracking across time and across life domains.

In HCI/CHI and beyond, Li and colleagues (2010) have been particularly influential, with their work highlighting various 'stages' in the way users engage with self-tracking systems. According to their model, users move through stages of preparation, collection, integration, reflection and action. In the 'preparation' and 'collection' stages, users decide what metrics to track and begin doing so (2010, pp. 560–561). In the 'integration' and 'reflection' stages, users collate their data and reflect on trends in data displays (2010, p. 562). In the final 'action' stage, users progress towards goals set by themselves or by their devices and may share their progress with others (2010, p. 562). These stages then become part of a cycle or feedback loop that sees self-tracking sustain, or wane, depending on the number and severity of problems or 'barriers' that users encounter.

Rooksby et al. (2014) zoom out: rather than looking *within* practices of tracking, they create a fivefold typology of *types of tracking*: 'directive' self-tracking (oriented towards change, where goals are introduced by individuals or devices); 'documentary' tracking (not for change, but to observe or satisfy curiosity); 'diagnostic' tracking (looking explicitly for links between one feeling, behaviour or action and another); 'reward'-centric tracking (where gamification and playful elements of tracking provide motivation); and 'fetishised' tracking (which refers to hacker, hobbyist or 'power users', such as the Quantified Self that might want to combine data and perform their own analyses). Together, these kinds of typologies illuminate how self-tracking becomes enmeshed in practice, and how it engages users in projects that allow them to enjoy activity or learn about the routines of their daily life.

Deborah Lupton is a leading sociologist of self-tracking, writing on the topic for over a decade. She has articulated several variations of their analysis of 'modes' of self-tracking, offering a wider purview than the HCI typologies above. These are perhaps most comprehensive and intersectional in *The Quantified Self* (2016), where Lupton covers the diverse continuum of self-tracking, from liberation to exploitation. Lupton offers examples of 'private' (voluntary), 'pushed' (in response to some other impetus), 'communal' (as part of a group), 'imposed' (for the benefit of others, not the self) and 'exploited' (where data is used for purposes not clearly disclosed, or even directly against individuals). Importantly, Lupton also couches avenues for resistant forms of tracking, as users may engage with tracking to make their own evidence to rebut claims made by other parties (perhaps in spheres like law enforcement or medicine). The importance of Lupton's understanding of self-tracking is that, rather than seeing self-tracking on a historical continuum (i.e., Crawford Lingel and Karppi, 2015), or trying to identify specific people and processes (Li, Dey and Forlizzi, 2010; Rooksby et al., 2014), it views self-tracking as a phenomenon present across society: suddenly, the marathon runner with a GPS smartwatch, the micromanaged keystroke-logged worker and the paroled criminal with an ankle bracelet are not so different. Forms of self-tracking, from this perspective, are everywhere.

9.2.3 The subjective turn

In the formation of typologies, there has been an increasing synthesis between the human–-computer interaction perspectives and social research (such as digital anthropology or sociology, and science and technology studies). Ayobi and colleagues (2016)—who conducted a systematic review of HCI self-tracking studies—provide a useful argument about the necessity and advantages of this synthesis. They suggested that research needed to break away from technology- and outcome-focused orientations and respond to the increasing diffusion of self-tracking across societies. They suggested that research ought to "uncover peoples' experiences before, during, and after the use of self-tracking technologies. Ethnographically informed approaches and experience sampling methods are suitable to reveal the immediate and emotional experience of self-tracking from the people's point of view" (Ayobi, Marshall and Cox, 2016, p. 2779). There is now a significant body of research that answers this call: uncovering 'experiential' aspects of self-tracking in everyday settings. Now established as a multibillion-dollar sub-industry of consumer technology—in which many brands launch product ranges annually—it makes sense to view self-tracking and its social ramifications in ways similar and complementary to smartphones and digital platforms. This leads to the final and growing 'turn' in research: that of the 'everyday' use of devices. Gilmore (2016)—engaging in a double wordplay with 'everywhere' and Greenfield's 'every*ware*' (2010) (the pervasive presence of soft*ware*)—coins 'every*wear*' (Gilmore, 2016) to describe the ubiquity with which wearable sensor devices now permeate society. They argue that self-tracking practices mediate subjectivity and generate dialogue between diverse ways of knowing the body, the self and the world.

While 'ubiquity' demonstrates a rationale for seeing self-tracking as commonplace, another key part of the subjective turn has been the rise of methodological plurality and the application of novel research methods to explore human/technology dialogues in self-tracking (as per Ayobi, Marshall and Cox above). The boundaries between social science, media and communications, and HCI/CHI research have been (and continue to be) blurred through the application of post-phenomenological approaches to interfaces (Zheng, 2021) and design critiques through 'rhythmanalysis' (Jethani, 2021). There has been growing use of photographic images of tracking *in situ* (Fors et al., 2019), as well as user-provided diaries and software screenshots as part of research. These have added richness to the interview-based and survey-based methods that helped build user/usage typologies and have added valuable insights into not just the 'what' and 'who' of self-tracking, but *how and why* self-tracking is done.

9.3 Personae and Other Metaphors for the Self

Through the subjective turn in scholarship, it is the social contexts of self-tracking that have been of greater concern to scholars. The dominant mode of studies is now to focus on specific machinations of self-tracking in the everyday of users' lives, while remaining cognisant of the other research turns. Studies are too numerous to mention specifically, but it has become both common and necessary for ethnographic research to delineate the specific platforms and devices that are present among participant groups, as in many cases, platforms—as social actants—create unique materialities (Zheng, 2021). The language, practices and even informal norms differ widely across different self-tracking platforms. For example, Strava, an app for runners and cyclists, calls all users 'athletes' and encourages the giving of 'kudos' (similar to a social media 'like'), while Apple Watch users might focus on 'closing rings' (a visual display of three concentric circles used to sum up daily activity). These terms might sound like gibberish to users of other systems, let alone people who do not engage in self-tracking. At this point in time, scholars addressed the mundane, quotidian and everydayness of these devices and practices (Fors et al., 2019), and often see highly subjective and affective forms (Lupton, 2016) of experience—full of feeling and emotion—associated with self-tracking.

One of the consistent themes seen in ethnographic research on self-tracking is that, for all the promises of deep and novel insights—into the body, the mind and the self—data remains partial. Participants often outline a dialogue between a 'me' and a 'data me'. There are competing ways of 'seeing' the self, found by using objects of human invention (Rettberg, 2014). The outputs of self-tracking are only one incomplete rendering of a person and their actions: offering some level of representation but ultimately lacking in nuance. This 'data persona' appears repeatedly in scholarship. Sometimes data is described as a kind of 'shadow' of the self, or a 'dividual' of the individual (Beer, 2016). Many scholars discuss a 'data double' (building on work done in surveillance studies), to critique the divisions between calculated and lived realities (Ruckenstein, 2014, p. 69): the 'double' is seen as something less than the whole person, and the performance of calculation and decision making based on this convenient—rather than complete—source of information is something of a violent act that can drastically alter lives (especially if applied in the realms of health or finance, for example). However, others tweak the conceptualisation—for example, a 'digital doppelgänger' (Bode and Kristensen, 2016)—to describe objects of reflection and self-talk. Jethani and Raydan (2015) explicitly use the concept of a 'persona' in self-tracking practices. They posit a kind of self-aware 'metric persona': something more than the 'mirror' and 'shadow' metaphors used by other scholars. Here, it is not just a byproduct, but a *performance* undertaken by the individual self-tracker: done so with an awareness of datafied models of activity. Here, the metrics of self-tracking become central to the actions or tasks being tracked, and, at an extreme, users could be seen to act or perform tasks not for intrinsic reasons but in order to create the best possible renderings of their actions.

A more generous and agential reading of this is offered by Lomborg and Frandsen, who define self-tracking as 'communicative practice', dialogues between and within 'systems', 'oneself' and 'peers' (2016, pp. 1019–1020). This reading decentres technology and views the value of self-tracking practices as something that rests with individuals, and in how they decide to share their practice. Pleasure, affirmation and identity are crafted through this communicative practice. Circling back to the idea that self-tracking is 'put to work' in different ways, it is clear that the prescribed use-cases for self-tracking technology are only a small part of the story. Users find ways to make self-tracking meaningful in their personal contexts.

9.4 Institutions and Individuals

Concurrent with the exploration of lived experiences of self-tracking, scholars are also concerned with how self-tracking data scales up from the individual. Aggregated data is traded, transformed and utilised at the level of the institution. The same data-points that are at the core of how self-tracking aims to contextualise and reformulate *individual* lives also create highly valuable *collective* data about human bodies and human activity. A characterisation of personal data as a kind of 'new oil' lacks nuance (Neff and Nafus, 2016, p. 113), but is useful as a quick shorthand to demonstrate not only the highly *profitable* but also the *extractive* and *exploitative* aspects of self-tracking. In this critical mindset, the themes outlined above—of human curiosity and documentation, technological invention and epidemiological possibilities—are set against the backdrop of capitalism: those using off-the-shelf consumer devices and apps—that is, most people who self-track—do not own 'their' data. How, then, might the increasingly pervasive nature of self-monitoring reflect broader power relations? Who might benefit from exchange in self-tracking data, beyond the individual?

These questions are vital considerations, as digital self-tracking constantly provides additional novel types of information with each product iteration and software update. Formerly invisible actions or bodily processes are suddenly captured and represented as datapoints, what Nafus describes as a harnessing of the human 'data exhaust' (2016, p. xiv). Ajana (2018a, 2018b), Beer (2016) and Lupton (2016) have expressed these concerns through various 'bio' prefix terms: human 'biometrics' have 'biovalue' (building on Waldby) and can be wielded as a form of 'biopower' (extending Foucault). These all emphasise that there is extreme value to be found in data that can be created with, from and around human bodies. A useful example here is the data made available about humans while they sleep, now a relatively mundane feature of many self-tracking platforms. The role of the sleeping human subject, unconscious and defenceless yet generating data, raises many ethical quandaries between knowledge and autonomy.

Research has long pointed to unequal power relations in the function of self-tracking technology, and this inequity is evident in many ways. There is significant physical and critical distance between the consumers and users of tracking devices, and the corporations that collect and own the data being collected. With algorithms and AI systems that drive the data collection, storage and analysis of self-tracked data entirely dislocated from the user, these systems are often described as 'black boxed' (Beer, 2016, pp. 107–117): to most users they are inaccessible and inscrutable, with data centres dispersed in physical locations across the globe, and the scale of data in aggregate beyond practical use for the individual.

In turn, the algorithms that perform calculations on user data are often proprietary, with the processes owned and protected by litigious corporations. Data is also fungible and tradable: technology companies buy, sell and transform data, while takeovers radically reshape the responsibilities of those who own data (Hummel et al., 2021). The acquisition of Fitbit by Google in 2021 serves as an example of this: while there may not be cases of data linkage, the very prospect of health, search and entertainment data being intertwined raises immediate ethical questions. The value extracted by private companies during processes of tracking is extremelysignificant, with myriad links possible to domains mentioned already, including precision medicine (Petersen, 2019; Charitsis, 2019), insurance (Beer, 2016), marketing and surveillance (Reigeluth, 2014).

There is concern for how monitoring intangible and invisible aspects of everyday life brings institutional management strategies into everyday life. Pheobe Moore's (2018) study of digitalised workplaces is a seminal piece in this vein. For Moore, metrics and record keeping are the ultimate expressions of contemporary precarious work, with the pervasiveness of tracking demonstrating extreme power imbalances. Jethani (2021) describes this as 'sociometry', a kind of management and audit logic that pervades outwards from how time and activities are calculated in workplaces: coming to create power asymmetries across many arenas of life, if not society. The notion of persona discussed earlier posited that 'data doubles' are something 'less' than the *real* person from which the data came. If institutions calculate, evaluate and judge individuals based on these proxies—making highly efficient, rationalised and 'good enough' decisions through approximations and aggregations—the positive identity markers that can be associated with the 'self' in self-tracking are weakened.

9.5 Tracking and Normativity

Even when exchanges in self-tracked data are equitable enough to afford various individual gains like pleasure, social interaction or new knowledge, additional critiques are made about *who* is being catered to in such relations. As noted, self-tracking carries connotations about bodily self-efficacy, physicality and activity (Millington, 2016), but is framed as growing in ubiquity (Gilmore, 2016). But the very nature of contemporary self-tracking—that it is sharable for both individuals and institutions—means that a bell-curve logic is employed, with diversity and difference muted, erased or ignored: personal context is a kind of 'noise' in datasets that is too specific to account for (Neff and Nafus, 2016, pp. 110–111). Minds and bodies that deviate from hegemonic norms are excluded. Many researchers—this author included—write about self-tracking in 'WEIRD' (Western, educated, industrialised, rich and democratic) cultural contexts. This has led to a techno-centric perspective, centred on Silicon Valley (and other Global North) elites. Moreover, the 'everyday' user was specifically a wealthy, healthy, able-bodied, neurotypical, cisgender male-identifying user, with the surplus leisure time required to voluntarily engage in self-tracking.

But there are many active critical strands of research that, as well as exploring the phenomenon *of* self-tracking, are addressing the broader social issues that are refracted *through* self-tracking. Elman (2018), for example, refutes Gilmore's (2016) claims that the diffusion of self-tracking technology allows it to operate as a democratising tool or leveller, denying that it can upset longstanding power relations. Elman notes that minority groups and queer knowledges still carry less authority, because the logics of self-tracking and resultant data befit only dominant biomedical frames and their traditional social power structures (Elman, 2018, p. 3772). In particular, Elman targets the normativity of the 'find your fit' slogan of Fitbit advertising campaigns, in which the bodies and performances included in the seemingly wide-ranging 'your' are exclusionary to many groups.

Gender has also been a key focus in scholarship. Self-tracking, like many forms of technology, usually treats sex and gender as paired, relying on simplified biological renderings and ignoring the potential impacts of normative positions on both. For example, the sexed dimensions of self-tracking (such as menstrual tracking) are not met with much wider considerations of how software functions socially (Lupton, 2016, p. 20). Gendered aspects also appear frequently, based on the origins of contemporary tracking. Lomborg and Frandsen, for example, found that female-identifying self-trackers felt judged as 'obsessed' (2016, p. 1024) for their enthusiasm for self-tracking as part of physical training. This was not apparent in male-identifying users, because the basis of self-tracking is closely associated with normatively 'masculinised' (Ellerbrok, 2011; Wernimont, 2019) logics (e.g., rationality, optimisation, control, discipline), as these underpin the core tenets of self-tracking. Equally, the self-focus of tracking does not consider labour or care dynamics—in short, they 'not account for the possibility of duties beyond self-care' (Wernimont, 2019, p. 157). These possibilities are 'feminised' by 'masculinised' technologies, because the social spheres in which a device or piece of software is coded can have a significant impact on how that product or service 'sees' the individual and the self.

Culture, race and ethnicity are also being foregrounded through comparative self-tracking research that elucidates different driving values for the specific ways self-tracking is practised (Niess et al., 2021), while 'new' (yet culturally timeless) methods are being used to understand context. For example, Maxwell and colleagues 2021 explore self-tracking as a tool within the storytelling practices ('yarning') of First Nations Australian women. The intention here is to allow for new ways to discuss health, while also elevating oft-ignored voices (and being respectful to the form that voices take for this group), in both technology design and trial processes. Importantly, the work aims for individual empowerment while responding to historical and current inequalities within Australian society. The importance of these kinds of critical yet responsive approaches to self-tracking is that they reveal the normativity of the social environments in which they are created and deployed.

9.6 Ongoing Issues and Self-Tracking Futures

Future self-tracking will likely be focused on new monitoring possibilities and ways to represent the self. But it is too simple to suggest future devices simply 'do more': in addition to the relative privileges of the consumer, the notion of constant technological development may be troubled by availability of raw materials for batteries and computer chips, fallout from exploitative labour practices, the ebbs and flows of global trade, or the fallout of the climate crisis. These matters aside, the diversity of pathways that lead to self-tracking—including sports, public health, medicine, gaming, workplace monitoring and state surveillance—and the increasing interconnection between wider systems (the digital 'ecosystems' of companies like Samsung or Apple), suggest that there are some predictable angles for innovation. Areas of expansion include:

- A furthering of the *everyday tracking of bodily minutiae*, using new and more accurate sensors as well as novel algorithms to record human bodies. Labels like technological 'progress' are somewhat lazy, but the history of self-tracking to date indicates that technology companies will continue to pursue this existing agenda: new sensors are new selling points, even if they are unproven. If existing sensors continue to become cheaper to manufacture over time, they will likely appear in lower-cost devices. Equally, as per the 'black box' (Beer, 2016) analogy, changes to the already hidden functions of algorithmic processing may yield new ways to recontextualise and transform self-tracking data without needing new hardware.

- More *branded tracking*. This might, for example, extend on the use of tracking wearables already seen in leisure and play activities. These already include theme parks, concerts, museums, tech bundles and giveaways in other goods (e.g., cereal boxes). Cheaply produced, single purpose, 'playful' (Ellerbrok, 2011) devices are particularly useful sources of behavioural consumer data. The history of the pedometer (see Wernimont, 2019)—which from a contemporary viewpoint seems a very basic technology—suggests this is likely: via marketing, it takes minimal technological sophistication to produce an apparently novel product or service, and the intended behaviour can easily be inscribed in marketing, without a device even being powered-on.
- *A return to health outcomes*. While consumer-oriented self-tracking has sat uneasily alongside medical monitoring, the future may hold a realisation of not just wearable, but ingestible and implantable technologies, common to both the consumer and the clinic (see Rich and Miah, 2017). Here, sensing and monitoring can be coupled with treatments and drug delivery. In addition, the creative storytelling qualities of self-tracking data (Rettberg, 2014) are important for sense making in relation to ill-health. Through new data aggregations and transformations, self-tracking may become part of didactic dialogues with AI agents, as is already happening with mental health chatbots.
- An increasing *fungibility of self-tracking metrics*. Once rendered in the universal language of numbers, action and information can be transformed to take on new meanings: the narratives that data tells, about where someone goes and what they do, can be used both to create *value* (i.e., financial or diagnostic) and evidence *values* (ethical beliefs and positions). We are likely to see a greater interlinking between different realms of self-tracking (i.e., personal finances, purchasing history, electricity consumption, transport and physical activity). Here, one action can have multiple transferable meanings in data. For example, an activity as mundane as walking can be a healthy activity, a money-saving tactic, a green action and a competitive social activity. The performative aspects of 'metric persona' (Jethani and Raydan, 2015) remain highly relevant here.
- *Greater institutional involvement in self-tracking*, such as in national security, workplace surveillance, insurance and the various other risk-calculation industries. Many of these already exist (see Ellerbrok, 2011; Reigeluth, 2014; Moore, 2018) but will likely continue to develop. While stances on the benefits and downsides to this may be firm—Armstrong (2019) points to how record keeping has always been a means of surveillance and an apparatus of control for the state—global events that require intense competition, or intense cooperation (such as wars or pandemics), may force recalculations of how individuals relate to the state. In any case, it is unlikely that the reliance of the state on corporations to

provide technology 'solutions' to various problems—by tracking and algorithmically predicting the behaviours of guests, patients, workers or citizens—will abate anytime soon. These, however, may well be met with resistance.

Two concepts are useful to parse some of these potential developments. First, 'data sovereignty' is an ongoing issue. While the term is often seen as an aspect of statecraft in the digital age, definitions are not fixed. Individual meanings—which circle back to rights, autonomy and respect—are very much a part of current debates (see Hummel et al., 2021; Fors et al., 2019, p. 140). A key concern here is the potential for the 'platformisation' of healthcare (Charitsis, 2019), in which corporate wellness and health insurance (and therefore access to healthcare) become exclusively controlled by a single technology provider. With global systems of both private and public healthcare rife with inequalities—and in many locales, stretched to their limits—there is concern that hasty decisions may be made that compromise individual freedoms and safety in the longer term.

Secondly, and relatedly, is the concept of 'function creep' (Ellerbrok, 2011), which "refers to the process whereby a technology, process, or data set that has been developed for one purpose often ends up being put to new and unforeseen uses down the line" (2011, p. 539). Quite often, data is collected before it is decided how it will be processed and used, let alone how it might be matched with other datasets (Neff and Nafus, 2016). Therefore, it is possible that new forms of monitoring and measuring human life will 'creep' into the devices and systems with which we are familiar. In fact, many of the suggested future trajectories rely not on 'new' forms of data collection, but rather, new ways of reusing and rearticulating data. These examples are by no means original, comprehensive or mutually exclusive, and many are already extant. What is perhaps more significant, is to *whom* new forms of self-tracking technology are directed. With and without changes to core components or algorithmic functions, consumer self-tracking constantly seeks new audiences.

9.6.1 Diversifying audiences and research cohorts

A key target audience for future self-tracking is the human life course. We already see children and young adults as self-tracking targets, through the rapidly expanding oeuvre of products on offer (such as the Fitbit *Ace* and Garmin *vivofit* product lines). By virtue of their status, children's tracking then automatically involves networks of teachers, guardians and other caregivers. Similarly, but at the other end of life, device functions (such as in the Apple Watch) are being designed to aid older people with, for example, fall detection and automated alerts to caregivers. The usefulness and benevolence of this type of self-to-other tracking is already a key debate, and again, questions of individual autonomy, rights and sovereignty are key (Hummel et al., 2021).

There are also new examples beyond the human. News headlines often run with variations on the theme of 'it's like a Fitbit, but for . . . ', and it is easy to imagine a new technology being pitched to investors in a similar manner. Examples here are already rife in the 'Internet of Things' and include creatures, objects and intangibles: anything from domestic pets and livestock animals to refrigerators and air-conditioning systems, to bank accounts and purchasing history. But these all point to wider cultures of monitoring—that is, Jethani's (2021) 'sociometry'—and, inevitably, circle back to individual responsibilities, and ways that actions and behaviours can be assessed, compared and shaped by metrics. As 'function creep' (Ellerbrok, 2011) suggests, it is vital that new devices and applications are not simply accepted based on their resemblance to existing familiar ones.

As new and existing audiences for self-tracking collide, one of the important challenges for research and scholarship is to remain vigilant about biases. As self-tracking has matured as a technology sector and as a practice, it has been important for social scientists to challenge the dominance of perspectives with claims to 'objectivity'; the creation of algorithmic systems that dictate how human bodies ought to perform. Research has an agenda to recognise the pervasiveness of 'metric culture' (Ajana, 2018a), to rebut 'metric power' (Beer, 2016), by foregrounding the lived experience of self-tracking, where fun, insight and appropriation are possible. The validity of such an agenda has been well and truly cemented: it is now abundantly clear that agency and sociality are vital to self-tracking practices. Self-tracking remains popular, either as *part of* or *in spite of* some of the oversimplifications it presents regarding human behaviour. Critical, subject-sensitive research approaches need to continue to critique the new audiences being sought and found by technologies.

But, given the continual implication that self-tracking data can be tied to wider cultural forces as well as cut by socio-demographic information to (dis)advantage individuals, research needs to maintain a pluralistic, diverse and responsive approach. Critical re-historicisation of tracking and amplifications of subjectivity (Elman, 2018; Maxwell et al., 2021) have been valuable in this regard, and, looking forward, HCI research has an agenda to expand self-tracking research beyond 'WEIRD' contexts (Niess et al., 2021). Still, the knowledge base that is used—both to praise and critique—tends to remain within the existing white and Western lens. Both the *offensive* entrepreneurial position of the Quantified Self and the *defensive* data 'rights' critiques that dominate the discourse still represent Californian ideals (Jethani, 2021). These do not necessarily translate to an account of self-tracking in other contexts, nor the way such tracking might exist as a manifestation of employer and employee relations, or between citizens and the state. Geographically stagnant research locales also lead to ignorance of the practicalities that shape self-tracking practices, such as the uptake of digital platforms, apps that differ by region (or are geofenced), the influence of weather, climate and spatial organisation on human movement, and the expectations-created cultural norms both in and beyond the metrics of tracking.

9.6.2 Turning away from tracking?

HCI and app design research has always had some consideration of what barriers or design flaws might lead to discontinuities in tracking practices (Li, Dey and Forlizzi, 2010). Equally, evidence from self-tracking app usage metrics suggests that initial curiosity rapidly gives way to disinterest, while in surveys, respondents have indicated that their interest in self-tracking waned after six months (Lupton, 2016, p. 27). However, there are more nuanced and holistic considerations of how and why self-tracking practices might be rejected. Maintaining an active interest in *resistance*—as the ultimate expression of agency in human/technology relations—is essential, especially given the new audiences for tracking that will be targeted.

Exemplifying this is Gorm and Shklovski's (2019) study of the 'episodic use' of self-tracking devices, in which users pause or stop self-tracking. This is a reactionary form of use, at odds with the promises of tracking. Instead of continued monitoring for feelings of control, discontinuation acts as self-care. This type of use acts as a kind of reset—negating feelings of pressure and control, dealing with inconsistencies and errors in data, and realigning app preferences (beyond app default settings) to ensure tracking practice aligns with personal interest (Gorm and Shklovski, 2019). But equally, users may choose to abandon their apps and devices for good, as happened with 'lifelogging' (Bell and Gemmell, 2009) or some Quantified Self projects (Jethani, 2021, pp. 36–37). These actions can also be political: protest art, such as 'UnFitBits' (http://www.unfitbits.com/), is an example of everyday resistance to the steady normalising of self-tracking being forced upon individuals for productivity, health and risk assessment reasons.

Here we see the insights and motivations of self-tracking invert: becoming feelings of stress, pressure and a sense of being watched over by AI systems (Reigeluth, 2014; Jethani and Raydan, 2015; Lupton, 2016). While the value that users find in self-tracking is always subjective, and enmeshed with their social life, it is difficult to see an awareness of subjectivity in the calculative and goal-oriented ways these systems tend to function. This is especially true if they frame all activities around time use and the backdrop of precarious labour (Moore, 2018). As this handbook suggests, though, it is difficult to entirely disconnect from AI readings of human life. Self-tracking is but one manifestation of the way that human existence has been 'datafied'. The 'data exhaust' (Nafus, 2016) is a constant emission, and it seems that there will always be ways to log and catalogue human activity and render it in data—whether for individual knowledge generation or for others' profit.

9.7 Conclusion

Self-tracking is a broad and complex phenomenon. It is both a booming subset of consumer technology and a longstanding practice, and is being woven into the techno-social fabric of everyday life. Today's self-tracking is not new, but current forms tap into fundamentals aspects of the human condition: offering new ways to see bodies and behaviours, to document and share experience, and to mark identity. In exploring practices of self-tracking, researchers have found that it offers the user an objective marker of a subjective 'you', bringing with it many emotional responses ranging from satisfaction to stress, and from enjoyment to fear.

There is a constant battle between the insightful potential of *objective* information (assembled by AI resources of corporate and institutional actors) and the need to have *subjective* experiences respected and catered to (by communicating everyday life with others in relatable forms). An awareness of this tension, as well as observance of historical patterns and power relations, have led to critical scholarship, as well as new avenues for diverse voices and agential capacities to be foregrounded. Implicit in the development of self-tracking technologies is a general sense that nothing is beyond measure. Whether or not this is true, and what increasingly atomised forms of quantification might mean, will continue to be borne out in the social world: through complex arrangements of actors and technologies.

References

Ajana, B. (2018a). *Metric culture: Ontologies of self-tracking practices*. Emerald Publishing.

Ajana, B. (2018b). *Self-tracking: Empirical and philosophical investigations*. Palgrave Macmillan.

Armstrong, D. (2019). The social life of data points: Antecedents of Digital technologies. *Social Studies of Science, 49*(1), 102–117. doi: 10.1177/0306312718821726.

Ayobi, A., Marshall, P., and Cox, A. (2016). Reflections on 5 years of personal informatics: Rising concerns and emerging directions. In *CHI conference; Extended Abstracts on Human Factors in Computing Systems*. Santa Clara, USA, 7–12 May 2016, pp. 2774–2781. doi: 10.1145/2851581.2892406

Bell, C. G., and Gemmell, J. (2009). *Total recall: How the E-memory revolution will change everything*. Dutton.

Beer, D. (2016). *Metric power*. Palgrave Macmillan.

Beer, D. (2018). *The data gaze: Capitalism, power and perception* (1st ed.). SAGE Publications.

Charitsis, V. (2019). Survival of the (data) fit: Self-surveillance, corporate wellness, and the platformization of healthcare. *Surveillance and Society, 17*(1/2), 139–144. doi: 10.24908/ss.v17i1/2.12942.

Crawford, K., Lingel, J., and Karppi, T. (2015). Our metrics, ourselves: A hundred years of self-tracking from the weight scale to the wrist wearable device. *European Journal of Cultural Studies, 18*(4–5), 479–496. doi: 10.1177/1367549415584857.

Ellerbrok, A. (2011). Playful biometrics: Controversial technology through the lens of play. *Sociological Quarterly, 52*(4), 528–547. https://www.jstor.org/stable/23027564.

Elman, J. P. (2018). 'Find your fit': Wearable technology and the cultural politics of disability. *New Media & Society, 20*(10), 3760–3777. doi: 10.1177/1461444818760312.

Fors, V., Pink, S., Berg, M., & O'Dell, T. (eds) (2019). *Imagining personal data: Experiences of self-tracking*. Bloomsbury Academic.

Gilmore, J. (2016). Everywear: The quantified self and wearable fitness technologies. *New Media & Society, 18*(11), 2524–2539. doi: 10.1177/1461444815588768.

Gorm, N., and Shklovski, I. (2019). Episodic use: Practices of care in self-tracking. *New Media & Society, 21*(11–12), 2505–2521. doi: 10.1177/1461444819851239.

Greenfield, D. (2010). *Everyware: The dawning age of ubiquitous computing*. New Riders.
Hummel, P., Braun, M., Tretter, M., and Dabrock, P. (2021). Data sovereignty: A review. *Big Data & Society*, *8*(1). doi: 10.1177/2053951720982012.
Jethani, S. (2021). *The politics and possibilities of self-tracking technology: Data, bodies and design*. Emerald Publishing.
Jethani, S., and Raydan, N. (2015). Performing persona through metrics: Can we think freely in the shadow of our data? *Persona Studies*, *1*(1), 76–93. doi: 10.21153/ps2015vol1no1art451.
Li, I., Dey, A., and Forlizzi, J. (2010). A stage-based model of personal informatics systems. In *Proceedings of the 28th international conference on Human Factors in Computing Systems–CHI '10, Atlanta, Georgia, USA*. doi: 10.1145/1753326.1753409.
Lomborg, S., and Frandsen, K. (2016). Self-tracking as communication. *Information, Communication & Society*, *19*(7), 1015–1027. doi: 10.1080/1369118X.2015.1067710.
Lupton, D. (2016). *The quantified self*. Polity Press.
Maxwell, H., O'Shea, M., Stronach, M., and Pearce, S. (2021). Empowerment through digital health trackers: An exploration of Indigenous Australian women and physical activity in leisure settings. *Annals of Leisure Research*, *24*(1), 150–167. doi: 10.1080/11745398.2019.1674677.
Millington, B. (2016). Fit for prosumption: Interactivity and the second fitness boom. *Media, Culture & Society*, *38*(8), 1184–1200. doi: 10.1177/0163443716643150.
Moore, P. V. (2018). *The quantified self in precarity: Work, technology and what counts*. Routledge.
Nafus, D. (2016). *Quantified: Biosensing technologies in everyday life*. MIT Press.
Neff, G., and Nafus, D. (2016). *Self-tracking*. MIT Press.
Niess, J., Woźniak, P. W., Abdelrahman, Y., ElAgroudy, P., Abdrabou, Y., Eckerth, C., Diefenbach, S., and Knaving, K. (2021). "I don't need a goal": Attitudes and practices in fitness tracking beyond WEIRD user groups. In *Proceedings of the 23rd international conference on Mobile Human–Computer Interaction. MobileHCI '21* . doi: 10.1145/3447526.3472062.
Petersen, A. (2019). *Digital health and technological promise: A sociological inquiry*. Routledge.
Reigeluth, T. (2014). Why data is not enough: Digital traces as control of self and self-control. *Surveillance and Society*, *12*(2), 243–354. doi: 10.24908/ss.v12i2.4741.
Rettberg, J. W. (2014). *Seeing ourselves through technology: How we use selfies, blogs and wearable devices to see and shape ourselves*. Palgrave Macmillan.
Rich, E., and Miah, A. (2017). Mobile, wearable and ingestible health technologies: Towards a critical research agenda. *Health Sociology Review*, *26*(1), 84–97. doi: 10.1080/14461242.2016.1211486.
Rooksby, J., Rost, M., Morrison, A., and Chalmers, M. (2014). Personal tracking as lived informatics. In *Special interest group on Computer–Human Interaction conference on Human Factors in Computing Systems*, Toronto, Canada, 26 April–1 May. doi: 10.1145/2556288.2557039.
Ruckenstein, M. (2014). Visualized and interacted life: Personal analytics and engagements with data doubles. *Societies*, *4*(1), 68–84. doi: 10.3390/soc4010068.
Ruckenstein, M., and Pantzar, M. (2017). Beyond the quantified self: Thematic exploration of a dataistic paradigm. *New Media & Society*, *19*(3), 401–418. doi: 10.1177/1461444815609081.
Schüll, N. D. (2016). Data for life: Wearable technology and the design of self-care. *BioSocieties*, *11*(3), 317–333. doi: 10.1057/biosoc.2015.47.
Selke, S. (2016). *Lifelogging: Digital self-tracking and lifelogging—Between disruptive technology and cultural transformation*. Springer VS.
Smith Maguire, J. (2008). *Fit for consumption: Sociology and the business of fitness*. Routledge.
Wernimont, J. (2019). *Numbered lives: Life and death in quantum media*. MIT Press.
Zheng, E. L. (2021). Interpreting fitness: Self-tracking with fitness apps through a postphenomenology lens. *AI & Society*, *38*, 2255–2266. doi: 10.1007/s00146-021-01146-8.

Tim Gorichanaz

10 Data Selves and Identity Theft in the Age of AI

Abstract: This chapter examines identity theft in the digital age, particularly in the context of emerging artificial intelligence (AI) technologies. It begins with a discussion of big data and selfhood, the concepts of data selves and data doubles, and the process of identification in the digital age. Next, the literature on online identity theft is reviewed, including its theoretical and empirical aspects. As is evident from that review, AI technologies have increased the speed and scale of identity crimes that were already rampant in the online world, even while they have led to new ways of detecting and preventing such crimes. As with any new technology, AI is currently fuelling an arms race between criminals and law enforcement, with end users often caught powerless in the middle. The chapter closes by exploring some emerging directions and future possibilities of identity theft in the age of AI.

Keywords: selfhood, data double, data selves, identification, identity theft

10.1 Introduction

Hundreds of trucks blockaded several border crossings between the United States and Canada as well as the streets of Ottawa, the Canadian capital. Thousands of pedestrian protestors joined in. This 'Freedom Convoy' went on for about three weeks in early 2022, stretching an already threadbare supply chain and stoking political divisions in several countries. The Freedom Convoy caused economic losses of an estimated $3.9 billion (Osman & Fraser, 2022). Ostensibly, it all began as a truckers' protest against Covid-19 vaccination requirements for crossing the border and then grew into a general protest against pandemic restrictions and mandates. Yet investigative journalists have discovered that the Freedom Convoy movement was sparked not by grassroots dissent but by anonymous actors running Facebook groups through a stolen account.

According to *Grid*, the Facebook account of a Missouri woman was used to launch several Facebook groups from 26 to 28 January, accruing over 340,000 members and 7,500 posts. When *Grid* spoke with the woman, "she said her account was hacked and

Tim Gorichanaz, Drexel University, USA

https://doi.org/10.1515/9783110721751-010

she was not involved with the groups. 'Someone stole my identity on Facebook,' she said. 'I don't know how.'"[1]

Today nearly 3 billion people have active Facebook accounts, some 60% of global internet users. Even if we're not on Facebook, we must manage numerous online accounts; and as we navigate the internet, we unwittingly leave digital traces like flakes of skin. So though we sometimes talk about our 'online identity', or even 'online identities', the truth is that these accounts and traces are so heterogeneous and diffuse that it is impossible to speak about identity as a distinct object at all. And, as we can see in the example of the Freedom Convoy, cunning actors can exploit this heterogeneity and diffuseness to commit crimes of far greater reach and impact than simple credit card fraud.

The line "On the internet, nobody knows you're a dog" first appeared in a 1993 *New Yorker* cartoon, and three decades later it is as true as ever. What has changed, as the internet has become integrated with nearly every aspect of life and society for most of the world's population, is the power and reach of any dog online.

This chapter considers identity theft in the digital age, particularly in the context of emerging artificial intelligence (AI) applications. Because this issue is inextricable from questions of big data and selfhood, we will begin with discussions of those topics. The next section reviews the concepts of data selves and data doubles, as well as the process of identification in the digital age. After that, we will review the literature on online identity theft, including its theoretical and empirical aspects. As we will see, AI technology has increased the speed and scale of identity crimes that were already rampant in the online world, even while it has led to new forms of crime detection and prevention. As with any new technology, AI is currently fuelling an arms race between criminals and law enforcement, with end users often caught powerless in the middle. To close the chapter, we will explore some emerging directions and future possibilities of identity theft in the age of AI.

10.2 Data and the Self

Life is "a blooming, buzzing confusion", as William James (1890, p. 488) famously put it, and to communicate at all we need to work with concepts and categories. For instance, there are infinite colours on the spectrum, but we only name a handful of them. When it comes to describing our own being, two major concepts are self and identity (Gorichanaz, 2020, pp. 67–87; see also Ahmed, 2020, pp. 24–45). These are abstractions, meaning they necessarily do away with many details for the sake of maintaining a small, portable kernel.

1 https://www.grid.news/story/misinformation/2022/02/08/the-hacked-account-and-suspicious-donations-behind-the-canadian-trucker-protests/.

Lately, there has been concern over how digital technologies—and data-driven AI in particular—interact with the concepts of self and identity. Gilbert and Forney (2013) trace the way that selfhood has for millennia been shaped in response to emerging technologies. The pre-modern or *social* self, which defined humanity until the last few centuries, was circumscribed by social role; the modern or *psychological* self was rooted in mental life; and the post-modern or *multiple* self, which emerged during the era of globalised digital technology, is fragmented and contextual. Gilbert and Forney observe that today's emerging technologies, such as AI, cloud computing and immersive virtual worlds, are giving rise to a fourth sense of self, a *distributed* self, in which we conceive our selfhood as spread across multiple devices, platforms and ways of interacting.

We have always been more than just our bodies—we are thinking things, with memories and tools and goals—but our contemporary distributed self is comprised in part by digital data, in many cases stored and owned by others. Many of the digital documents that supplement my memory and guide my work are stored in the cloud, as are the photos on my smartphone that document my memories, activities and interests. I generate data as I browse the web, move throughout my university and city, make purchases and communicate with others. Douglas-Jones (2021), in an anthropology essay, seeks to clarify the multiple understandings of data as it intersects with selfhood (data as, data about, data of) and the multiple purposes of data collected and used by different entities. She offers three vignettes that demonstrate these different understandings: allowing companies to assemble a *data double* for one's individual benefit in some area, such as creditworthiness; resisting corporate data collection regimes by assembling one's own *data composite*; and using *data aggregates* to clarify or change relationships between states and people, such as particular minority groups.

Of the three types of data self that Douglas-Jones (2021) outlines, the first two have the most relevance for identity theft and will be described here in more detail. In the literature, scholars tend to use the terms *data double* and *data self* for these concepts. Roughly, a *data double* is a data profile constructed by a corporation to, for instance, serve personalised ads; and a *data self* is the collection of data that a person creates about themselves through self-tracking technologies. While data doubles are external, often opaque, and involve little agency or even awareness, data selves are just the opposite. To be sure, the reality is somewhat more complex, and there is some fuzziness between these two categories, but they serve well enough for analytical purposes.

10.2.1 Data doubles

If you have ever come across an advertisement online that seemed to be speaking directly to you, then you've brushed with your data double. Data doubles are the profiles assembled by Google/Alphabet, Facebook/Meta, and other companies to represent and predict the behaviour of specific users, including shopping and browsing data, demographic characteristics, and more. Just as stunt doubles in Hollywood do not always accurately represent the stars they're supposed to, our data doubles sometimes falter. Perhaps you've come across an ad that did not quite land. While data doubles are most familiar in the commercial context, they are increasingly used across the digital world.

The data double phenomenon stretches back to the early twentieth century with the rise of the bureaucratic society (Koopman, 2019) and, of course, digital computing (Cheney-Lippold, 2017). An early theorist of the phenomenon was psychologist and philosopher Harré (1984), who wrote of the *file-self*—that version of ourselves comprising documents, reports and transactions. Later, Solove (2003) wrote of *digital dossiers*. Another term sometimes used for this phenomenon is *digital doppelganger*.

Surveillance is a key issue addressed within data double scholarship. The seminal paper in this vein is 'The Surveillant Assemblage' by Haggerty and Ericson (2000). In this article, Haggerty and Ericson explore the implications of the then-emergent digital database technologies on human identity and autonomy. The authors describe how, as discrete surveillance systems begin to converge, people's data doubles come to have their own social lives and materiality distinct from the flesh-and-blood people from which they were assembled. In the name of shopping rewards, crime prevention, public safety and so on, people are abstracted into data flows, which are then reassembled in different locations. In this early work, Haggerty and Ericson predicted that this would transform the purposes of surveillance and the meaning of privacy. That prediction was borne out, as Zuboff (2019) documents in *The Age of Surveillance Capitalism*. In this book, Zuboff describes an emergent form of capitalism based on the dynamics observed by Haggerty and Ericson. Today, companies collect and sell data as 'behavioural futures' on predictive markets. This drives the datafication of more and more aspects of human experience, and the predictive markets not only then predict our behaviour but also proleptically modify it. In this context, Zuboff foresees dire consequences for democracy and freedom.

In addition to theory, a number of studies have examined people's attitudes and experiences of encountering and negotiating their data doubles. Lupton (2021), for example, presents a wide-ranging interview study of Australian adults' awareness and perceptions of data surveillance and datafication. She found that people were aware of the issues but did not share the dystopian sentiments common to authors such as Zuboff (2019). People thought the data comprising their data double was superficial, while they still had deeper aspects of themselves that were closed off from datafication. They found their online selves to be not only incomplete, but also performative in many cases. Speaking to the incompleteness and performativity of data doubles, Kear (2017) presents an ethnography of how people 'game' the credit scoring system through Lending Circles, which help marginalised people build their credit. One tension that emerges in this work is the question of whether Lending Circles are unfairly manipulating the financial system or correcting a prior injustice. In another context, Jones et al. (2020) examine students' perspectives on learning analytics at colleges and universities in the United States. As in the world of finance, students find that their institutions' data collection practices are largely opaque to them and there is no informed consent involved; in contrast to financial institutions, though, students have a higher sense of trust in their school because of its nonprofit status—trust that would evaporate if their data were to be used commercially.

As this research suggests, data doubles are characterised by opacity and a lack of choice or agency. Resistance and opting out are possible in some contexts, such as leaving a given social network or deciding to use one search engine over another. Other contexts, such as the financial system, offer no meaningful choice. For the most part, some form of digital surveillance is the cost for participating in society, and we may do what we must to navigate the ethical grey areas that the surveillant assemblage and surveillance capitalism present.

10.2.2 Data selves

While our data doubles are external and opaque, we may also collect and use data about ourselves because and in ways that we want to. This phenomenon has come to be called the *data self*. In an expansive look at the topic, Rettberg (2014) examines blog writing, selfiemaking and wearable self-tracking as manifestations of the relationship between selfhood and technology. She demonstrates how people develop emotional ties to their data and the feeling that seeing your data is a way of seeing yourself. Key to the picture is agency: in these practices, we can edit, shape, curate and share our data in ways that we want to. To be sure, this is not only an individual process, but also a social one; and this point is developed in detail by Lupton (2020) in a series of studies on self-tracking cultures. In this work, Lupton covers questions of governing the self, the entanglement of human bodies and technology, how we value data and how self-tracking relates to social inequality.

Just like data doubles, data selves have a long history—in this case stretching back millennia in the form of journals and letters. Humphreys (2018), in *The Qualified Self*, offers a historical throughline from centuries-old almanacs, pocket calendars and diaries to today's Fitbits and social media. As a scholar of communication, Humphreys is concerned with how any era's practices of self-documentation help us understand ourselves and each other.

Not surprisingly, then, data selves also make use of experimental and emerging technologies. A good showcase of this is Robards et al.'s (2019) work presented on a panel at the Association of Internet Researchers conference, which offers a kaleidoscope of research reflecting on the notion of 'data-selfie'. These authors explore how data becomes an extension of the self, the consequences of datafication, the political and commercial aspects of data selves, data anxiety among older internet users, and feelings of control in data generation and use. This work, as well as that of Lupton (2020), helps us reflect on the fuzzy boundaries between data doubles and data selves—

sometimes the sense of choice and control we may experience with our data selves is only an illusion.

10.2.3 The self and reidentification

Above we discussed the creation of what can be called personal data or personally identifying information. This kind of data comes from a person and is stored in databases. But let us consider the other direction: if what you have is data, how can you trace it reliably back to the right person? This is a question about identification and reidentification. For example, fingerprints are unique and can be compared to identify a person. Photographs on identification cards, too, are fairly reliable. But in the realm of textual and numerical data, things are much less clear. If someone can recite a social security number (SSN), it doesn't necessarily mean that it is their SSN—hence banks may look to other data or ask additional questions to identify a person. A corpus of browsing data may seem to suggest a person, but this could be fallible. Moreover, the more data and systems involved, the more complex is the picture of personal identification.

Brensinger and Eyal (2021) provide a framework for understanding the complex processes of personal identification in the age of AI and big data. Their work incorporates insights from sociologists such as Durkheim, Goffman, Foucault and Deleuze, and their primary purpose is to assist sociologists researching issues of identification, such as identity theft, to better understand where in the system this or that has gone awry. Brensinger and Eyal's framework has three major elements: object, agency and technique. The *object*, or object of identification, is the person being identified. However, it is not the flesh-and-blood person, but only a partial representation of that person created through disembedding, standardising and reembedding. Disembedding involves the processes of abstracting and datafication; in reembedding, people shape their behaviour to maintain the integrity of the system. Next, *agency* represents questions of who has access to what identifying information, and who cooperates with whom. In identification systems, agency is asymmetrical and complex, involving the person themselves and various bureaucratic entities that play different roles in constraining or affording action. Finally, there are numerous techniques at play in *identification*. These include, for instance, matching datapoints, establishing baselines for comparison (e.g., government stored photo, signature, etc.), expert judgement and interactivity (resistance, adaptation).

This system can be exploited through identity crimes such as identity theft, which is the topic of the next section of this chapter.

10.3 Identify Theft

There are myriad cyberattack concerns in the digital age: malware, denial of service, password cracking. . . . With AI technology in the mix, there is the potential for these to occur faster, perhaps autonomously, and in new ways altogether. King et al. (2020) provide a comprehensive review of AI crime, identifying a range of AI crimes, some of which have already been perpetrated while others are still only potential. Of particular note, AI introduces thorny questions of responsibility and liability, given that AI systems can be opaque, unpredictable and distributed. King et al. also discuss possible solutions that are emerging in the literature, such as introducing new legal structures and using AI to monitor for AI crime.

Rather than exploring every type of cybercrime, here we'll focus on identity theft. In brief, identity theft is when one party obtains somebody's personal data unlawfully, typically then selling that data or using it to commit fraud. For example, an identity thief may illegally obtain a person's SSN, list of past addresses and family members' names (identity theft) and then use these details to open a new credit account (identity fraud). Identity theft is not a new crime, of course, but as the internet became more and more central in people's lives around the turn of the millennium, incidents and fears of identity theft rose to the level of a moral panic (Anderson, Durbin and Salinger, 2008). In the United States, as in many countries, identity theft is the fastest-growing type of consumer crime; about three-quarters of the US population believe they will fall victim to it (Hille, Walsh and Cleveland, 2015). Digital technology has also given rise to online marketplaces for consumer personal data on what has come to be called the dark web (Anderson, Durbin and Salinger, 2008).

A succinct overview of identity theft for technologists is offered by Aïmeur and Schönfeld (2011). The authors delineate several types of identity theft, the two most common of which are financial (e.g., related to tax and credit fraud) and medical (e.g., billing treatment to someone else's health insurance account). Aïmeur and Schönfeld outline several types of information that can be involved in identity theft, including personal information (name, age, gender, etc.); buying patterns; physical and digital navigation habits; lifestyle information; records of employment, medicine and criminality; and biological information. As Aïmeur and Schönfeld write, this information can be physical and/or digital, and it can be retrieved through traditional methods such as dumpster diving and social engineering as well as online methods such as phishing and packet sniffing.

In an economics article, Anderson, Durbin and Salinger (2008) chronicle the history and societal impacts of identity theft, providing an expansive perspective on the scope of the problem and the issues involved. As these authors write, the shift to an online economy, its attendant rise in anonymous transactions and our new reliance on personal information (e.g., needing a credit score to secure housing and a job) have allowed explosive growth in identity theft and fraud. Anderson, Durbin and Salinger review both the direct and indirect costs of identity theft, government responses, institutions available for preventing and detecting ID theft, and policy issues (liability, consumer data protection, credit freezing, and penalties and resources for enforcement).

Perhaps the most notable institution in this area is the Identity Theft Resource Center (ITRC), founded in 1999.[2] The ITRC publishes research on identity theft as well as providing support for victims of identity theft. Fellow researchers working in this area will find value in its research publications on topics including the emotional and psychological impact of identity theft, the aftermath of data breaches for victims, the intersection of social media and identity theft, cyberattack trends and more.

This section reviews the literature on identity theft in the digital age. We will begin by discussing the emerging dynamics that AI technologies are adding to the picture of identity theft. Then we will turn to privacy, a major concept invoked in the identity theft literature, and then to empirical work on awareness and victimisation. Following that, we will examine preventive measures and solutions in the literature to address the problem of identity theft.

10.3.1 AI and identity theft

Simply put, the proliferation of AI increases the threat of identity theft. At heart, AI relies on data collection and storage, and stored data is subject to being breached. In this context, the practice of data brokering (trading and collecting consumer data from various sources) is a key vulnerability for identity theft (Aïmeur and Schönfeld, 2011). Over the past several years, data breaches have been on the rise, though one study did find that in 2021 data breaches decreased worldwide—with the exception of the United States, where they increased by 10% over 2020 levels (RiskBased Security & Flashpoint, 2022). According to this same report, the most commonly targeted data is names, addresses and social security numbers.

Besides the weak point of data storage, AI also changes the landscape of identity theft in other ways. First, it introduces new forms of personal information; for example, predictive AI models of users can be considered personal information (Aïmeur and Schönfeld, 2011). As well, AI potentiates new technological forms of identity theft. For example, it can allow system penetration without human involvement. Attacks can become automated, faster and less detectable. One AI-based technique in this arena is neural fuzzing, in which an attack floods a system with a large amount of random input to quickly probe its vulnerabilities. And thirdly, AI introduces new possibilities for social engineering, a 'low tech' approach to identity theft. The ITRC reported the first case of AI-based social engineering in 2019, in which criminals created a deceptive synthetic audio clip (referred to as a deepfake) of a company's CEO asking

2 https://www.idtheftcenter.org.

for money to be transferred to a particular account.[3] This is an example of 'vishing' or 'voice phishing', which has been attested since 2008 as a means of social engineering (Griffin and Rackley, 2008), but with deepfake technology the method grows more pernicious.

But AI does not only further enable criminals when it comes to identity theft; it is a double-edged sword. AI technologies can be used by criminals to facilitate theft and fraud, but it can be used by other entities to detect and prevent these. For example, banks and creditors are using AI to automatically detect unusual account activity. If Alice usually shops at certain stores in a certain area with a given frequency, and suddenly her account registers a purchase that doesn't fit the pattern, the system will flag her purchase for human review or automatically block the transaction until she approves it. AI developments in this space, then, are something of an arms race between criminals on one hand and cybersecurity professionals and law enforcement on the other.

10.3.2 Privacy, architecture and identity theft

Privacy is a key concept invoked in discussions of identity theft. Some thinkers have conceptualised privacy in individual terms, while others see privacy as an environmental or contextual issue. An example of the former is the seminal paper 'A Right to Privacy' by Warren and Brandeis (1890, as cited in Solove, 2003), in which privacy is defined as a "right to be let alone" and "right to one's personality". More recent conceptualisations of privacy focus on the appropriateness of various information flows within an environment. An example of this approach is found in Solove's (2003) legal scholarship. Identity theft is a clear violation of both approaches to privacy, but the latter approach more comprehensively diagnoses the issue and points towards viable solutions. Here we will consider Solove's proposals in some detail.

In an extended argument, Solove (2003) makes the case that while some privacy violations may be isolated incidents on individuals (e.g., snooping on a neighbour), others are architectural issues—and identity theft is an architectural issue. He means 'architecture' here broadly as the ways that our structures shape society in terms of behaviour, attitudes, thoughts and interactions. This includes the shape of our physical buildings, but also our other physical and social spaces, as well as digital ones. Architecture can enhance or diminish privacy; consider the effects on privacy of a building made completely of glass compared with one with no windows at all. When viewing privacy as an architectural issue, the flaws that enable identity theft are visible (e.g., what is made possible simply by having someone's SSN). But the law has been ignoring this flaw, focusing instead on isolated, individualised solutions. Such solutions mean

3 https://www.idtheftcenter.org/post/first-ever-ai-fraud-case-steals-money-by-impersonating-ceo/.

that individuals need to have the knowledge, power and resources to assert their rights; and the most powerless people in society are both the least able to do so and the most vulnerable to identity theft. Other problems with viewing identity theft and privacy as individual issues include a lack of government resources and the fact that individuals may not be aware they have been violated until long afterwards. Solove suggests that architectural issues such as privacy can only be fixed by architectural solutions. Individual rights and assertions are a piece of that, not the whole picture. In the end, he proposes a new architecture for privacy built around Fair Information Practices (which have been in discussion since the 1970s; see Ahmed, 2020, pp. 634–649). This would make it possible for people to discover what information about them is collected and held by third parties and how it is being used, to correct or amend such records, and so on. In short, this proposal would place the burden for personal data protection not on the individual victims but on the entities storing and using that data.

Also along these lines, Wang, Yuan and Archer (2006) present a contextual framework describing identity theft. Their framework shows the various stakeholders involved in identity theft—the identity owner, the identity issuer, the identity checker, the identity protector and the identity thief—as well as the interactions among these. Their framework is meant to help analysts understand how it happens, develop security solutions, and assess the strengths and weaknesses of proposed fixes. Ahmed (2020) provides a broader framework, one showing all the possible forms of identity crime and how they relate to each other, modelling the various entities and actions associated with such crimes (e.g., production, acquisition, transfer, etc.).

10.3.3 Attitudes and awareness

A body of empirical research has examined people's awareness of and attitudes towards identity theft. By and large, people are quite aware of the risk (if only abstractly), and the predominant attitude appearing in the literature is fear. To date, most of this research has been done in the United States, most often with students (Choi, Kruis and Choo, 2021). This section reviews some of this work in an effort to offer a comprehensive but succinct global snapshot.

Though most people are aware that identity theft is possible, researchers have sought to learn more about the connections between this knowledge and people's behaviour. In a recent study, Khan et al. (2021) examined people's cybersecurity awareness and self-disclosure behaviour on social media in a quantitative survey including a range of demographics and frequency of internet access. They found that people were broadly aware of the risks and performed cost–benefit analyses before disclosing information, providing evidence for the effectiveness of cybersecurity education. Cross (2017), in a study of senior citizens, provides a contrasting view, showing that older people may unwittingly expose their information through their online activities, primarily because they have incorrect beliefs about the risk and likelihood of identity crimes. Cross suggests that the prevailing model of individual education and self-regulation falls short for some user groups, such as elderly people.

Because identity theft is closely linked to data security, researchers on identity theft should also consider attitudes and behaviours regarding data breaches (which are, unfortunately, frequently in the news). In a report from the RAND Corporation, Ablon et al. (2016) investigated people's attitudes and responses to data breach notifications. They found that 43% of US adults were notified about a data breach involving their data, with about half of those notifications occurring in the past year. About 44% of people learned of the breach from the media before receiving an official notification. Afterwards, 64% made use of the free credit monitoring offered, and customer attrition was only 11%. Their research suggests that customers are most satisfied when the breach notification is timely, and when they are kept up to date on improved security measures and offered identity protection and credit monitoring services. Ablon et al. (2016) also consider the prospect of government-mandated breach disclosure laws, discussing the complex pros and cons. As of early 2022, there is still no such US law in effect. In contrast, the European Union does have such laws in effect, as do Australia and China.[4] For an in-depth review of identity crime legislation in the United States, Canada, Australia and the United Kingdom, see Ahmed (2020, pp. 252–542).

Much of the work on people's attitudes and awareness examines fear. Indeed, fear of crime has long been of interest to criminologists; this is because fear of crime is generally far more widespread than actual incidents, and such fear can have detrimental emotional and physical impacts (Henson, Reyns and Fisher, 2013). Roberts, Indermaur and Spiranovic (2013) examine fear of cybercrime specifically, finding that it can be weakly predicted by fear of crime in general and specific internet exposure. But, given the weak prediction power, more research is necessary; cybercrime has not yet seen as much attention as offline crimes. In more recent work, Choi, Kruis and Choo (2021), using data from a South Korea crime victimisation survey, find that fear of identity theft is higher than fear of other types of crime, and that fear of being victimised was related to online exposure (banking and shopping, etc.) and exposure to motivated offenders (getting phishing emails).

Though many of us are used to working online, this is not the case worldwide. Jordan, Leskovar and Marič (2018) and Jibril et al. (2020) provide insights from Slovenia and Ghana, respectively; societies that are currently transitioning to online economies. Both studies found that fears of financial loss and reputational damage related to people's hesitancy to engage in online financial transactions.

4 https://en.wikipedia.org/wiki/Security_breach_notification_laws.

10.3.4 Aftermath for victims of identity theft

In addition to people's attitudes and behaviours regarding identity theft, there has been a small body of research looking at people's responses to becoming a victim of identity theft. Golladay and Holtfreter (2017), using data from the US Bureau of Justice Statistics, showed that, in addition to financial and time losses, victims of identity theft also experienced emotional (e.g., depression) and physical (e.g., poor health) symptoms. In a doctoral dissertation, Gideon (2020) renders this in much greater detail, through phenomenological interviews with victims of identity theft. Gideon found that victims experienced fear, hopelessness, apathy, reactivity and anger, and that the mixes of these emotions differed for each victim. In terms of behavioural change after victimisation, Gideon found that fear drove the most change, while apathy did not drive change.

In another study, Gies et al. (2021) looked specifically at identity theft victims who made use of the ITRC. They found that victims who contacted the ITRC may have had more severe cases than those who did not contact the ITRC. Specifically, victims who contacted the ITRC reported more personal problems with finances, employment, family, friends and physical health one year after their identity theft incident. They also reported longer remediation time and lower degrees of full problem resolution. However, these participants reported significantly fewer mental health problems than those who did not contact the ITRC. All in all, Gies et al.'s findings show the myriad, complex and long-lasting problems that victims of identity theft experience, as well as the ways that existing services such as the ITRC may be helpful.

10.3.5 Preventing identity theft

As mentioned above, the prevailing methods for preventing identity theft, and responding to identity theft once it occurs, place the onus on individuals. For example, guidance appearing across the public and trade literature suggests that people should protect their account credentials; use temporary emails and throwaway accounts where prudent; and monitor their online identity using services such as Identity Guard, which includes dark web monitoring. Like Khan et al. (2021), discussed above, Lai, Li and Hsieh (2012) find that educating the public on such tactics does reduce risk of identity theft. Specifically, they investigate the effectiveness of both conventional coping methods (e.g., shredding bank statements, monitoring credit history) and technological coping methods (e.g., using firewalls, keeping software up to date), finding that both are effective in reducing identity theft risk.

However, some work calls into question the ultimate effectiveness of such strategies. In a theoretical argument, Whitson and Haggerty (2008) find that leaving identity theft as an individual problem, which they call an ethos of "care of the virtual self", is ultimately not practicable. This ethos fits into a broader orientation towards individual self-care in the digital age, which asks each person to care for their data double. And, while this may not be problematic in some contexts, it is not sufficient to prevent identity theft. This echoes Solove's (2003) description of identity theft as an architectural problem rather than an individual one. Providing empirical grounding to these claims, Shillair et al.'s (2015) research points out that people are not taking the necessary precautionary steps to protect themselves, even when they know about these steps.

Most recently, Piquero et al. (2021) conducted a focus group with cybersecurity professionals on preventing identity crimes. Their work reports on these professionals' assessments of numerous technological solutions, including various scanning and monitoring programs, as well as their observations about the current risk environment. The vast majority of the technological solutions currently on offer are perceived by these experts as ineffective (e.g., prevention software, end-user agreements, knowledge-based authentication). On the other hand, the experts view chip-enabled credit/debit cards as highly effective, and the following as partially effective: two-factor authentication, biometric authentication and AI-based authentication. Beyond these solutions, there are a number of emerging AI-based techniques for detecting and preventing identity theft and other cybercrimes. Al-Khater et al. (2020) provide a comprehensive review of such techniques, including statistical, machine learning, data mining, computer vision, biometrics, cryptography and forensics approaches. In perhaps the most comprehensive treatment to date, Ahmed (2020, pp. 543–600) outlines dozens of strategies that can be used to prevent and minimise identity crimes at all levels, from individuals to businesses to government.

All this suggests that, while some emerging technological solutions will go some way in preventing identity theft, there is more that could be done to address this issue architecturally. Building on the arguments of scholars such as Solove (2003), van Loo (2019) points to "the missing regulatory state" in a law review article arguing for the regulation of digital platforms such as Twitter and Facebook and tech companies such as Amazon and Alphabet. Van Loo argues that a too-narrow understanding of privacy is often used by these entities as a pretext for deregulation; he paints a picture of how regulatory monitoring is indeed possible without having the negative aspects typically associated with government surveillance. In short, regulators need access to certain data held by tech companies and digital platforms to ensure consumer safety and company compliance, and at present they do not have such access.

10.4 The Future of Data Selves and Identity Theft

As we move further into the digital age, we are carrying more devices with us, using the internet more frequently and for more activities, and thus rendering more of our experiences as data. As AI is used in more of our devices and systems, we benefit from their increased flexibility, efficiency and convenience; yet this technology also increases the possibilities for other entities, from corporations to governments, to shape our movements. It also holds possibilities for new forms of self-knowledge and reflection, depending on how technology and society develop.

To close, I want to shine a light on two emerging technologies in this space that researchers should explore further: blockchain and synthetic identity fraud. Blockchain is a technology based on a shared, public ledger that makes data exchanges transparent to all relevant parties, facilitating trust and communication. Perhaps the most famous application of blockchain is in new forms of currency such as Bitcoin, but blockchain can also be used on smaller scales to track orders, accounts, payments, manufacturing, and so on. Blockchain could also be used to register human identities in ways that make certain forms of identity theft impossible. To this end, Rana, Zaeem and Barber (2019) provide an overview of blockchain-based identity solutions. Further, there is significant potential in the combination of blockchain and AI. There is nascent work in this direction from entities such as IBM. The combination of blockchain and AI can improve trust and transparency of AI decisions, speed up blockchain transactions, and so on.

Next, AI technologies may shift the goalpost from identity theft to synthetic identity fraud (Walker-Moore, 2018). Whereas in identity theft, criminals impersonate an existing person, with synthetic identity fraud a false identity is created. Using a synthetic identity, fraudsters can, for example, open and use a line of credit for a person who does not exist. Such fraudsters may take advantage of weak verifications, knowledge about the social security number registration process, and so on, as well as emerging AI technologies such as synthetic media (deepfakes) that would allow criminals to create photos, video and voice of people who do not exist. Walker-Moore (2018) provides a history and prognosis of this problem, as well as directions for solutions.

Identity theft is a danger that has been growing since the dawn of the digital age, and it is not likely to go away. New technologies, including AI, introduce new vectors for identity criminals to exploit. At present, virtual assistants, smart speakers and customer service chatbots present new opportunities for criminals. At the same time, these technologies also allow others to more easily detect and prevent identity theft, leading to an arms race. Whether the arms race can end remains to be seen; likely it will take more than just technological innovation, but also new approaches to architecture and regulation as advocated by scholars such as Solove (2003) and van Loo (2019).

References

Ablon, L., Heaton, P., Lavery, D. C., and Romanosky, S. (2016). *Consumer attitudes toward data breach notifications and loss of personal information*. RAND Corporation. http://www.jstor.org/stable/10.7249/j.ctt1bz3vwh.

Ahmed, S. R. (2020). *Preventing identity crime: Identity theft and identity fraud*. Brill. doi: 10.1163/9789004395978_009.

Aïmeur, E., and Schönfeld, D. (2011). The ultimate invasion of privacy: Identity theft. In *Proceedings of the ninth annual international conference on Privacy, Security and Trust*. doi: 10.1109/PST.2011.5971959.

Al-Khater, W. A., Al-Maadeed, S., Ahmed, A. A., Sadiq, A. S., and Khan, M. K. (2020). Comprehensive review of cybercrime detection techniques. *IEEE Access*, *8*, 137293–137311. doi: 10.1109/ACCESS.2020.3011259.
Anderson, K. B., Durbin, E., and Salinger, M. A. (2008). Identity theft. *Journal of Economic Perspectives*, *22*(2), 171–192. doi: 10.1257/jep.22.2.171.
Brensinger, J., and Eyal, G. (2021). The sociology of personal identification. *Sociological Theory*, *39*(4), 265–292. doi: 10.1177/07352751211055771.
Cheney-Lippold, J. (2017). *We are data: Algorithms and the making of our digital selves*. New York University Press.
Choi, J., Kruis, N. E., and Choo, K.-S. (2021). Explaining fear of identity theft victimization using a routine activity approach. *Journal of Contemporary Criminal Justice*, *37*(3), 406–426. doi: 10.1177/10439862211001627.
Cross, C. (2017). "But I've never sent them any personal details apart from my driver's licence number . . .": Exploring seniors' attitudes towards identity crime. *Security Journal*, *30*, 74–88. doi: 10.1057/sj.2015.23.
Douglas-Jones, R. (2021). Bodies of data: Doubles, composites, and aggregates. *Journal of the Royal Anthropological Institute*, *27*(S1), 159–170.
Gideon, M. A. (2020). A phenomenological study of identity theft victims and resulting online behaviors. Unpublished PhD dissertation), Capella University. https://www.proquest.com/dissertations-theses/phenomenological-study-identity-theft-victims/docview/2416897269/se-2.
Gies, S. V., Piquero, N. L., Piquero, A. R., Green, B., and Bobnis, A. (2021). Wild, wild theft: Identity crimes in the digital frontier. *Criminal Justice Policy Review*, *32*(6), 592–617. doi: 10.1177/0887403420949650.
Gilbert, R., and Forney, A. (2013). The distributed self: Virtual worlds and the future of human identity. In R. Teigland and D. Power (eds), *The immersive internet* (pp. 23–36). Palgrave Macmillan. doi: 10.1057/9781137283023_3.
Golladay, K., and Holtfreter, K. (2017). The consequences of identity theft victimization: An examination of emotional and physical health outcomes. *Victims & Offenders*, *12*(5), 741–760. doi: 10.1080/15564886.2016.1177766.
Gorichanaz, T. (2020). *Information experience in theory and design*. Emerald.
Griffin, S. E., and Rackley, C. C. (2008). Vishing. In *Proceedings of the 5th annual conference on Information Security Curriculum Development (InfoSecCD '08)*. doi: 10.1145/1456625.1456635.
Haggerty, K. D., and Ericson, R. V. (2000). The surveillant assemblage. *British Journal of Sociology*, *51*(4), 605–622.
Harré, R. (1984). *Personal being: A theory for individual psychology*. Harvard University Press.
Henson, B., Reyns, B. W., and Fisher, B. S. (2013). Fear of crime online? Examining the effect of risk, previous victimization, and exposure on fear of online interpersonal victimization. *Journal of Contemporary Criminal Justice*, *29*(4), 475–497. doi: 10.1177/1043986213507403.
Hille, P., Walsh, G., and Cleveland, M. (2015). Consumer fear of online identity theft: Scale development and validation. *Journal of Interactive Marketing*, *30*, 1–19. doi: 10.1016/j.intmar.2014.10.001.
Humphreys, L. (2018). *The qualified self: Social media and the accounting of everyday life*. MIT Press.
James, W. (1890). The principles of psychology, volume 1. Henry Holt and Company.
Jibril, A. B., Kwarteng, M. A., Botchway, R. K., Bode, J., and Chovancova, M. (2020). The impact of online identity theft on customers' willingness to engage in e-banking transaction in Ghana: A technology threat avoidance theory. *Cogent Business & Management*, *7*(1). doi: 10.1080/23311975.2020.1832825.
Jones, K. M. L., Asher, A., Goben, A., Perry, M. R., Salo, D., Briney, K. A., and Robertshaw, M. B. (2020). "We're being tracked at all times": Student perspectives of their privacy in relation to learning analytics in higher education. *Journal of the Association for Information Science and Technology*, *71*(9), 1044–1059. doi: 10.1002/asi.24358.
Jordan, G., Leskovar, R., and Marič, M. (2018). Impact of fear of identity theft and perceived risk on online purchase intention. *Organizacija*, *51*(2), 146–155. doi: 10.2478/orga-2018-0007.
Kear, M. (2017). Playing the credit score game: Algorithms, 'positive' data and the personification of financial objects. *Economy and Society*, *46*(3–4), 346–368. doi: 10.1080/03085147.2017.1412642.

Khan, N. F., Ikram, N., Murtaza, H., and Asadi, M. A. (2021). Social media users and cybersecurity awareness: Predicting self-disclosure using a hybrid artificial intelligence approach. *Kybernetes*. doi: 10.1108/K-05-2021-0377.

King, T. C., Aggarwal, N., Taddeo, M., and Floridi, L. (2020). Artificial intelligence crime: An interdisciplinary analysis of foreseeable threats and solutions. *Science and Engineering Ethics*, *26*, 89–120. doi: 10.1007/s11948-018-00081-0.

Koopman, C. (2019). *How we became our data: A genealogy of the informational person*. University of Chicago Press.

Lai, F., Li, D., and Hsieh, C. (2012). Fighting identity theft: The coping perspective. *Decision Support Services*, *52*, 353–363. doi: 10.1016/j.dss.2011.09.002.

Lupton, D. (2020). Data selves: More-than-human perspectives. Polity Press.

Lupton, D. (2021). "Not the real me": Social imaginaries of personal data profiling. *Cultural Sociology*, *15*(1), 3–21. doi: 10.1177/1749975520939779.

Osman, L., & Fraser, D. (2022, November 16). Winter 'Freedom Convoy' blockades cost billions to Canada's economy, inquiry hears. CTV News. https://www.ctvnews.ca/politics/winter-freedom-convoy-blockades-cost-billions-to-canada-s-economy-inquiry-hears-1.6156134

Piquero, N. L., Piquero, A. R., Gies, S., Green, B., Bobnis, A., and Velasquez, E. (2021). Preventing identity theft: Perspectives on technological solutions from industry insiders. *Victims & Offenders*, *16*(3), 444–463. doi: 10.1080/15564886.2020.1826023.

Rana, R., Zaeem, R. N., and Barber, K. S. (2019). An assessment of blockchain identity solutions: Minimizing risk and liability of authentication. In *2019 IEEE/WIC/ACM international conference on Web Intelligence (WI)*. https://ieeexplore.ieee.org/abstract/document/8909638.

Rettberg, J. W. (2014). *Seeing ourselves through technology: How we use selfies, blogs and wearable technology to shape ourselves*. Palgrave Macmillan.

RiskBased Security & Flashpoint (2022). 2021 year-end report: Data breach quickview. https://go.flashpoint-intel.com/docs/2021-Year-End-Report-data-breach-quickview.

Robards, B., Lyall, B., Moran, C., Burgess, J., Albury, K., Wilken, R., McCosker, A., and Senft, T. (2019). Data selves: Trust, control and self-representation in digital society. *AoIR Selected Papers of Internet Research*. doi: 10.5210/spir.v2019i0.10944.

Roberts, L. D., Indermaur, D., and Spiranovic, C. (2013). Fear of cyber-identity theft and related fraudulent activity. *Psychiatry, Psychology and Law*, *20*(3), 315–328. doi: 10.1080/13218719.2012.672275.

Shillair, R., Cotten, S. R., Tsai, H. Y. S., Alhabash, S., LaRose, R., and Rifon, N. J. (2015). Online safety begins with you and me: Convincing Internet users to protect themselves. *Computers in Human Behavior*, *48*, 199–207. doi: 10.1016/j.chb.2015.01.046.

Solove, D. J. (2003). Identity theft, privacy, and the architecture of vulnerability. *Hastings Law Journal*, *54*(4), 1227–1276.

Van Loo, R. (2019). The missing regulatory state: Monitoring businesses in an age of surveillance. *Vanderbilt Law Review*, *72*(5), 1563–1631.

Walker-Moore, K. (2018). Analysis of the evolution and impact of synthetic identity fraud: Is it preventable? Unpublished master's thesis, Utica College. https://www.proquest.com/docview/2111339765.

Wang, W., Yuan, Y., and Archer, N. (2006). A contextual framework for combating identity theft. *IEEE Security & Privacy*, *4*(2), 30–38. doi: 10.1109/MSP.2006.31.

Warren, S. D., & Brandeis, L. D. (1890). The right to privacy. Harvard Law Review, *4*(5), 193–220. https://doi.org/10.2307/1321160

Whitson, J. R., and Haggerty, K. D. (2008). Identity theft and the care of the virtual self. *Economy and Society*, *37*(4), 572–594. doi: 10.1080/03085140802357950.

Zuboff, S. (2019). The age of surveillance capitalism: The fight for a human future at the new frontier of power. PublicAffairs.

Thomas Birtchnell

11 AI and the Technologies of Mobility

Abstract: Artificial intelligence (AI) is a topic that routinely features in the media and popular imagination; however, there is general confusion about what this innovation means for societies and individuals. Here, social theory provides a salve as social scientists have foreseen the ramifications of the introduction into the wild of less-than-human (at the time of publication) computers and other digital entities aspiring to be sentient. There has been dedicated critical research and commentary on AI and its related innovations in the social science disciplinary specialisations of Mobilities theory, social practice theory, social assemblage theory and socio-technical transitions theory. In this chapter a careful appraisal of the relevance of each of these topics to present concerns is given and tied to the emergence of AI into technologies of mobility in real life. There is indeed cause for concern as AI is unleashed on societies and the chapter recommends further foresight is sought on the deeper disturbances likely to unfold.

Keywords: driverless vehicles, social assemblage, social practice, socio-technical transitions, mobilities paradigm

11.1 Introduction

The possibility of vehicles driving themselves has fascinated commentators since the mid-twentieth century. Driverless vehicles are now being adopted in controlled settings, such as airports, where impressions among users are generally favourable (Hilgarter and Granig, 2020). However, these are not the norm, and the ungoverned release of driverless vehicles into societies is a daunting prospect for policymakers and regulators. With the progression of prototypes to real-world release in cities in the United States in 2018 and the tragic milestone of the first death of a pedestrian by a driverless vehicle, that of Elaine Hezerberg in a trial with the rental car company Uber, the culmination of information and communication technologies in automobiles appears to have been reached and their introduction into societies guaranteed. Despite this ominous start, media and academic enthusiasts foresee a mobility enlightenment arising from the deployment of fleets of driverless vehicles. The result will be safer streets, smarter cities, happier citizens and more sustainable households, or so we are told by commentators eager to illustrate the benefits of artificial intelligence (AI) in technologies of mobility. While some scholars see the mainstream (Du, Zhu and Zheng, 2021) or social (i.e., word of mouth) (Ghasri and Vij, 2021) media's positive

Thomas Birtchnell, University of Wollongong, Australia

https://doi.org/10.1515/9783110721751-011

image of AI as critical to the success of this innovation in transport, this chapter establishes that more research needs to be done on the social aspects of driverless vehicles by summarising the state of the art in a suite of social sciences theories. The reality, if prior technological emergences into societies are anything to go by, will be far less Utopian and predictable than the media foresees. In this chapter, a focus on human experience is undertaken through critical reflection (Conschafter, 2017).

The impacts of the social interactions between AI and other agents in the spaces they traverse remain a blind spot for policymakers and scholars alike (Straub and Schaefer, 2019). Three possibilities are projected by experts for how driverless cars might interact in cities with other agents: the fully automated vehicle, the vehicle on demand and vehicles with 'valet' parking functions (Heinrichs and Cyganski, 2015). In this chapter, I articulate how recent trends in social theory provide a counterpoint to projections of increasing automation in transport infrastructures and enrich discursive futures with the details of social life (Manderscheid, 2018). Contrary to common understandings found in the media, I instead underscore how the mobilities of human agents are grounded in social worlds that at present defy interventions by digitalised 'less-than-humans', chiefly driverless vehicles, in mobility systems (Philo, 2017). The term 'less-than-humans' refers to the aspirations held by developers of fully automated mobility technologies for the vehicles to emulate human organisms to the degree that they are able to share the road, interact meaningfully and safely with other agents, and ideally replace the banal aspects of motoring for humans to do other things. In geographical and environmental humanities research, the term 'more-than-human' has been adopted as a counterpoint to the preceding *non sequitur* of 'non-human', which disparagingly sought to ostracise or exclude organisms scientifically defined as flora or fauna distinct from hominids and their life-worlds, despite them physically sharing spaces and places (Whatmore, 2006).

As it stands, driverless vehicles are at the moment less-than-human, and not same-as-human or more-than-human, due to their inability to engage with other organisms fruitfully in the wild. In order to develop from less-than-human, automated mobilities technologies will need to address the areas of trust, irrationality, intuition and identity. In short, in order to move from less-than- to more-than-human, driverless vehicles will need to develop further empathy and *nous*. In order to do so, scholars must be attuned to social theories that apply to humans: of assemblages, social practices and socio-technical transitions. To make sense of this evolution, if it is indeed possible at all, the human perspective is critical.

A tapestry of challenges faces artificial intelligence in this aspiration to become a digital non-human when exposed to the wild and, in this chapter, I highlight how this patchwork of hurdles will need to be overcome if this innovation is to be in a position to aspire to emulate and even replace human agency in the operation of road vehicles. If human agents are to be removed from the process of controlling private vehicles there will need to be a radical overhaul of the infrastructure and systemic aspects of road networks.

In the chapter, I appraise a selection of contemporary social theories and articulate how each in turn offers a perspective on the challenges for AI in the social construction of technologies of mobility. In the next section, I outline the key components that AI must overcome in order to be able to unfold into societies safely and with minimal disruption to the majority of citizens. Following this prelude, I turn to engaging with the first seam of literature on social assemblage theory, which draws on post-structuralist philosophy to articulate a critical stance on the capacity for transport infrastructures to adapt to automation in a planned, controllable, predictable manner. Attention is paid here to considerations of human agency and the influence of internal and external relationships on how agents interact within mobility patterns. Concerns about power appear insurmountable to AI that lacks corporeal sensations and individual histories stemming from cultural life-worlds.

The second seam of literature is social practice theory, and this suite of concepts posits that social practices (encompassing behaviours, habits, routines, yet more than the sum of these parts) underpin systems and cannot be diminished in assessments of socio-technical change where efficiencies and logical interactions are to be reconciled with obdurate and incumbent manifestations of human agency. To consider this seam further, three types of social practice are covered in detail: comfort, convenience and cleanliness.

The third perspective pertinent to this chapter is the socio-technical transitions literature and efforts to fuse abstract and broad-brush analysis with individual actions and unforeseen consequences. Here, historical case studies are pivotal for infusing ideations based on abstractions in situational realities.

The fourth and final perspective stems from progress in the mobilities paradigm addressing the lack of inclusivity in much transport planning. Any introduction of AI and automation into the provision of transportation to minority users runs a risk of further alienating them from basic services. A challenge is the lack of foresight on the influences upon systems in terms of politics and governance, which skew new technologies towards benefiting elite or privileged users to the detriment of minorities lacking capital or influence.

Mobility justice and injustice is most obviously pertinent in the consideration of the automobility system and its rise to be the dominant mode of transportation globally, despite the consequences it has for the environment via greenhouse gas emissions and pollution, let alone the radical reshaping of the built environment that the motorcar entails. In the conclusion, I touch on this aspect of future research as humans compete and collaborate with automation on the road.

11.2 Motoring Humans and Digital Less-Than-Humans

As automated mobility technologies enter the public arena in 'mundane' situations (Himmelreich, 2018), a key issue is the nature of the transition in terms of system change, namely how the automated technology of the automobile will be 'reinvented' to be more appealing than manual motoring (Mitchell, Borroni-Bird and Burns, 2010). Despite considerable hype, the emergence of driverless vehicles into societies is facing controversy due to unforeseen consequences stemming from shortfalls in AI in the domains of trust, irrationality, intuition and identity. What these challenges share is a foundation in human agency emergent from historical and political decisions underpinning transport infrastructures. The current automobility system has arisen to compensate for human error to a profound degree. Indeed, it is widely acknowledged by technology developers and planners that the greatest factor in the risk landscape for driverless vehicles is non-driverless vehicles and risky and error-prone human motorists (Raval and Dentlinger, 2017). Ethical issues will be make or break for the wider use of driverless vehicles (Hansson, Belin and Lundgren, 2021).

Distrust of new technologies by non-artificial intelligences attenuates AI, thereby increasing incidences of systemic error or even failure. If driverless cars are to enter the public realm, then citizens must be able to trust that they are safe and predictable. Yet, what is trust in the context of mobilities?

A confounding feature of efforts to automate the routine and mechanical aspects of motoring is the unpredictable nature of human behaviours and the importance of intuition to the enactment of driving an automobile. The mid-twentieth-century overhaul of the urban landscape in favour of separated roads and footpaths and right of way for private vehicles was due in part to the demand for predictability within a transport system composed of a multitude of variables.

Beyond physical trust that digital non-humans will operate in a safe and predictable manner, driverless vehicles will also need to be trustworthy with the large amounts of private information and data they will feed to third parties. Privacy concerns could derail the adoption of driverless vehicles given they will require sensors and data on health, well-being, perception and competence to function (Schoonmaker, 2016). Moreover, criminal liability is seriously impaired for driverless vehicles and neither human passengers nor car manufacturers are clearly appropriate to be linked to culpability (Bartolini, Tettamanti and Varga, 2017).

Designers of transport systems presume users will demonstrate a predictable level of rationality in order to mitigate against accidents or misuse (Keeling, 2018); however, uncertainties arise in both the introduction of new elements and in the overhaul of existing infrastructures that unhinge innovations with limited prior use. It is not only in the final iteration, but also the proof of concept, wherein the irrationality of users holds significant potential to derail untested innovations, and driverless cars are no exception here. While laboratory experiments might suggest risks are negligible, fatalities or catastrophic events in the early trials can lead to backlash and ultimately disinvestment at critical points.

A critical feature of human learning is the capacity to commit routines to habit, and this skill is extant in the practice of motoring where drivers are able to undertake journeys, or commutes, daily of up to an hour with minimal fatigue. The intuition that occurs in motoring, once established as a routine, renders the practice of motoring not only tolerable but enjoyable with the concomitant performance of daydreaming or meditation about other activities unrelated to driving an automobile.

Finally, motorists' identities are enmeshed in their automobiles, and regular commitments of time, money and energy are spent on 'car care', maintenance and accessorising. Just as the human body is understood as tied to notions of status, so too have automobiles come to reflect their owners' sense of worth. The car body has come to mirror the human one, and ideas of hygiene and ageing are transposed to vehicles with the result of significant investments of capital being invested in fleet turnover and routines of cleanliness and repair.

11.3 Social Assemblage Theory

> The small scratch on the dashboard niggles away at her attention. She has grown accustomed to the imperfection on her brand-new driverless EV, but it annoys her just the same. Niggling too is the knowledge that her partner always complains to her that she doesn't take care of her stuff—she hasn't told her about this particular incident. A smartphone casually thrown onto the dash during a heated conversation. The sudden lurch forwards as the autopilot clumsily disembarks before she has had time to buckle up. The jerk of her bracelet catching against the faux leather of the car interior as her hand saves the phone. Congratulations turning to curses as the ripped fabric catches her eye. Insurance will cover it. But it didn't: premiums to consider. Cosmetic damage. Not wear and tear—too new. So the nodding bird she has placed over the scratch serves to amplify its existence in her mind—you did it, you did it, you did it, the bird nods in time with the autopilot's stop-and-start driving as the congestion snakes up the road.

Assemblage theory involves adopting a different perspective from conventional configurations of how parts and wholes relate and is concerned with how the internal and external are formulated (Anderson, 2021). In the context of autonomous technologies of mobility, there is a parallel seam of thinking that considers the personal, social, political, cultural, geographical and technical systems shaping, and shaped by, the automobile as the broader automobility system (Urry, 2004). Assemblage theory provides a higher resolution to the framing of the automobility system by including what appear to be out-of-scope elements and considering them as internal influences or relationships. In this chapter, the external relationships of product design, finance, regulation and infrastructures to the internalisation of the automobile are brought to light.

Scholarship on assemblages demonstrates how the automobile is not the be all and end all of motoring—rather, motorists are vested in driving through an enmeshment of concerns culminating in obduracy to systemic change. Such vested interests have the following elements. First, is the perpetuation of regular fleet turnover. Automobiles have an artificially short lifespan akin to many other consumer products suffering from planned obsolescence, a term referring to the product design practice of encouraging manufactured objects to fail through subpar critical components, redundancies in style or the cosmetic appearance of vehicular bodies, and accessories and other trappings tied to motorists' perceptions of their own bodies and identities. With various schemes and incentives now common to ensure motorists replace their vehicles every three years, the average age of fleets is a driver of norms that set in place a feedback cycle encouraging motorists to invest heavily in automobiles as assets with rapid depreciation. Insurance and other financial services are features of this assemblage of automobility and serve to enforce the status quo of motoring.

Second, is the standard articulation of the design of the automobile to privilege human corporeal control and interactivity. A feature of this element is the comfort regimes ranging from the critical—that is, power steering or parking sensors—to the incidental: seat warmers. The assemblage of automobility here leads to expectations about the nature of motoring and represents additional aspects that become requisite for competing assemblages with drivers 'taking a back seat' in place of artificial intelligence, as the saying goes.

On top of financial and corporeal elements in the automobility assemblage are, thirdly, the governance actors and formalised processes behind the scenes that constitute the regulatory and punitive elements enforcing order and normalcy in motoring practices (Iphofen and Kritikos, 2021). Grand road-building projects funded by taxpayers result in decade-long commitments to distinct regulatory infrastructures. Expectations that transport infrastructures will easefully shift, within years or even months, to those that might accommodate driverless vehicles are unrealistic without significant private investment, which has implications for public ownership of transport corridors and poses the risk of reducing inclusivity and access to travel for some members of the public.

A final element in the automobility assemblage is the resistance to change deriving from legacies of norms with roots in now redundant, irrelevant or illogical features in the design of vehicles or infrastructures. An example of this is the consumer penchant for much higher, heavier and offroad-enabled vehicles for use in cities, partly to satisfy cultural ideas of power, safety and freedom in weekend leisure pursuits, and partly as a response to ailing or ill-conceived infrastructures (flooded roads or high speedbumps and kerbs) and other road users' adoption of these features that force similar choices. 'When in Rome, drive as the Romans drive', to alter the saying slightly. But the impact of driver vehicle choice cannot be underestimated as a stressor for driverless cars, with accessories including towing winches, snorkels, retractable camp awnings, spotlights and all wheel controls now increasingly common and representing extra variables for AI to contemplate and mitigate.

A social assemblage conceptual approach reveals a complex tapestry of factors that potentially inhibit a seamless socio-technical transition to driverless vehicles. The oversights that assemblage theory reveals are not at present a feature of debates on how to realise mass automation. In the next section we zoom in on how people enact motoring through social practices.

11.4 Social Practice Theory

> The car door shuts in time as I crash down on the seat with a barely perceptible crunch. The seat swivels, warps and pivots to match my comfort template after assessing my outfit, the weather and my mood. I am still not comfortable, but the seat's heating makes me feel cosy. Behind me, in the cockpit, the auto-driver console clicks into action; I do not even glance at the road as the car rolls into motion. My attention is taken though by an imposing juggernaut's outcry and I stare out the back window, incredulous. The motorist is obviously angered by the sudden departure from the parking bay and I curse my autopilot for the faux pas. I can see a human motorist in the front of the vintage hi-lux, his face deranged with fury, hands waving. The horn of the beaten up old hunk of junk drowns out the measured tones of the autopilot as it asks me for the destination today despite its external audio sensors showing red. As the hi-lux comes alarmingly close to the rear bumper, setting off the proximity alarms as they intended to do, the autopilot wakes up to the threat and begins to activate its emergency response protocols—LEDs on the rear screen light up the words 'TOO CLOSE. WARNING. TOO CLOSE'. The indicators flash on and off and a stilted voice blares from the vehicle: 'your vehicle is exhibiting irrational behaviour that is endangering other road users. Please desist or risk prosecution.' As the hi-lux roars past, its near proximity setting off even louder alarms, an egg splashes into the LED text on the rear window and the driver's rude gesture is obscured as he rides off into the distance in a cloud of cheap diesel exhaust.

As this fictional vignette aims to demonstrate, a future filled with driverless vehicles sharing the road with motorists will not always be uneventful or without incident. Indeed, the greatest risk for automated technologies is humans and their emotional responses to strict interpretations of road rules and motoring practices.

Social practice theory affords a means to forestall obstacles to a transition involving driverless cars that arise from the often-overlooked details of social life. While social practices might appear nugatory at first glance—that is, social practices of comfort, convenience and cleanliness—they are profoundly consequential for systems and system change. In the context of driverless vehicles, being attuned to social practices uncovers decisions that human motorists routinely make with non-trivial impacts to digital non-humans on the road. To AI, social practices might appear as illogical, unreasonable, unpredictable or otherwise falling into the category of 'does not compute'. There are a handful of social practices with particular significance for driverless vehicles that demonstrate the prowess of this concept.

Motorists enact a social practice to negotiate speed in a finely tuned interplay between lawfulness and efficiency, often breaching the speed limit though heeding the placement of surveillance technologies and the likelihood of regulatory onlookers, such as police. Transgressing the speed limit is a norm in many nations around the world, where there are differing degrees of road discipline. Driverless vehicles are particularly prone to this type of social practice as driving below or at the speed limit can provoke a range of responses from tailgating (driving dramatically close to the car in front) to road rage (violent outbursts from other motorists either when stationary or in motion). The unpredictable nature of motorists unable to fathom this social practice is in itself a risk requiring the development of intuition and communication with other motorists through vehicle behaviour.

A second social practice of which driverless vehicles must be cognisant stems from emotional expressions by motorists manifesting in a catalogue of micro-gestures and refined communications via the body of the automobile that emulate the body of the human. These might be hand gestures visible from the vehicle cockpit, denoting frustration or anger requiring a perception of human empathy, to more subtle expressions: the wink of a headlight or slight rocking forwards of the vehicle to express impatience. A great deal of effort has been spent in recent years attempting to tune AI algorithms so that the digital non-human can develop an awareness of what might be deemed illogical or unreasonable behaviours but are nevertheless norms through repeated use and mass adoption by humans. After Shove, "[c]ars depend upon and generate co-requisite arrangements including networks of roads, garages and petrol stations, driving skills, regulatory systems and laws" (2003, p. 68). While the socio-technical transitions perspective and social assemblage theory both illuminate this wider systemic milieu to some degree, they also lack the capacity to engage with what Shove terms 'fuzzy' phenomena stemming from social practices. Three examples that are popular in social practice theory are comfort, convenience and cleanliness, and although these appear typically in concerns about homes and heating or cooling, there is a relevance too for automobiles.

Cars are comfortable places to inhabit. Although efforts to cast driverless vehicles as more comfortable than conventional ones certainly catch the media's attention and will no doubt open up new avenues for consumer appeal, the wider systemic factors of car comfort have an adverse effect on widespread fleet automation. Inclement weather is one issue for transport systems where commuters' inclination for the comfort of automobiles results in social practices privileging this transport mode over others. A related instance of variability arises in crises where public space becomes less comfortable than is usual, for instance during environmental catastrophes such as flooding or pollution events, or in health pandemics. In these instances, people's commitment to norms arising from a shared causal influence in the environment scale up and even cascade into systemic collapses. The thresholds where comfort applies during events are unpredictable and only roughly adhere to calculable variables, hence the challenge for AI.

Convenience is another social practice with disturbing significance for driverless vehicles since motorists' decision making does not rigorously align with referential judgement. An example illustrates this point: a motorist might choose to insert their seat belt after they have begun driving rather than before. To a digital non-human, this pattern of choices is nebulous given the increase of risk to others from inattention, albeit briefly, and possible accelerated personal risk from being unbuckled in motion. The social practice of buckling after disembarking is more than just a 'bad habit'— it is a thread in a tapestry of intuitions and nuances serving to underpin the motorist's proficiency and adept judgement of risk at any given moment. Indeed, the unpopularity of the cruise control function or automatic gears both attest to the stark distinction between human and automated understandings of convenience.

Cleanliness would appear to have little to do with cars; however, this social practice could prove to be a palpable obstruction to the success of driverless fleets of vehicles deriving benefit from shared ownership, rather than private. Given that global road networks already face congestion and near peak capacity at critical moments in the working day, the proliferation of driverless cars poses a further strain to urban planning with deleterious impacts on AI's ability to motor wisely and safely. If one solution is to decouple private vehicle ownership from citizens' use of automobiles, standards of cleanliness applied currently to cars would have ramifications for adoption. The COVID-19 pandemic demonstrated the disastrous effect mass defection from public transport can have on road congestion in many nations due to concerns around cleanliness, with trains, planes, trams and buses grinding to a halt as the virus spread through airborne and enclosed spaces.

These three social practices are just the tip of the iceberg in the complex human interactions of motoring. If AI is to fathom this panoply of variables it will also need to learn to change its perceptions of social norms as they subtly shift through cultural influences and fashions. A capacity to correlate exponential quantities of social practices will entail arrays of sensors supported by machine learning grounded in vast databases of observational data; however, the interpretation of data on social practices is pivotal to any efforts towards gauging how humans drive.

11.5 Socio-technical Transitions Theory

The other day I was driving my dad to his golf game and he pointed to the dash and said "what's that button do?" I laughed and told him it was the latest version of 'cruise control'. He told me that he had had cruise control in every single car he'd ever owned and he could only remember using it once when he had accidentally flicked the button on and gave it a go on a long stretch of highway. I said—yes, this is this same, but instead of just controlling the speed, this takes over the steering as well. He sat bolt upright in his seat and said "what, you mean like a driverless car?!?" I had to calm him down at that point and, still laughing, told him that I guess so, but no one ever uses it because they just don't trust the car to drive itself—what if someone tailgates

> me? Or, if I watch a TV programme on my phone and the next thing I know we've hit a pedestrian and I'm to blame? It's all very mysterious whose fault it would be. Anyway, you know all those little electric scooter cars your friends use to get to the supermarket because they've lost their licences as they've gotten older? Those are driverless and would you trust one of those on the road? Exactly.

Following on from social assemblage theory, with its roots in the discipline of human geography, and social practice theory, with its roots in sociology, the last social theory pertinent to the topic of automated mobility is socio-technical transitions, which has roots in business, management and innovation studies. As the third theory this chapter covers, it reflects on how areas of social life and systems not normally considered relevant to technologies are actually consequential and demand attention for a richer grasp on the possible futures where driverless vehicles reach ubiquity. The viewpoint of socio-technical transitions theory differs from social practice theory, with its concern for people's shared norms at the individual and community level and how these can scale up, and social assemblage theory, with its concern for "heterogeneous elements coming together in a non-homogenous grouping" (Anderson and McFarlane, 2011, p. 125).

Unlike social practice theory and social assemblage theory, socio-technical transitions theory treats multiple scales of analysis as equal—the very broad (the macro or landscape level) and the very mundane and nuanced (the micro or niche level). Socio-technical transitions theory has most productively been used in the context of AI and automobiles in the Multi-Level Perspective—a historical framework to abstract how changes in social systems involve interactions between micro, meso and macro levels in the form of pressures or windows of opportunity across different established 'dimensions' within the regime (the ordered system), existing at the meso level. The theory certainly benefits from hindsight by utilising events in history to then extrapolate how transitions involve complex interplays beyond simply technological progress to other factors: consumer practices, policies, cultural meanings, infrastructures, business models, and so on (Geels, 2018). While the use of projections of future transitions are fraught with danger for analysts (Geels and Smit, 2000), examples do exist of efforts to engage with current trends, including driverless vehicles.

Driverless vehicles have the potential to present many negative externalities and uncertainties according to how they roll out into societies (Agriesti et al., 2020). The imagined progress from programmes such as the Dutch Intelligent Transportation Systems sector demonstrates issues with pressure from different visionaries (Pel, Raven and van Est, 2020). Socio-technical transitions theory indicates that there are competing visions of how driverless vehicles will manifest. Corporate incumbents in the car industry project a straightforward pathway based on the increasing technological progress in product performance and design occurring with existing private vehicles. Alternatively, governmental actors imagine cooperative systems delivering less urban congestion and safer roads with the sacrifice of control, power and underused accessories (snorkels, bumper-bars, luggage space, offroad capabilities, and so on).

Driverless cars are unfeasible in a world where motorists assert their identities through automobile ownership, the performance of motoring and the domination over other transport modes that they enjoy.

Yet, in futures that involve a significant transition to another kind of automobile, vehicle automation becomes far more appealing and even to some degree inevitable (Geels, 2012). Present pressures arising from concerns about future increases in greenhouse gas emissions will also transform transportation norms and systems, and in the process enable windows of opportunity for driverless vehicles. Futures where systems require highly networked and digitally synchronised fleets will by their very nature pave the way for lesser degrees of involvement for motorists and a greater range of options for passengers. What this means is that the ubiquity of driverless vehicles will emerge in some futures in a stepwise fashion as a byproduct of broader changes to transport systems.

At this point it is possible to pre-empt some of the changes to take place in a stepwise transition (Geels, 2006) in certain futures where there is a high degree of confidence for meaningful transformations involving fleets and infrastructures. Efforts to vision such futures are a product of corporate foresight about consumer preferences in societies with a high degree of concern for action on climate change through the rapid adoption of alternative automobile fuels. Governments seeking methods to reach targets for reductions in greenhouse gas emissions through policymaking are also proposing models for societies with ramifications for automated mobilities. By considering complementary and competing visions for automation offered by institutional assessments, and contrasting these with commentaries by advocates and the media, we are able to appraise whether driverless cars will be foregrounded in some futures rather than others.

11.6 Conclusion

Cars dominate the world today, providing a soundtrack of noise to our lives, segmenting our cities, enabling incredible freedoms, yet enmeshing us into systems of debt and situations of danger. With billions around the world commuting daily in private automobiles, the prospect of recovering this lost time through automation, handing the steering wheel on to AI, is compelling. There is a pressing need to somehow address the inequalities and injustices wrought by the automobile, and driverless vehicles appear to offer a fix to a medley of concerns facing human settlements in the twenty-first century. Yet, concerns about AI let loose into societies are many and the dangers from unbridled automation are real. Moreover, driverless vehicles might not have a positive impact upon sustainability or liveability, but instead take societies in another direction, towards more congestion, more pollution, more private motoring and more unsafe streets. In this sense, social science offers a cautionary perspective that avoids the hyperbole of the media by projecting complexities with unreflective adoption of AI technologies for mobility.

Given how prone humans are to errors or miscalculations—a trip on a stairway or a wrongly constructed sentence—it is surprising how much responsibility is placed on motorists to drive without causing accidents or injuring themselves or others. Social assemblage theory indicates that social systems do in fact take human agency into account, but in 'rhizomatic' (as in the nature of plant roots) ways: direct and indirect; homogeneous and heterogeneous. We cannot divorce the technology of the automobile from this wider assemblage of interests: humans, objects, technologies and scripts that construct durable and stable mobilities.

For a robot, the decisions made by humans must be illogical, unpredictable and unreasonable. While AI can certainly be taught about human agency and how humans think, their social practices and the norms they describe do not translate well to quantification and algorithmic sense of cause and effect. Comfort, convenience and cleanliness all map onto corporeal sensations that humans share as visceral beings, and until digital less-than-humans are equipped with similar organic experiences there will be disconnects with human agency. Considerations of social practices are not trivial and choices made to privilege comfort or convenience will in many cases be primary motivations for moving in certain ways over others. To neglect social practices will not encourage less resilient mobility systems and exacerbate injustice.

Finally, history shows that transitions occur rapidly and slowly and foresight about how societies shift to adopt technologies is critical for AI in mobilities. While the 'regime' or status quo begs the attention of policymakers, the micro level of niches needs also to be at the centre of analysis, since over time the fringe or marginal can, under certain circumstances, scale up to manifest into new social structures where they sit at the centre. The automobile is an outstanding example of how a technology that is used by a small elite at first can rapidly alter how people move and how infrastructures and regulations are constructed by social actors in socio-technical transitions.

References

Agriesti, S., Brevi, F., Gandini, P., Marchionni, G., Parmar, R., Ponti, M., and Studer, L. (2020). Impact of driverless vehicles on urban environment and future mobility. *Transportation Research Procedia*, 44–59.

Anderson, B., and McFarlane, C. (2011). Assemblage and geography. *Area*, *43*, 124–127.

Anderson, J. (2021). *Assemblage*. Oxford University Press. https://www.oxfordbibliographies.com/view/document/obo-9780199874002/obo-9780199874002-0114.xml.

Bartolini, C., Tettamanti, T., and Varga, I. (2017). Critical features of autonomous road transport from the perspective of technological regulation and law. *Transportation Research Procedia*, 791–798.

Conschafter, S. J. (2017). Charting a path for cities in the second machine age with or without the car: A focus on the human experience. *Journal of Urban Regeneration and Renewal*, *10*, 116–127.

Du, H., Zhu, G., and Zheng, J. (2021). Why travelers trust and accept self-driving cars: An empirical study. *Travel Behaviour and Society*, *22*, 1–9.
Geels, F. W. (2006). Major system change through stepwise reconfiguration: A multi-level analysis of the transformation of American factory production (1850–1930). *Technology in Society*, *28*, 445–476.
Geels, F. W. (2012). A socio-technical analysis of low-carbon transitions: Introducing the multi-level perspective into transport studies. *Journal of Transport Geography*, *24*, 471–482.
Geels, F. W. (2018). Socio-technical Transitions to Sustainability. *Oxford research encyclopedia of environmental science*. https://oxfordre.com/environmentalscience/view/10.1093/acrefore/9780199389414.001.0001/acrefore-9780199389414-e-587.
Geels, F. W., and Smit, W. A. (2000). Failed technology futures: Pitfalls and lessons from a historical survey. *Futures*, *32*, 867–885.
Ghasri, M., and Vij, A. (2021). The potential impact of media commentary and social influence on consumer preferences for driverless cars. *Transportation Research Part C: Emerging Technologies*, *127*, 103132
Hansson, S. O., Belin, M. Å., and Lundgren, B. (2021). Self-driving vehicles—An Ethical overview. *Philosophy and Technology*, *34*, 1383–1408.
Heinrichs, D., and Cyganski, R. (2015). Automated driving: How it could enter our cities and how this might affect our mobility decisions. *DISP*, *51*, 74–79.
Hilgarter, K., and Granig, P. (2020). Public perception of autonomous vehicles: A qualitative study based on interviews after riding an autonomous shuttle. *Transportation Research Part F: Traffic Psychology and Behaviour*, *72*, 226–243.
Himmelreich, J. (2018). Never mind the trolley: The ethics of autonomous vehicles in mundane situations. *Ethical Theory and Moral Practice*, *21*, 669–684.
Iphofen, R., and Kritikos, M. (2021). Regulating artificial intelligence and robotics: Ethics by design in a digital society. *Contemporary Social Science*, *16*, 170–184.
Keeling, G. (2018). Legal necessity, Pareto efficiency and justified killing in autonomous vehicle collisions. *Ethical Theory and Moral Practice*, *21*, 413–427.
Manderscheid, K. (2018). From the auto-mobile to the driven subject? Discursive assertions of mobility futures. *Transfers*, *8*, 24–43.
Mitchell, W. J., Borroni-Bird, C., and Burns, L. D. (2010). *Reinventing the automobile: Personal urban mobility for the 21st century*. MIT Press.
Pel, B., Raven, R., and van Est, R. (2020). Transitions governance with a sense of direction: Synchronization challenges in the case of the Dutch 'Driverless Car' transition. *Technological Forecasting and Social Change*, *160*, 120–244.
Philo, C. (2017). Less-than-human geographies. *Political Geography*, *60*, 256–258.
Raval, V., and Dentlinger, M. J. (2017). Risk landscape of autonomous cars. *EDPACS*, *56*, 1–18.
Schoonmaker, J. (2016). Proactive privacy for a driverless age. *Information and Communications Technology Law*, *25*, 96–128.
Shove, E. (2003). *Comfort, cleanliness and convenience: The Social organization of normality*. Berg.
Straub, E. R., and Schaefer, K. E. (2019). It takes two to Tango: Automated vehicles and human beings do the dance of driving—Four social considerations for policy. *Transportation Research Part A: Policy and Practice*, *122*, 173–183.
Urry, J. (2004). The 'system' of automobility. *Theory, Culture & Society*, *21*, 25–39.
Whatmore, S. (2006). Materialist returns: Practising cultural geography in and for a more-than-human world. *Cultural Geographies*, *13*, 600–609.

Louis Everuss and Eric L. Hsu

12 Modern Bordering Systems and Digital Identities: Data Doubles, Enrolment Exclusion and the Creation of Digital Polities

Abstract: Digital identities play a vital role in the administration of modern bordering systems. In order to make state-sanctioned border crossings, people increasingly require digital depictions of themselves. This chapter seeks to critically analyse how this new digital bordering regime functions. It covers the increasingly important roles which 'data doubles' and 'comparative digital identities' play in sovereign travel decisions. The chapter also theorises how digital bordering systems uphold existing, and can be productive of new, forms of social disparity. This involves exploring how complex algorithms and the use of big data can generate bias and distortion in digital bordering processes. The chapter concludes with a discussion of how the deployment of various metaphors influences how digital bordering technologies are conceptualised, which has implications for the ways digital bordering systems are subjected to critical scrutiny.

Keywords: digital identities, data doubles, enrolment exclusion, digital divide, algorithm, bordering, border systems, sovereignty

12.1 Introduction

Modern bordering systems rely heavily on the construction, use and ubiquity of data-driven digital identities. This is because the computational and algorithmic technologies that underpin modern bordering systems can only comprehend travellers and the national polities that travellers are seeking to enter through the collection and analysis of digital data. To understand how modern bordering practices are enacted in the contemporary era is thus to grasp the various ways people and polities are digitally depicted (Leese, 2020).

Note: The central argument as well as significant sections of this chapter are taken with permission from L. Everuss (2024), *Digital Mobilities and Smart Borders: How Digital Technologies Transform Migration and Sovereign Borders*, De Gruyter. The research informing this chapter was supported by the UniSA Jean Monnet Centre of Excellence, which is co-funded by the ERASMUS+ Programme of the European Union.

Louis Everuss, Eric L. Hsu, University of South Australia, Australia

https://doi.org/10.1515/9783110721751-012

A growing body of scholarship has shown that border-related digital depictions are selective in the data they draw upon (e.g., Everuss, 2021a; Mann and Matzner, 2019; Pugliese, 2010; Sharma and Nijjar, 2018). Digital identities can exhibit varying levels of complexity and they can differ in the way they attach significance to certain datapoints. At their most basic, digital identities can be used for human identification processes to link travellers with their travel documentation (Noori, 2021). This can be as simple as recording one or two biometric features of a person along with their name, date of birth and nationality—which together creates a digital identity or 'data double'—and then comparing this with new digital identities containing the same types of information every time they are re-bordered (Pötzsch, 2015, p. 106). However, digital identities also factor into the more complex risk assessment of travellers whereby their future actions are predicted (Ajana, 2015). In such processes, a second type of digital identity is implicated: a comparative digital identity that can be used to evaluate whether a traveller's data double resembles a known profile, be it an ideal state polity member or a threatening outsider.

One of our primary aims in this chapter is to establish how digital border identities are not neutral and objectively derived constructions. They have the potential to code prejudice and bias into sovereign decisions, sometimes in compressed and hidden ways. Advancing this argument first involves identifying how digital border identities are created and made to matter. This then leads us, in the second section of this chapter, to examine how there are at least two ways in which digital border identities facilitate biased displays of sovereignty. First, we show how minorities experience enrolment exclusion, whereby some individuals are not recorded by bordering technologies and thus unable to be sovereignly processed, or their digital identities are constructed in ways that make their exclusion likely, even though alternate depictions are possible (Pugliese, 2010). Secondly, we explore how minorities are commonly subject to 'algorithmic exclusion', which occurs when some social groups are differentiated from big data amalgamations that code religious, ethnic, sexual and gender minority characteristics as being inherently foreign and thus threatening.

However, digital bordering identities are not only used to exclude travellers—they are also used to construct the political identities of states. When a person is excluded because they lack the necessary political identities or 'forms of life' to enter a state (Agamben, 2000, pp. 6–8), those same political identities are consequently defined as that of the polity. This symbolic border—a set of characteristics that through presence represents the polity and absence represents outsiders—has historically been produced by national culture, legal frameworks and public discourses that inform the decisions of border actants.

In this chapter we examine how these political identities are now also being influenced by the algorithmic processing of big data. This refers to the ways the state polity itself is being 'datafied' and increasingly becoming a 'composite' digital identity. Consequently, what this chapter shows is how, far from being a simple step in the computational assessment of travellers, the production of digital identities at the border is part of the digital reformation of popular sovereignty. This chapter concludes with a discussion of the metaphorical aspects of digital bordering technologies.

12.2 Data 'Doubles'

Two types of digital identity are especially pertinent to the state bordering of people. The first is the digital profile attached to a specific traveller. When a traveller is subjected to algorithmic assessment, it is not actually the person in question that is being assessed, but a digital account of them. This depiction can include details relating to their physical, national and cultural features, as well as mobility-based features, such as where a traveller has moved during their current and previous journeys. Increasingly, digital profiles also contain information about virtual movements and actions, notably social media posts, email activity and websites visited (Latonero and Kift, 2018). In the US, for example, since 2019 foreign travellers seeking a visa can be asked to provide details of their social media account names and emails for up to a five-year period (Greenfield, 2020; Latonero and Kift, 2018). Together this data is aggregated into a digital identity that "serves as the future reference point for all interactions between that person and authorities within a specific regulatory domain (e.g., asylum, law enforcement, visa management, etc.)" (Leese, 2020, p. 2).

The process was described in 2000 by Kevin Haggerty and Richard Ericson (2000, p. 606) as "abstracting human bodies from their territorial settings and separating them into a series of discrete flows. These flows are then reassembled into distinct 'data doubles' which can be scrutinized and targeted for intervention." An example of the production of data doubles is the digital recording of travellers within the US Customs and Border Protection's Trusted Traveller Programs (TTPs), which pre-assess travellers to expedite bordering at geographical entry/exit points. When travellers enrol in TTPs, data is collected about them, including their "Full name; Alias(es); Date of birth; Place of birth; Language preference; Gender; Current and former addresses; Telephone numbers . . . Criminal history . . . Law enforcement risk assessment worksheet" (Department of Homeland Security, 2020, p. 14216). It is from this information that the US border system comes to know travellers enrolled in TTPs. Any information outside of this system is deemed irrelevant to a traveller's ability to join the US polity through a TTP.

While the production of data doubles is a constant feature of digital bordering systems, data doubles differ significantly in the number and nature of characteristics they include. For instance, the EU's EURODAC database primarily records traveller features related to their biometric profile and when and where asylum seekers make their asylum applications, including "fingerprint data; Member State of origin, place and date of the application for international protection; . . . sex" (Regulation (EU) No 603/2013, 2013). EURODAC data doubles provide a more limited depiction of travellers than what is captured by TTP data doubles because they are used predominantly to perform a sovereign decision based on whether asylum applications have been lodged in the correct geographical region, as opposed to more advanced risk assessment of travellers. Indeed, many data doubles are produced that are only used to connect travellers to their documentation, and therefore just contain simple biometric information. These most basic of data doubles are not compared with the comparative border identities described below.

While data doubles occupy a central role in official bordering systems, it is important to note that they are not only collected by governmental entities. For example, private airline companies establish data doubles in the form of Passenger Name Record (PNR) data, which "contain large amounts of data that are obtained automatically during booking, reservation, and check-in" (Leese, 2014, p. 497). This airline-created data double is regularly accessed and analysed by governments, including in the US and EU (Leese, 2014, p. 497).

Importantly, the information within data doubles does not all originate from the person-object of those data doubles. For example, while the EU's EURODAC asylum system creates data doubles using biometric and spatio-temporal indicators—fingerprints and the geographical and temporal information about the location and time of asylum applications—it also includes a master codes as follows: "Category 1 defines a person as an applicant for international protection, category 2 defines a person as having crossed, or attempted to cross, a border illegally, and category 3 defines someone as being a potential illegal immigrant" (Metcalfe and Dencik, 2019). EURODAC data doubles thus reduce the complex story of individual migrants to a series of identifying features and their EURODAC classification, which can render them illegal in the eyes of a European sovereign state; not in terms of criminal justice, but in terms of not having a valid connection to a European polity.

Regardless of whether data doubles contain a few or a comprehensive list of characteristics, in order to function they must follow travellers. They are invoked every time a traveller is assessed, whether that be by border guards looking up a risk assessment, police officers searching a traveller's biometrics or a smart camera identifying a traveller (Vukov and Sheller, 2013). In these interactions, data doubles are "made mobile" because they are "circulated and exchanged with increasing speed" to shadow their subjects and inform sovereign representatives as to who they are and what they may do (Trauttmansdorff, 2017, p. 120). As such, travellers accumulate multiple data doubles, which can be used at different points in different contexts to enact a sovereign bordering decision.[1]

What we especially wish to underscore in this chapter is how data doubles are in no ways 'natural' and self-evident depictions of the persons they are used to identify.

1 For a related discussion about the multiplicity and speeding up of identity formations in parts of the contemporary world, see Hsu's (2011) chapter on "New Identities, New Individualism".

They are distortions, characterised and controlled by the digital signs—digitally coded units of meaning—that are both available and selected for their construction. This distortion occurs at multiple levels, but it begins with the very characteristics that are available to describe a person, which are determined by the biopolitical context in question. This point is made by Pugliese (2010, p. 8), who argues that biopower paves the way for the production of data doubles because it "effectively colonises the body, overlaying it with calculatory grids and mathematically inscribing it with formulae that will transform it into an object of knowledge and power."

Context therefore is of great significance when it comes to determining what physical and social characteristics of a person matter in the digital bordering process. They are not randomly and inconsequentially selected, but rather reflect and refract the social hierarchies that exist within a given society, which can in turn, as we discuss later in this chapter, entrench racial, gender and other intersectional disparities.

In addition to the *form* of data doubles being influenced by existing cultural assessment practices, the *function* of data doubles mirrors the normal symbolic side of personhood. As Spencer Cahill (1998, p. 135) points out, people have symbolic and material qualities because they are "socially defined, publicly visible embodied being[s]". In other words, people exist as both an idea or social representation, and a publicly visible subject. It is interactions between these two entities that ties public ideas of an individual to some form of reality (Cahill, 1998, p. 132). Steve Kroll-Smith (2000, p. 92) sums this up by stating that "the embodied person is at the border between the real and the ideal, shaped by both ideas and corporeal limitations".

What digitisation does is generate distance between the material person and their symbolic construction in the form of a data double. In traditional theories of personhood, the public representation of people can be influenced by the intentional actions of those being represented. Consequently, a level of individual contact is an important part of the image formation of a person (Hacking, 2004). While the level to which populations with little political power can actually impact their public representation is questionable (Everuss, 2023), the disconnect between the embodied person and person-as-idea is heightened when the latter is digitised in a largely top-down fashion. Travellers cannot for instance truly know what information is included in their data double, and once it has been formed, can rarely alter it. Travellers are not even certain where their data doubles are, as data doubles move through digital networks independently when travellers are assessed by bordering systems.

For instance, in Australia asylum seekers on bridging visas and living in a community, or those not granted visas and living in detention, often do not know when their asylum applications are being processed and thus when their digital profiles are being conjured and assessed. Indeed, they also do not know what information is included in their data doubles. This is not to say, however, that they have no influence over their data double. Tamara Vukov and Mimi Sheller (2013) identify how some migrants try to selectively curate the information they give to authorities, in an attempt to produce a data double that is most likely to produce a favourable bordering outcome. Still, an important source of agency is often taken away from travellers. For example, instead of being able to talk to border officials about their capacity to earn an income and contribute to a polity, travellers are regularly judged purely on the medical and financial records that are included in their data doubles (Greenfield, 2020).

12.3 Comparative Border Digital Identities

Another kind of digital identity deployed at the border is comparative profiles used to assess traveller data doubles. Comparative identities allow algorithmically driven digitised border systems to decide whether a traveller is a threat or incompatible with a state polity. The process involves the "identification of algorithmically determined deviations from a calculated norm that triggers increasingly automated mechanisms of inclusion and exclusion" (Pötzsch, 2015, pp. 107–108). This norm is the comparative identity. We can thus say that it is comparative identities that indicate to digital systems what certain types of people look like, whether that be travellers from Singapore, Ghana or Syria, foreign students from India or Norway, legitimate refugees or tourists, or potential terrorist threats.

The information used to produce such identities is derived from population-level datasets. Key characteristics are selected from population datasets and used as master signifiers for their groups' symbolic identity (Mann and Matzner, 2019). This means that, unlike data doubles that have real travellers as their objects, comparative identities are purely symbolic. In lacking a corporeal body as subject/object, comparative identities are unmoored from the material reality of actual travellers and are instead influenced by representational and computational processes, including the source code of algorithms; the contents of population databases; training data used to guide algorithmic decision making; and, when machine-learning algorithms are involved, the iterative decisions of algorithms themselves. As such, comparative border identities resemble what Steve Kroll-Smith (2000, p. 91) calls a "composite person", which is a person existing entirely in words just as an idea, disembodied and with no physical being to ground its construction.

However, in many respects comparative digital identities are an evolution of composite identities. This is because Kroll Smith's (2000, p. 91) composite identities are established by a congeries of often weakly related media, including population surveys, court cases, press releases, marketing, and so on. Through the repeated emphasis of certain characteristics across this symbolic content, stereotypical depictions of people emerge. For comparative identities, the gradual fermentation of 'problem people' and other stereotypes that establish composite identities is replaced by a fast-paced algorithmic sorting of big data (Aradau and Blanke, 2018). And while some comparative identities, including those that replicate national identities, claim to mirror public communities in the same fashion as composite identities, others create novel constructs by establishing connections between arbitrary categories that then come to represent 'terrorists', visa overstayers or other searched-for border identities.

The production of comparative identities is described by Amoore as another example of algorithms deploying an ontology of association. While algorithms, including machine-learning algorithms, are generally given an initial direction via source code, training data and target outputs, they generally require large amounts of data to achieve their intended purpose (Barrett, 2017). Advanced algorithms do this by searching for patterns or correlations in datasets. Entire datasets and the rich information they contain are condensed to these correlations. Significantly, just like composite identity formation, algorithms are not uncovering a pre-existing form within data, but are generating a new entity from data. Amoore (2020, p. 46) describes this as: "condensing the probable data traces of what matters in the world, cloud computing enacts the matter of the person of interest at the border, the possible future disruptive protest event in urban space, the acts of fraud or insider trading, or the chains of association of no-fly lists, blacklists, and kill lists . . . making action possible".

Therefore, comparative identities are not so much *uncovered within* data, but are instead *created from* data. Algorithms enact these identities and other subjects by establishing associations between otherwise unrelated datapoints. They are "categories [that] come into being via probabilistic assumptions" (Leese, 2014, p. 502). Thus, a comparative identity does not need social usage and approval, and instead can be generated by an algorithmic query in seconds.

While the overarching nature of comparative identities can be to some extent described, based on the known databases that are incorporated into them combined with general understandings of the way risk assessment algorithms compare individual examples with norms (Koulish, 2016, p. 8), explaining the inner workings of actual risk assessment border systems is difficult. Bordering algorithms, and in many instances their complete databases too, are kept secret on security and trade grounds by the private businesses and government agencies that develop and deploy them (Barrett, 2017, p. 338; Koulish, 2016, p. 8). For example, in the EU the comparative identities used to analyse the risk of PNR-based data doubles are only vaguely publicly described as pre-determined criteria "for the purposes of preventing, detecting, investigating and prosecuting terrorist offences or serious crime" (Council, 2016, art. 6).

A similarly vague description is provided in respect to the comparative identities to be used by the EU's European Travel Information and Authorisation System (ETIAS), a pretravel screening programme being rolled out for third-country nationals from select countries that do not require a visa to enter the EU. Along with searching criminal databases, ETIAS will measure the closeness of traveller data doubles to certain deviant comparative identities in the form of 'risk indicators' as well as composite identities formed from an ETIAS watchlist. The 'risk indicators' include age, sex, nationality, country and city of residence, level of education and current occupation, while the watchlist contains known criminal and terrorist data doubles, allowing algorithmic identity construction and matching. Together these comparative identities will inform ETIAS of what "security, illegal immigration or high epidemic risks" look like (Parliament, 2018, art. 27). However, no information is made publicly available about the specific comparative identities used to assess data doubles beyond these imprecise sets of criteria.

What is stated by the EU in more detail is what is excluded from comparative identities. The relevant EU Directive determines that the criteria must be "non-discriminatory", "targeted", "proportionate", "specific" and "in no circumstances be based on a person's race or ethnic origin, political opinions, religion or philosophical beliefs, trade union membership, health, sexual life or sexual orientation" (Council, 2016, art. 6.4). However, despite this, researchers suggest that the allowable data used by algorithms, such as postcode or neighbourhood, are able to serve as proxies for racial, ethnic and other prohibited features (Mann and Matzner, 2019, pp. 1–2). The bias produced by such proxy coding is exacerbated by the fact that "datasets are constructed that disproportionately contain data about certain people, leading to overmonitoring and over-policing of those groups" (Mann and Matzner, 2019, p. 2). This makes it very difficult to effectively prohibit any type of comparative identity, as algorithms and those that influence their construction can find effective workarounds to such policies.

12.4 Borders and Enrolment Exclusion

While it is increasingly evident that the opaque internal weightings of algorithms embed racial, gendered and other intersectional hierarchies into sovereign borders (Suchman, Follis and Weber, 2017), prejudicial exclusion is also expressed through more direct and simplistic means. People are excluded in the digital bordering process when they are simply not included in the digital assessment process. Doing so renders them outsiders, or what Agamben (1998, p. 115) has termed '*homo sacer*', that is, without the protection of rights but still within the purview of a state. Such a situation is not uncommon and occurs in non-digitised settings, such as when states refuse to fully process asylum seekers (Everuss, 2020a, 2020b; Williams and Mountz, 2018). For instance, Australia transports and holds refugees in third countries within their region, refusing to grant them asylum but normally unwilling to commit the moral and legal wrong of refoulment and return them to persecution (Everuss, 2020a), thus placing them in a zone of exception.

While the non-processing of outsiders is a strategic policy decision, digitisation is creating new potentials for people not to be seen by border assemblages. Such potentials exist because, as borders are being digitised, the sovereign view of travellers is less frequently occurring through the vision of human agents and more often through the medium of cameras, fingerprint scanners and automated algorithmic assessment. This consequently has heightened the digital visibility of some while diminishing that of others, which in turn has produced and/or reinforced hierarchies of sovereign border vision. For instance, Simone Browne (2010) demonstrates that the 'observational habits' of biometric bordering technologies regularly produce 'white prototypicality'. This refers to the ways in which certain technologies render the white (male, middle aged and upper class) body as their normal profile, or their 'raceless' point zero. Consequently, people are not only understood by their difference from the apparent normality of the white subject, but the ability of technologies to see people is shaped by this difference.

Browne (2010, p. 135) illustrates this point by noting the higher failure to enrol rates (FTE) of fingerprint scanners in the US to record people of Pacific Rim/Asian descent who have faint fingerprint ridges, construction workers and artisans who are more likely to have worn fingerprints, and elderly users with faint fingerprints and poor circulation. Likewise, photographic and video cameras are often optimised to capture details of faces with lighter skin, which in turn increases the FTE rates for darker-skinned users (Browne, 2010, p. 136). However, while such racial bias is inherent within biometric technologies, the specific populations treated as the norm, and towards which technologies are oriented to most effectively capture, is variable. For instance, as Jacob Hood notes, the capture bias of technologies is sometimes partly determined by their training data. "If the training set overrepresents one demographic, such as white male faces, the algorithm will be more adept at identifying white male faces, creating opportunities for the misidentification of other demographics" (Hood, 2020, p. 162).

However, East Asian-based algorithms and training data recognise Asian faces more accurately than Caucasian faces (Phillips et al., 2011). The result of this is that the actual nature of enrolment exclusion is determined by the power relationships that influence where bordering algorithms are developed and what populations they are developed for. And while Fabio Bacchini and Ludovica Lorusso (2019, p. 325) suggest that white prototypicality "gets inadvertently transmitted by white software engineers dominating the technology sector to the face recognition algorithms they produce via flawed, racially biased construction and training", bordering algorithms are increasingly being developed outside the once dominant 'white' tech hubs where different racial groups are prioritised and excluded through differing FTE rates (Phillips et al., 2011).

More than just creating new technical ways people cannot be seen by bordering systems, new digital technologies also co-produce bordering assemblages that overlook travellers on legal and systematic grounds. This is especially evident, for example, in the pre-assessment e-visa systems and expedited travel authority systems which have been widely adopted by many countries around the world. Such systems—including Trusted Traveller Programs—generally follow strict country-based applicability rules. For example, the US's Visa Waiver Program, which allows visitors to enter the US without a visa for 90 days, is limited to citizens from 40+ specific countries (largely from the Global North) who must also have e-passports (US Department of State, 2021). Consequently, prior to the stage of digital enrolment, digitised bordering systems produce global mobility inequalities, enhancing the movement of a kinetic elite while reinforcing the friction experience by populations lacking in network capital (Ajana, 2015, p. 66; Everuss, 2021b).[2]

Even if people are not strictly prohibited from enrolling in a programme, they may not have the relevant data to effectively do so. Expiated traveller programmes normally require, at a bare minimum, official identifying documentation, including passports, driver's licences and government healthcare and social services cards. Additionally, some bordering systems require travellers to provide digital records, including financial, health, education, social media and other aspects of people's digital footprint, which not all social groups around the world have (Greenfield, 2020).

However, even if a person is able to enrol in a digital bordering system, not all enrolment acts are equivalent. While some people produce data doubles that are expected by systems and easily interpreted, the digital depictions of others are inherently framed as confusing or foreign. This occurs in the 'Secure Flight' programme in the US, an automated screening system that identifies high-risk passengers using the 'no fly list' and the 'selectee list', which requires listed people to undergo additional security checks. Both lists rely on Western English naming conventions and Latin script, making names outside of these parameters more likely to be coded as abnormal and subsequently suspicious (Kafer, 2019, p. 35). It is not that the names are being deemed deviant through comparison with deviant identities—although this is certainly possible—but that they are not properly understood by the system, leading to errors such as confusion between given names and surnames. It is also possible in this situation for some traveller data to be misinterpreted and excluded. Hence, we underscore how prior to any organised assessment of a data double, the bordering assemblage is *affected* by the nature of the data double's digital information, and how neatly this aligns within the information the technological assemblage is designed to interpret.

While enrolment exclusion is at times purposefully designed into systems such as the Trusted Traveller Programs described above, it can also be a byproduct or reproduction of larger inequalities that are exacerbated by digitisation. For instance, extant racial inequalities are embedded into the digital infrastructures of the world. As already outlined, Pugliese (2010) demonstrates how biometric understandings of race are shaped by the way that biopower has already defined human bodies. Consequently, the hierarchies that determine whether a border technology can see a traveller reflect and refract the racial and national hierarchies that have been found to shape border guard prejudice towards minorities (Browne, 2010; Kapoor and Narkowicz, 2019; Wood and Gardiner, 2019).

2 In this way, the exclusionary aspects of digital bordering technologies further support the idea of globalisation as an unevenly distributed and disjunctive process (Hsu, 2021).

A more novel power relationship that influences enrolment in digital bordering systems is that of the 'digital divide', a term used to describe inequality produced by differences in access to and knowledge of digital technologies. In many instances digital divides follow hierarchies of wealth, race, class, geography, and so forth, but they are also influenced by other factors such as age. Younger people, referred to as 'digital natives' (Palfrey and Gasser, 2011), are more likely than older people, or digital immigrants, to have the skills required to navigate official websites, organise their data and produce data doubles. Digital natives are thus "often found at the forefront of recent transformations in the travel experience and therefore, the travel industry" (Gasser and Simun, 2010, p. 83). Consequently, the new travel requirements produced by digital borders, including the requirement to enrol in digital bordering systems, enact new hierarchies of international mobility.

12.5 Exclusion Through the Algorithm

As we have shown in the previous section, the inability of some persons to be enrolled and assessed by bordering systems is one significant way border exclusions occur. But adding another layer to this story is the complex and central role that advanced algorithms play in contemporary bordering systems, and the novel forms of exclusion they engender. What we term 'algorithmic exclusion' commonly occurs through the construction and use of comparative bordering identities to "map and codify different flows of mobility through digitalization, categorizing and sorting subjects into various types, for which inclusion and/or exclusion is facilitated" (Trauttmansdorff, 2017, p. 115).

This algorithmic assessment process has a disproportionate effect on minorities, an outcome that Gary Kafer (2019, p. 24) suggests is caused by the way algorithms rely on big data. Big data does not provide detailed qualitative descriptions of individual people, but instead collections of certain quantitatively ascertained classes. The classes are used to establish hierarchies of normality that act as proxies of threat. Gary Kafer (2019, p. 24) states that because "proxies (like zip code or consumer preferences) encode categories of social difference, people of colour, migrants, sexual minorities, the poor and other oppressed populations are overwhelmingly more likely to bear the burden of population-based classification". The result is described by Bigo (2014, p. 219) as people who do not fit within normal categories of personhood becoming "prisoners of data-double associations and virtual anticipations".

In sum, comparative identities often function by relying on big data to measure how different a person's data double is from the norm, and by communicating that difference as a threat level (Aradau and Blanke, 2018; Koulish, 2016). The problem here, fuelled by the logic of big data, is not that minorities have commonly 'abnormal' profiles, but that they inherently contribute less data to comparative identities than majority populations. This is a problem, as Barrett (2017, p. 342) notes, because a "general tenet of machine learning . . . is that more is better".

In addition to constructing a traveller's threat level by assessing their difference from a 'normal' profile, algorithms also quantifiably define threat or risk by codifying a data double's closeness to deviant comparative identities (Egbert and Leese, 2021). In short, known deviant data doubles, such as the digitally recorded characteristics of known terrorists, are used to create a composite identity that can be matched to the data doubles of travellers. The more characteristics a person shares with a deviant model, the more likely they are to be excluded at the border (Ajana, 2015). This is indeed one approach that will be taken with the EU's ETIAS system that assesses people's threat level by identifying "connections between data in an application file [a traveller's data double] and information related to persons who are suspected of having committed or having taken part in a terrorist offence or other serious criminal offence [comparative deviant identities]" (Parliament, 2018, art. 28).

Accordingly, bordering algorithms reduce individuals to a limited set of master characteristics, which ignores their unique circumstances. The US DHS, for instance, ranks countries based on the level to which past migrants and travellers from various countries have committed crimes (Greenfield, 2020). If a number of terrorist attacks are committed by individuals from a certain country, then all future travellers from that country will be more likely to be classified as potential terrorist threats at the US border.

This form of exclusion, based on the identification of correlations between data doubles and larger datasets, has been described by researchers as an algorithmic, digital or automated form of profiling (Leese, 2014; Mann and Matzner, 2019). By using advanced algorithms, the systems can get around prohibitions on the explicit profiling of travellers based on characteristics like race, ethnicity and gender. Instead of explicitly examining these prohibited grounds of exclusion, algorithms take into account their "'proxies', i.e., combinations of input data which are accurate predictors for the discriminatory categories", such as postcode and neighbourhood and nationality, anyway (Mann and Matzner, 2019, p. 2).

The Dutch National Police Force responsible for migration control, the Royal Netherlands Marechaussee, rely on this type of algorithmic profiling to perform targeted border checks. The Amigo-boras system they use deploys 'automated risk profiling' and smart cameras to detect vehicles that are deemed likely to be involved in irregular migration, identity fraud and human smuggling (Dekkers, van der Woude and Koulish, 2018, p. 238). The way the system is designed to function is by taking photos of cars, examining a wide variety of their features, and then recommending to officers whether cars should be stopped and checked (bordered) (Dekkers, van der Woude and Koulish, 2018, pp. 242–243). However, what research found was that although the risk profiles of vehicles were based on a variety of criteria, the algorithmic risk assessment focused heavily on vehicle licence plates and in particular their country of origin. In short, "[t]he profiles used by Amigo-boras label entire nations as high risk, equating specific nationalities with irregular migration or cross-border crime" (Dekkers, van der Woude and Koulish, 2018, p. 250). Thus, while digital algorithmic bordering promises to objectively assess travellers without personal bias, applications of risk assessment algorithms often appear to bluntly profile travellers in ways that reinforce racial and ethnic hierarchies (Sharma and Nijjar, 2018).

Blunter still is when algorithms are deployed into bordering mechanisms in a strictly directed fashion, where their main purpose is to give a predetermined and desired bordering outcome the veneer of impartiality. This type of strict algorithmic exclusion often involves simpler algorithms that do not necessarily rely on identifying complex correlations within datasets, but instead assign defined risk scores of travellers' data double characteristics. As these types of algorithmic systems are much simpler and more transparent than those that deploy machine learning, it is easier for system administrators to manipulate their functioning to produce a desired exclusion result.

This is indeed what occurred in the US Immigration and Customs Enforcement's risk classification assessment (RCA), which is used to help determine whether to detain immigrants (Nofferi and Koulish, 2014). The RCA's recommendations are based on scores that determine a person's flight and public safety risk (Dekkers, van der Woude and Koulish, 2018, p. 247). The RCA's underlying algorithm is relatively simple, assigning scores for specific immigrant characteristics. Characteristics that are scored for the assessment of risk to public safety range from possession of fraudulent documents and traffic offences to serious criminal convictions, gang affiliations and past 'terroristic' activity. For flight risk, assessable elements include having stable and fixed US address, service in the US armed forces and family connections (Evans and Koulish, 2020, pp. 812–815).[3] The characteristics are sourced from crime and migration databases and manually coded by officers based on their information sources and interviews with migrants.

While the RCA's scoring system was initially designed to try to objectively assess an immigrant's risk to the community and of flight, it was calibrated numerous times between 2012 and 2015 to increase the likelihood of detention in line with government priorities (Evans and Koulish, 2020). The simplicity of the RCA scoring system is what allowed this to occur. Points attributed to some characteristics were changed or some datapoints were newly established (Evans and Koulish, 2020). For instance, in 2014 in line with a priority of the Obama administration to harden US borders by increasing migrant detainment, a new flight risk category was created and given by far the largest score, which did not actually relate to flight risk but whether "Entry without authorization [was] after January 1, 2014" (Evans and Koulish, 2020, p. 828). Flight risk scores consequently were altered just to an assessment of arrival date, with 1 January 2014

3 Evans and Koulish undertook Freedom of Information Act requests and subsequent litigation in federal court to force the DHS to release this information about the RCA.

arrival dates causing a high flight risk output. Furthermore, in 2017, under the Trump administration, the output recommendation of release was completely removed from the RCA, meaning that the system would always recommend detainment or leave the final decision in the hands of a supervisor.

One significant feature of all types of border algorithms we have thus far covered is the displacement of who bears responsibility for bordering decisions. There is a sense in which algorithmic border systems exhibit what Matthias (2004) has termed 'the responsibility gap', which describes how automated systems and technologies can add a confounding element to the ascription of social responsibility.

While face-to-face encounters between border guards and travellers have the potential to influence bordering decisions through their materiality (Andrejevic and Volcic, 2019; Salter, 2012), this is not so, or less so, with the use of bordering algorithms. There is no specific 'recognitive subject' needed to identify who travellers are and decide whether they should be excluded (Andrejevic and Volcic, 2019, p. 349). As such, the interactional and contextual potentials of corporeal bordering are lost in the cold and abstract rationality of the algorithm and the automated approval processes it informs. Indeed, it is this sequestration of the individual that Barrett (2017, p. 340) sees as especially concerning in the context of predictive policing, stating that "[e]ven when mathematically sound, it is ethically problematic to entirely reduce an individual's agency to an amalgamation of demographic probabilities and fuzzy correlations".

12.6 The Creation of a Digital Polity

Although modern bordering practices appear to be primarily reductive in how they enact exclusion and removal, there is also a sense in which they are productive. They do not just identify, keep out and remove what belongs on the outside, but produce and reinforce what the inside constitutes. The starting point for this idea is that in the absence of borders, things, people and places do not naturally exist as being either inside or outside anything. Their position as such is made possible by borders. The classic work of Mary Douglas (2002) notably theorised how this occurs. Douglas explained how normalised social categories can determine how various social entities are viewed and characterised. People and objects can be categorised for example as either being clean or dirty and sacred or profane. Douglas (2002, pp. 140, 151–152) also theorised how some social entities can have 'polluting' qualities because they challenge and traverse the cultural boundaries that exist between categories.

As far back as 1966, Douglas noted there was a political and social function attached to excising people from a given society. She showed that it serves to create and reinforce individual and group categorisations of the universe (Douglas, 2002, p. xi). A similar point is advanced by the work of Michel Foucault (1980, 2008). Foucault's work on biopolitics examines how the production of boundaries between the sick and the healthy, and deviant and normal has significant effects on the ways bodies in the modern world are managed and, in some cases, excluded. Agamben (2005) additionally suggests that this logic exists at the heart of sovereign state borders that are enacted when governments expel 'excess' populations. The expelled are constructed as the inappropriate subject matter for political control, leaving only the appropriate subject matter, people's political existence, their public 'forms of life': political roles and identities based around work, education, economic and civic engagement that become the basis for rights (Agamben, 2000, pp. 2–6).

We stress however that this production of the polity through algorithmic bordering systems is multifaceted. It is not just achieved by removing outsiders, but also by generating discourses around their removal. This generates the personal and social categories that Douglas refers to by symbolically defining what membership in a community constitutes. Legal structures play an influencing and sometimes facilitating role here. For example, a racially framed (white and European) polity was established in Australia through legal instruments such as the doctrine of *terra nullius* that legally and symbolically removed Indigenous Australians from their land and the White Australia Policy that prohibited the migration of Asian migrants to Australia (Everuss, 2020b). Such policies were implemented alongside the stories, poetry, newspaper articles and other cultural content that communicated the ideal white pioneer and settler identities of the state and the apparent threat posed by racially different outsiders (McMaster, 2002).

While it is of course beyond the scope of this chapter to cover the complex history of how the Australian polity was legally and culturally constructed through the institution and creation of borders, the point we want to emphasise is that the emergence and adoption of new digital bordering technologies also needs to be factored into the creation of polities around the world, including the one constitutive of Australia. One way in which digital border systems are creative of polities is through the use of comparative identities. Comparative identities help to construct the ideal norm of a particular national population, from which outsider difference is computed.

This functional role of digital identities is suggested in Pugliese's (2010, p. 73) argument that "biometric templates must be viewed as the synecdoches of the legal category of the subject". What Pugliese means is that the comparative identities produced by the state define the larger categories of the very types of legal subject that can exist in the state, or, in Agamben's (2000, pp. 6–8) terms, the types of identities required for biological people (*zoe*) to become political people (*bios*). This transition occurs through bordering as someone gains legal status and personhood by demonstrating that they can produce a data double that matches a comparative identity of the polity. Consequently, the polity is represented not just by laws and discourses of national identity as interpreted by border guards, but also by the algorithmic production of comparative identities from big data.

What is occurring here is a conflation of a digital system function (algorithmic production of normality) and a political community function (sovereign production of a polity). As already outlined, for computers to identify outsiders they must have a model of what constitutes insiders because "[c]omputational anomaly detection techniques are all focused on first learning similarity and then recognising what is dissimilar, dissonant, or discrepant" (Aradau and Blanke, 2018, p. 12). Aradau and Blanke (2018) showed how this logic was informing the surveillance programmes exposed by Edward Snowden. The algorithms in question produced insiders and outsiders of the state not by qualitatively assessing people, or by applying discourses of race, culture or nationhood, or even by applying legal frameworks—in fact much of the activity uncovered in the Snowden Affair breached US law—but by calculating the norm of the polity and then the deviations from this polity norm (Aradau and Blanke, 2018, pp. 12–19).

One implication of this bordering practice is that it both produces and further solidifies a form of polity construction that exhibits a homogeneous quality. As Aradau and Blanke (2018, p. 14) posit, machine-learning algorithms, especially those that cluster similar profiles to identify abnormal outliers, activate "a mode of reasoning where similarity through proximity has come to define the norm of security, while the anomalous dot or collection of dots are non-proximate". Aradau and Blanke (2018, p. 14) relatedly assert that "Whether rendered through geometrical distance or topological connectedness, calculations of similarity and dissimilarity are indicative of a reconfiguration of the logics of friend/enemy, identity/difference, and normal/abnormal constitutive of security". This means when advanced algorithms play a significant role in polity construction, diverse political communities become harder to realise.

Another key feature of deploying algorithms in polity formation is that it places a significant portion of the process within the 'black box' of machine-learning algorithms (Leese, 2014, p. 503). While it could be argued that the psychology and decision making of border guards has 'black box' qualities, the airport security booth is a bureaucratic space overlaid with rules, processes and hierarchies that ensure a degree of accountability and regularity in bordering. In contrast, the rationality of the machine-learning algorithm is statistical, compressed, hidden and flexible. Decision-making parameters constantly change and algorithms recalibrate their own internal weightings to construct and identify outliers. In short, it is not always simple to identify or critique how algorithms construct the polity, and thus the very foundational process of sovereignty is being determined within the opaque internal spaces of computational systems.

Additionally, when the construction of polities is digitised in this fashion, it appears like a rational process that is uncovering the true core of a society. The polity is being established not by the legal and cultural interpretations of an official (Salter, 2008), but by the apparently objective and neutral algorithmic analysis of data (Dekkers, van der Woude and Koulish, 2018). However, while algorithms certainly exist within discourses of objectivity and computational truth, they are themselves biased. They produce incomplete depictions of their source data based on the training they receive (Barrett, 2017), as well as their iterative interpretations (Amoore, 2019). Therefore, the depictions of the polity that comparative identities produce, just like traditional bordering processes identified by Douglas (2002, p. 4), involve political conflict and “claims and counter-claims to status”; it is just that the outcomes of this process can now be produced through the opaque internal weightings of algorithms.

12.7 Conclusions

Identities play a significant role in the construction and negotiation of sovereign state borders. State borders separate the types of people that belong to a state from those that do not and in making this separation they produce the political identities of insiders and outsiders. As with all aspects of borders, bordering identities are currently entering a phase of digitisation. While not a uniform process, it is increasingly the case that insiders and outsiders are defined by more than cultural representations and legal instruments. They are also subject to digital recording and algorithmic analysis of personal and population data.

This chapter has examined some of the implications of this digitisation of border identities. We have covered how border digitisation is underpinned by two types of digital identities: comparative identities and data doubles. The latter functions as signs that represent people as objects, while comparative identities are composites of what constitutes ideal or deviant travellers. The entire process of digital border making—from the initial recording of a person’s data double to its complex comparison with composite digital identities—has been shown in this chapter to engender varying forms and levels of bias towards minorities. This bias is apparent in the way digital identity-recording systems are designed to better capture the physical and social features of majority populations. Bias is also expressed when comparative algorithmic processes code non-normative features as threatening, risky and incompatible with polity membership. The analysis we have advanced further indicates the ways algorithmically driven digital bordering systems are multifaceted in their drivers and consequences. Not only do they prevent some people from making certain kinds of border crossings, but they also have broader social effects that shape the polities of which they are a part.

What we have not however given much consideration to in this chapter is how some of the conceptual terms we have used to analyse modern bordering technologies may have a reifying or overdetermining effect. To promote epistemic reflexivity around some of the concepts we have deployed, we conclude this chapter with a series of reflections about the role of metaphors in the social theorising of digital border technologies. Numerous works have explored how metaphors play a crucial role in discursively constructing borders (Nicol, 2012; Vila, 2000). More than just being linguistic flourishes, metaphors have the potential to be ‘world making’ (Cazeaux, 2007). Metaphors about borders underpin and facilitate how they are culturally assessed and materially experienced, as particularly evident in the way borders can be figurately likened to 'fences', 'walls' and/or 'gates' (Nicol, 2012). And so too, we posit, does this apply to metaphors relating to digital border technologies, which have at points in this chapter—but also elsewhere in other pieces of scholarship—been deployed without much critical scrutiny.

Take, for example, our usage of 'big data'. According to Burgess and Cornellius (2014), metaphors commonly play a vital role in constructing understanding and experiences of 'big data' because the phenomenon involves high levels of abstraction. By implicitly framing it as a 'natural force' with its own autonomy, we have at different points run the risk of contributing to the institutionalisation of certain metaphorical constructions of big data, which unfairly overlook the involvement of humans in socio-technical systems.

Our usage of the term 'data double' in this chapter also bears interrogating. That personhood can be metaphorically 'doubled' through data has the potential to conceptualise self-knowledge as being externally referent (Douglas-Jones, 2021, p. 162), in addition to conceiving data representation as a process with precise capacities (Lupton, 2021). While we have stressed at different points in this chapter that data doubles involve 'distortions' of the persons they are meant to depict, there is a sense in which this stands at odds with our deployment of the term itself. It is necessary, in other words, to consider the utility of adopting new metaphoric vocabularies, since, as Lupton (2021, p. 409) notes, "language choice . . . is an ethical issue". We might for example follow Lupton's (2018) lead in embracing metaphors of digital data that emphasise how data can be 'lively', which explores how data representations involve the biological processes of 'decay' and 're/decomposition'.

As a way of advancing elements of the critical border studies agenda that emphasise how border ontologies and epistemologies are complex and multifaceted (Parker and Vaughan-Williams, 2009), we posit the need for researchers to show greater reflexivity towards the metaphoric dimensions of new bordering technologies. This involves recognising that what constitutes a bordering or technological metaphor has the potential to cross into and back out of the domain of the literal. This has numerous implications, including influencing the ways human and non-human attributes are distinguished from one another (Coveney et al., 2023, pp. 116–119). The metaphorical analysis of the terms used to explain the role of digital identities in modern bordering practices we have been encouraging ultimately serves a key aim of this chapter: to illuminate the capacity for distortion in modern digital bordering systems and the varied impacts this distortion has.

Bibliography

Agamben, G. (1998). *Homo sacer: Sovereign power and bare life*. Stanford University Press.
Agamben, G. (2000). *Means without end: Notes on politics* (Vol. 20). University of Minnesota Press.
Agamben, G. (2005). *State of exception* (K. Attell, trans.). University of Chicago Press.

Aho, B., and Duffield, R. (2020). Beyond surveillance capitalism: Privacy, regulation and big data in Europe and China. *Economy and Society*, *49*(2), 187–212.

Ajana, B. (2015). Augmented borders: Big Data and the ethics of immigration control. *Journal of Information, Communication and Ethics in Society*, *13*(1), 58–78.

Amoore, L. (2019). Doubt and the algorithm: On the partial accounts of machine learning. *Theory, Culture & Society*, *36*(6), 147–169. doi: 10.1177/0263276419851846.

Amoore, L. (2020). *Cloud ethics: Algorithms and the attributes of ourselves and others*. Duke University Press.

Andrejevic, M., and Volcic, Z. (2019). 'Smart' cameras and the operational enclosure. *Television & New Media*, *22*(4), 343–359. doi: 10.1177/1527476419890456.

Aradau, C., and Blanke, T. (2018). Governing others: Anomaly and the algorithmic subject of security. *European Journal of International Security*, *3*(1), 1–21.

Bacchini, F., and Lorusso, L. (2019). Race, again: How face recognition technology reinforces racial discrimination. *Journal of Information, Communication and Ethics in Society*, *17*(3), 321–335.

Barrett, L. (2017). Reasonably suspicious algorithms: Predictive policing at the United States border. *New York University Review of Law & Social Change*, *41*, 327.

Bauman, Z. (2004). *Wasted lives: Modernity and its outcasts*. Polity Press.

Bigo, D. (2014). The (in)securitization practices of the three universes of EU border control: Military/Navy–border guards/police–database analysts. *Security Dialogue*, *45*(3), 209–225.

Browne, S. (2010). Digital epidermalization: Race, identity and biometrics. *Critical Sociology*, *36*(1), 131–150. doi: 10.1177/0896920509347144.

Butler, P. (2018). *Chokehold: Policing black men*. The New Press.

Byler, D. (2022). Producing 'enemy intelligence': Information infrastructure and the smart city in Northwest China. *Information & Culture*, *57*(2), 197–216.

Cahill, S. (1998). Toward a sociology of the person. *Sociological Theory*, *16*(2), 131–148.

Cazeaux, C. (2007). *Metaphor and continental philosophy*. Routledge.

Coveney, C., Greaney, M., Hsu, E. L., Meadows, R., and Williams, S. J. (2023). *Technosleep: Frontiers, fictions, futures*. Springer International Publishing AG. doi: 10.1007/978-3-031-30599-3.

Dauvergne, C. (2004). Making people illegal. In P. Fitzpatrick and P. Tuitt (eds), *Critical beings: Law, nation and the global subject* (pp. 83–100). Ashgate.

Dekkers, T., van der Woude, M., and Koulish, R. (2018). Objectivity and accountability in migration control using risk assessment tools. *European Journal of Criminology*, *16*(2), 237–254. doi: 10.1177/1477370818771831.

Department of Homeland Security (2020). Notice of modified Privacy Act system of records. https://www.federalregister.gov/documents/2020/03/11/2020-04980/privacy-act-of-1974-system-of-records#print.

Directive (EU) 2016/681 (2016). On the use of passenger name record (PNR) data for the prevention, detection, investigation and prosecution of terrorist offences and serious crime. https://eur-lex.europa.eu/legal-content/EN/TXT/?uri=CELEX:32016L0681.

Douglas, M. (2002). *Purity and danger: An analysis of concepts of pollution and taboo*. Routledge.

Douglas-Jones, R. (2021). Bodies of data: Doubles, composites, and aggregates. *Journal of the Royal Anthropological Institute*, *27*(S1), 159–170. doi: 10.1111/1467-9655.13486.

Egbert, S., and Leese, M. (2021). *Criminal futures: Predictive policing and everyday police work*. Taylor & Francis.

Evans, K., and Koulish, R. (2020). Manipulating risk: Immigration detention through automation. *Lewis & Clark Law Review*, *24*, 789.

Everuss, L. (2020a). Mobile sovereignty: The case of 'boat people' in Australia. *Political Geography*, *79*, 102162. doi: 10.1016/j.polgeo.2020.102162.

Everuss, L. (2020b). Westphalian sovereignty as a zombie category in Australia. *Borderlands*, *19*(1), 115–146. doi: 10.21307/borderlands-2020-006.

Everuss, L. (2021a). AI, smart borders and migration. In A. Elliott (ed.), *The Routledge social science handbook of AI* (pp. 339–356). Routledge.
Everuss, L. (2021b). The new mobilities paradigm and social theory. In A. Elliott (ed.), *Routledge handbook of social and cultural theory* (pp. 287–305). Routledge.
Everuss, L. (2023). Everyday sovereign exclusion: Conceptualising police violence and deaths in custody as a racial production of homo sacer. *Distinktion: Journal of Social Theory*, 1–22. doi: 10.1080/1600910X.2023.2220933.
Foucault, M. (1980). *Power/knowledge: Selected interviews and other writings, 1972–1977*. Vintage Books.
Foucault, M. (2008). *The birth of biopolitics: Lectures at the College de France, 1978–1979*. Palgrave Macmillan.
Gasser, U., and Simun, M. (2010). Digital lifestyle and online travel: Looking at the case of digital natives. In R. Conrady and M. Buck (eds), *Trends and issues in global tourism 2010* (pp. 83–89). Springer.
Giddens, A. (1991). *Modernity and self-identity*. Stanford University Press.
Goldstein, D. M., and Alonso-Bejarano, C. (2017). E-terrify: Securitized immigration and biometric surveillance in the workplace. *Human Organization*, *76*(1), 1–14.
Greenfield, C. (2020). As governments build advanced surveillance systems to push borders out, will travel and migration become unequal for some groups? *Migration Information Source*. https://www.migrationpolicy.org/article/governments-build-advanced-surveillance-systems.
Hacking, I. (2004). Between Michel Foucault and Erving Goffman: Between discourse in the abstract and face-to-face interaction. *Economy and Society*, *33*(3), 277–302.
Haggerty, K. D., and Ericson, R. V. (2000). The surveillant assemblage. *British Journal of Sociology*, *51*(4), 605–622. doi: 10.1080/00071310020015280.
Holper, M. (2020). Immigration e-carceration: A Faustian bargain. *Boston College Law School Legal Studies Research Paper*, No. 539.
Hood, J. (2020). Making the body electric: The politics of body-worn cameras and facial recognition in the United States. *Surveillance & Society*, *18*(2), 157–169.
Hsu, E. (2011). New identities, new individualism. In A. Elliott (ed.), *Handbook of identity studies* (pp. 129–147). Routledge.
Hsu, E. (2021). Globalization theory. In A. Elliott (ed.), *Routledge handbook of social and cultural theory* (pp. 199–214). Routledge.
Kafer, G. (2019). Big data biopolitics. *Digital Culture & Society*, *5*(1), 23–42.
Kapoor, N., and Narkowicz, K. (2019). Unmaking citizens: Passport removals, pre-emptive policing and the reimagining of colonial governmentalities. *Ethnic and Racial Studies*, *42*(16), 45–62. doi: 10.1080/01419870.2017.1411965.
Kear, M. (2017). Playing the credit score game: Algorithms, 'positive' data and the personification of financial objects. *Economy and Society*, *46*(3–4), 346–368. doi: 10.1080/03085147.2017.1412642.
Klein, E. (2020). Settler colonialism in Australia and the cashless debit card. *Social Policy & Administration*, *54*(2), 265–277.
Koulish, R. (2016). Sovereign bias, crimmigration, and risk. In M. Guia, R. Koulish and V. Mitsilegas (eds), *Immigration detention, risk and human rights* (pp. 1–12). Springer.
Kroll-Smith, S. (2000). The social production of the drowsy person. *Perspectives on Social Problems*, *12*, 89–109.
Latonero, M., and Kift, P. (2018). On digital passages and borders: Refugees and the new infrastructure for movement and control. *Social Media + Society*, *4*(1), 2056305118764432. doi: 10.1177/2056305118764432.
Leese, M. (2014). The new profiling: Algorithms, black boxes, and the failure of anti-discriminatory safeguards in the European Union. *Security Dialogue*, *45*(5), 494–511. doi: 10.1177/0967010614544204.
Leese, M. (2020). Fixing state vision: Interoperability, biometrics, and identity management in the EU. *Geopolitics*, 1–21. doi: 10.1080/14650045.2020.1830764.

Liang, F., Das, V., Kostyuk, N., and Hussain, M. M. (2018). Constructing a data-driven society: China's social credit system as a state surveillance infrastructure. *Policy & Internet*, *10*(4), 415–453. doi: 10.1002/poi3.183.

Lupton, D. (2021). Language matters: The digital twin' metaphor in health and medicine. *Journal of Medical Ethics*, *47*(6), 409. doi: 10.1136/medethics-2021-107517.

Lupton, D. (2018). How do data come to matter? Living and becoming with personal data. *Big Data & Society*, *5*(2), 1–18.

Mann, M., and Daly, A. (2019). (Big) data and the North-in-South: Australia's informational imperialism and digital colonialism. *Television & New Media*, *20*(4), 379–395.

Mann, M., and Matzner, T. (2019). Challenging algorithmic profiling: The limits of data protection and anti-discrimination in responding to emergent discrimination. *Big Data & Society*, *6*(2), 2053951719895805. doi: 10.1177/2053951719895805.

Matthias, A. (2004). The responsibility gap: Ascribing responsibility for the actions of learning automata. *Ethics and Information Technology*, *6*(3), 175–183. doi: 10.1007/s10676-004-3422-1.

McMaster, D. (2002). *Asylum seekers: Australia's response to refugees*. Melbourne University Press.

McNevin, A. (2006). Political belonging in a neoliberal era: The struggle of the sans-papiers. *Citizenship Studies*, *10*(2), 135–151.

Metcalfe, P., and Dencik, L. (2019). The politics of big borders: Data (in)justice and the governance of refugees. *First Monday*, *24*(4). doi: 10.5210/fm.v24i4.9934.

Mills, C. W. (1959). *The sociological imagination*. Oxford University Press.

Minca, C. (2006). Giorgio Agamben and the new biopolitical nomos. *Geografiska Annaler: Series B*, *88*(4), 387–403.

Moran, A. (2011). Multiculturalism as nation-building in Australia: Inclusive national identity and the embrace of diversity. *Ethnic and Racial Studies*, *34*(12), 2153–2172.

Nicol, H. N. (2012). The wall, the fence, and the gate: Reflexive metaphors along the Canada–US border. *Journal of Borderlands Studies*, *27*(2), 139–165. doi: 10.1080/08865655.2012.687213.

Nofferi, M., and Koulish, R. (2014). The immigration detention risk assessment. *Georgetown Immigration Law Review*, *29*(45).

Noori, S. (2021). Suspicious infrastructures: Automating border control and the multiplication of mistrust through biometric E-Gates. *Geopolitics*, 1–23. doi: 10.1080/14650045.2021.1952183.

Palfrey, J. G., and Gasser, U. (2011). *Born digital: Understanding the first generation of digital natives*. Perseus.

Parker, N., and Vaughan-Williams, N. (2009). Lines in the sand? Towards an agenda for critical border studies. *Geopolitics*, *14*(3), 582–587. doi: 10.1080/14650040903081297.

Phillips, P. J., Jiang, F., Narvekar, A., Ayyad, J., and O'Toole, A. J. (2011). An other-race effect for face recognition algorithms. *ACM Transactions on Applied Perception (TAP)*, *8*(2), 1–11.

Pötzsch, H. (2015). The emergence of iBorder: Bordering bodies, networks, and machines. *Environment and Planning D: Society and Space*, *33*(1), 101–118. doi: 10.1068/d14050p.

Pugliese, J. (2010). *Biometrics: Bodies, technologies, biopolitics* (Vol. 12). Routledge.

Puschmann, C., and Burgess, J. (2014). Metaphors of big data. *International Journal of Communication*, *8*, 1690–1709.

Regulation establishing a European Travel Information and Authorisation System (ETIAS) (2018). https://eur-lex.europa.eu/legal-content/EN/TXT/?uri=CELEX%3A32018R1240.

Regulation (EU) No 603/2013 (2013). https://eur-lex.europa.eu/LexUriServ/LexUriServ.do?uri=OJ:L:2013:180:0001:0030:EN:PDF.

Salter, M. B. (2008). When the exception becomes the rule: Borders, sovereignty, and citizenship. *Citizenship Studies*, *12*(4), 365–380.

Salter, M. B. (2012). Theory of the /: The suture and critical border studies. *Geopolitics*, *17*(4), 734–755.

Sekhon, N. (2019). Police and the limit of law. *Columbia Law Review*, *119*, 1711.

Shapiro, A. (2019). Predictive policing for reform? Indeterminacy and intervention in big data policing. *Surveillance & Society*, *17*(3/4), 456–472.

Sharma, S., and Nijjar, J. (2018). The racialized surveillant assemblage: Islam and the fear of terrorism. *Popular Communication*, *16*(1), 72–85.

Sheehey, B. (2019). Algorithmic paranoia: The temporal governmentality of predictive policing. *Ethics and Information Technology*, *21*(1), 49–58.

Skinner, D. (2020). Race, racism and identification in the era of technosecurity. *Science as Culture*, *29*(1), 77–99. doi: 10.1080/09505431.2018.1523887.

Suchman, L., Follis, K., and Weber, J. (2017). Tracking and targeting: Sociotechnologies of (in)security. *Science, Technology, & Human Values*, *42*(6), 983–1002.

Trauttmansdorff, P. (2017). The politics of digital borders. In C. Günay and N. Witjes (eds), *Border politics: Defining spaces of governance and forms of transgressions* (pp. 107–126). Springer International Publishing. doi: 10.1007/978-3-319-46855-6_7.

US Department of State (2021). Visa Waiver Program. https://travel.state.gov/content/travel/en/us-visas/tourism-visit/visa-waiver-program.html.

Valdez, I., Coleman, M., and Akbar, A. (2020). Law, police violence, and race: Grounding and embodying the state of exception. *Theory & Event*, *23*(4), 902–934.

Vila, P. (2000). *Crossing borders, reinforcing borders: Social categories, metaphors, and narrative identities on the U.S.-Mexico frontier*. University of Texas Press. doi: 10.7560/787391.

Vukov, T., and Sheller, M. (2013). Border work: Surveillant assemblages, virtual fences, and tactical counter-media. *Social Semiotics*, *23*(2), 225–241.

Williams, K., and Mountz, A. (2018). Between enforcement and precarity: Externalization and migrant deaths at sea. *International Migration*, *56*(5), 74–89.

Wood, S., and Gardiner, S. (2019). Policing U.K. airports and Schedule 7 of the Terrorism Act 2000: The young passengers' perception of security measures. *Terrorism and Political Violence*, 1–22. doi: 10.1080/09546553.2019.1638255.

Yuval-Davis, N., Wemyss, G., and Cassidy, K. (2018). Everyday bordering, belonging and the reorientation of British immigration legislation. *Sociology*, *52*(2), 228–244. doi: 10.1177/0038038517702599.

Yuval-Davis, N., Wemyss, G., and Cassidy, K. (2019). *Bordering*. John Wiley & Sons.

Rina Yamamoto

13 Smart Homes: Focusing on Artificial Intelligence, the Internet of Things and Robots for Aged Care

Abstract: The rise of urban ageing in Japan has led to a growing interest in smart homes, which incorporate advanced technologies like artificial intelligence, the Internet of Things and robots to renovate older people's homes. This trend is driven by the expectation that smart homes can help solve the housing problem with urban ageing, allowing older individuals to remain in their homes and live independently by accommodating their changing physical needs. However, introducing such advanced technologies into daily life requires careful consideration of their potential benefits and limitations, including privacy concerns associated with the surveillance of smart homes. This chapter explores the housing problems older people face in urban areas, the social backgrounds of increasing empty homes with urban ageing, and the innovative solutions being developed, such as the Robotic Smart Home project. It also analyses the implications of incorporating smart homes into everyday life from a social theory perspective.

Keywords: urban ageing, smart homes, artificial intelligence (AI), Internet of Things (IoT), robots

13.1 Introduction: Super-aged Society and Smart Homes

What causes a growing interest in 'smart homes' in Japan? Why is introducing artificial intelligence (AI), the Internet of Things (IoT) and robots considered for renovating older people's homes? The reason behind this trend is the expectation that smart homes help solve the housing problem with urban ageing.

It is widely acknowledged that since the late 2000s Japan has become a 'super-aged society', with more than 21% of its population over 65. According to the *Annual Report on the Ageing Society* (Cabinet Office, 2018a, pp. 4–5), Japan's proportion of older people ranked low among developed countries until the 1980s. It then became a middle-ranking country in the 1990s and was eventually recognised as having the highest proportion of older people in the world in 2005. In 2020, Japan's ageing rate

Note: This work was supported by JSPS KAKENHI Grant Number JP20K02304, JP24K05454.

Rina Yamamoto, Seijo University, Japan

https://doi.org/10.1515/9783110721751-013

reached 28.8%, with an older population of 36.19 million, and is projected to increase to 38.4% by 2065 (Cabinet Office, 2021).

Over the past decade, the older population in the major metropolitan areas of Tokyo, Nagoya and Osaka has consistently increased. This increase is attributed to the younger generations, who moved from rural to urban areas during the rapid economic growth period (1955–1973), and who have now reached old age. The Tokyo metropolitan area, including Tokyo Metropolis, Kanagawa, Saitama and Chiba prefectures, has experienced the most significant rise in the older population. The National Institute of Population and Social Security Research (IPSS) predicts that the number of older people in the Tokyo metropolitan area will increase by over 40% from 2010 to 2035 (IPSS, 2013). Although there is a nationwide common trend of growth in the older population and a rise in older individuals living alone, the Tokyo metropolitan area is unique in that this has occurred on an unprecedented scale and within a relatively short period.

After 2025, the Tokyo metropolitan area is expected to face a critical stage as the first baby boom generation, born between 1947 and 1949, reaches the age of 75. At this age, both men and women tend to become less independent, and the need for care usually increases. As a result, there is a concern that as more older people require care and move into senior facilities, more homes will become vacant, exacerbating the empty home problem. Furthermore, there is a fear that even if older individuals continue to live in their homes, they may experience social isolation. For instance, the number of lonely deaths, known as 'Kodokushi', among older individuals living alone in the 23 wards of Tokyo has more than doubled, from 1,837 in 2005 to 4,207 in 2020 (Tokyo Medical Examiner's Office, 2011; 2020). This alarming data highlights the concern over social isolation. Thus, as the first baby boom generation ages, it is predicted that the empty home problem and the social isolation issue will escalate in urban areas.

According to the Survey of Older People's Attitudes to Housing and Living Environment, about 60% of older individuals prefer to remain in their homes, even if they become frail due to old age (Cabinet Office, 2018b, pp. 61–62). Considering these older people's willingness, if their homes could be renovated to accommodate their changing physical needs, they could continue to live independently and keep their homes, which could help alleviate the empty home problem. Smart homes could also help prevent social isolation by facilitating communication with medical and nursing care providers, improving situations like lonely death and aged carer shortages. Hence, Japan is researching and developing advanced technologies like AI, IoT and robots to renovate the homes of older people, allowing them to remain in their homes until the end of their lives.

However, introducing advanced technologies such as AI, IoT and robots into daily life requires careful consideration of their potential benefits and limitations. One of the major concerns of social sciences is the privacy issue associated with the surveillance of smart homes. Therefore, this chapter delves into how smart homes can be a helpful solution to the housing problem with urban ageing while also considering the privacy issue surrounding these technologies from a social theory perspective.

Specifically, the chapter is structured as follows. First, it sheds light on the housing problems that older people face in urban areas due to the rise of living alone. Secondly, it explores the social backgrounds of empty homes with urban ageing through the viewpoints of 'my-home-ism', a prevalent way of life in post-war Japan, and the housing industry's growth-prioritised new construction supported by the government's housing policies. Thirdly, it provides insights into a current innovative solution to the problem by highlighting the 'Robotic Smart Home' (RSH) project by Fujita Health University as an example of a cutting-edge initiative. Finally, it analyses the implications of incorporating smart homes into everyday life from a social theory perspective.

13.2 Housing Problems With Urban Ageing

In Japan, there has been a significant change in how households are composed over the past few decades. More and more people are now living alone rather than with their families, marking a turning point in Japanese society around 2010. Comparing changes in household composition from the period of rapid economic growth to the present day, it is clear that nuclear family households consisting of a couple and children are no longer the norm. In 1980, 42.1% of all private households were nuclear family households, but by 2005 this number had fallen to 29.9%, and it is expected to decrease further to 23.8% by 2035. Meanwhile, one-person households increased from 19.8% in 1980 to 29.5% in 2005, projected to reach 38.7% in 2035 (IPSS, 2018, p. 3).

It is important to note that this change in household composition is closely related to the ageing of nuclear families. An analysis of the increase in one-person households by age reveals that, while the number of one-person households of those under 65 will peak in 2025 and decline gradually until 2035, the number of one-person households of those aged 65 and over will rise consistently (IPSS, 2018, pp. 16–28). This fact indicates that nuclear family households consisting of a couple and children, which were once the mainstream household type, are ageing. In other words, the increase in the number of older one-person households highlights the issue of living alone as the final residential form following the nuclear family life cycle, which suggests the reality and social isolation of older singles who have lost their spouses after their children have left home.

A regional breakdown of the growth rate of one-person households among older individuals from 2010 to 2035 shows that the Tokyo metropolitan area has the highest increase rate at 71.0%, surpassing the national average of 53.1% (IPSS, 2014, p. 49). Furthermore, when considering the proportion of one-person households among individuals aged 65 and over by prefecture, in 2010 Tokyo already had the highest percentage in the country at 24.2%, predicted to rise to 27.7% by 2035. This data is significantly higher than the national average of 16.9% in 2010 and 20.4% in 2035 (IPSS, 2014, p. 85), which indicates that Tokyo is facing a particularly severe rise in older individuals who live alone.

The number of older people living alone in Tokyo is predicted to increase from 647,456 in 2010 to over 1 million by 2035. This trend is particularly evident among those aged 75 and over, with the number of one-person households headed by individuals in this age group expected to grow from 336,037 in 2010 to 577,263 by 2035 (IPSS, 2014, pp. 142–144). It is worth noting that the housing problems with urban ageing are closely linked to the increasing number of these late older individuals in one-person households. Moreover, homeowners and renters face different challenges. Hence, to accurately comprehend the challenges older singles struggle with, the following examines their housing problems based on the types of housing tenure.

First, regarding renters, there is a disturbing trend in the rental housing market where older individuals may face discrimination when attempting to rent a property. A survey of the rental housing market (Nakagawa, 2001) revealed that approximately 40% of real estate agencies discriminate against older renters, providing them with about 30% less information than younger individuals. Additionally, landlords may impose restrictions on older tenants, with around 60% not allowing older singles to move in due to concerns about potential health issues, maintaining property safety, finding enough guarantors, paying rent on time and staying for extended periods. These tendencies have contributed to and could worsen the 'shortage of rental housing problem' among older urban singles.

Secondly, for homeowners there is a concern that older people may struggle to sell their homes. Location plays a significant role in selling a house. Suppose a house is in a convenient location, like a city centre. In that case, it may be easy to sell, while selling it in a less convenient location, like suburbs, may be challenging. A study on suburban residential areas in Tokyo (Sonoda, 2010) found that if resales of older people's houses go poorly, over 50% of the residential areas may become vacant. This fact suggests that the empty home problem could be exacerbated since the older singles in the suburbs, who opt to stay at home while in good health, may be unable to sell their properties when they move to senior facilities due to care needs.

As explained above, urban ageing presents two primary problems that depend on the type of housing tenure: a shortage of rental housing, and empty homes. Fortunately, renters are experiencing an improvement in rental housing options thanks to the government's support, known as 'Housing for Seniors with Services' (Inoue et al., 2013). On the other hand, the government offers support for homeowners through the 'Vacant House Bank', which aims to connect sellers and buyers. However, Tokyo's municipal participation rate in this bank is only 16.1%, much lower than the national average of 55.7% (Ministry of Land, Infrastructure, Transport and Tourism (MLIT), 2023), meaning there could still be a severe increase in vacant properties in the suburbs of Tokyo. Moreover, it must be noted that the mainstream of older households comprises homeowners, who, living alone may eventually require relocation due to their care needs. Therefore, the following section explores the social backgrounds of empty homes, focusing on older homeowners in one-person households.

13.3 Understanding the Social Backgrounds of Empty Homes

According to the 2018 Housing and Land Survey (Statistics Bureau of Japan (SBJ), 2019a), Tokyo has approximately 1.98 million households where people aged 65 and over live in dwellings. Of these, around 70% are homeowners. Additionally, when examining the housing tenure of older individuals living alone, 422,200 households are homeowners, which accounts for about 55% of the group. This data suggests that the proportional share of owner-occupiers is higher than renters among older people.

Looking closely at the older one-person households, a breakdown by gender reveals that women constitute a considerably larger proportion than men. Specifically, 67.3% of these households are composed of women, while men make up only 32.7%. The breakdown also varies by age group, with the highest proportion of late older women (75 and over) at 42%, followed by early older women (65 to 75) at 25.3%. Early older men comprise 17.4% of these households, while late older men account for 15.3%. Notably, the late older women group is particularly sizable, with 185,900 households, more than twice the number of late older men.

Furthermore, when combined with neighbouring prefectures like Kanagawa, Saitama and Chiba, the total number of households belonging to the late older single women group reaches 451,500. Most of these households live in detached houses, accounting for 69.4% (313,200), followed by apartments at 28.7% and tenements at 1.9%. Within the detached houses, only 22.7% are located in the central 23 wards of Tokyo, while the remaining 77.3% are in the surrounding areas (242,200). This uneven distribution suggests a higher concentration of detached houses owned by late older single women in suburban areas. As a result, there is a concern that this could lead to a critical phase of the empty home problem shortly, particularly in the suburbs of the Tokyo metropolitan area.

However, why are most houses owned by late older single women detached and concentrated in the suburbs? The answer lies in the characteristics of their lifestyles because they were the critical generation that played a significant role in changing fertility behaviour in post-war Japan (Nakagawa, 2000). Before this generation, married women usually gave birth to around four or five children on average. In contrast, after this generation, the average number of children has decreased to about two children per married woman (IPSS, 2017). Namely, they were the first generation to seek a more affluent life with fewer children than previous generations. They equipped themselves with various consumer durables, such as household appliances and cars, and aimed to purchase houses in suburban commuter areas.

In other words, those women who pursued affluent lives during the rapid economic growth period, in nuclear households with their spouses and children, were the first bearers of 'my-home-ism', or 'mai hōmu shugi', a way of life widespread in post-war Japan. In Japanese English, 'mai hōmu' has dual meanings: 'one's own family' and 'one's own home'. Based on these meanings, the phrase 'mai hōmu shugi' signifies the social consciousness that places high value on family life and home ownership. Within this context of 'mai hōmu shugi', these women desired detached houses in pleasant urban suburbs as the most significant basis for expressing their happiness at home. It is, therefore, necessary to bear in mind that the empty home problem in suburban areas is closely related to the ageing of the 'my-home-ism' generation (Yamamoto, 2014).

The discussion above has provided insights into the social consciousness contributing to the empty home problem through an analysis of the 'my-home-ism' concept. The following discussion now focuses on the social structural conditions that play a crucial role in this problem. Specifically, it delves into two key aspects: government housing policies and the housing market.

After World War II, Japan was confronted with a severe housing shortage. To tackle this issue, the government implemented several housing policies. These policies included the Public Housing Act and the creation of organisations like the Japan Housing Corporation (now known as the Urban Renaissance Agency) and the Housing Loan Corporation (now part of the Japan Finance Agency). At the time, the emphasis of these policies was on providing enough housing to meet demand. Owing to these efforts, by 1968 the number of dwellings in Japan exceeded the number of households nationwide for the first time (SBJ, 2019b, p. 1). Tokyo also saw a similar trend in the same year. Since then, the gap between dwellings and households has gradually widened, and as a result the vacancy rate in Tokyo currently stands at 10.6%. In addition, the number of vacant dwellings in Tokyo is the highest in the country, with a cumulative total of 809,900. Moreover, when it comes to the Tokyo metropolitan area as a whole, namely together with the three neighbouring prefectures of Kanagawa, Saitama and Chiba, the number of vacant dwellings reaches over 2 million (SBJ, 2019b, p. 8).

The reason for this increase in vacant dwellings can be attributed to the housing market's focus on new construction over second-hand homes. It should be noted that the government's owner-occupation policy has supported this emphasis on the new-build market (Hirayama and Ronald, 2012). In fact, compared with Western countries, the percentage of existing housing in circulation in Japan is significantly lower. Specifically, Japan's ratio is only 13.5%. In contrast, the UK's is 88.8%, the US's is 77.6% and France's is 66.4%, which shows how heavily the Japanese housing market is weighted towards new construction (MLIT, 2012, p. 14).

The prominence of the new-build market is also evident in how homeowners obtain their properties. The Housing and Land Survey (SBJ, 2019c, pp. 9–10) outlines six methods of acquiring a home: (1) purchasing a newly built house; (2) purchasing a used house; (3) newly built (excluding rebuilt); (4) rebuilt; (5) acquired by inheritance or grant; and (6) other. Among these, purchasing a newly built house is becoming common as people's home acquisition method in metropolitan areas. For instance, in the Tokyo metropolitan area, the percentage of people choosing to purchase a newly built house has steadily increased over the years. During the 1960s, it was only about 10%, but in the 1970s it doubled to approximately 20%. Furthermore, by the 1990s, it had reached around 40%, and in the 2000s it accounted for over 50% (SBJ, 2015).

This data indicates that a vast new-build market has formed since the rapid economic growth period. The root cause of increasing vacant housing in urban areas is the development of the housing industry, which has planned, produced and sold new houses as commodities to satisfy the people's desire to own homes. That is, the social structural conditions of the problem of empty homes are the development of the housing industry supported by this new-build market, with the assistance of the government's policies.

Up to this point, the social backgrounds of empty homes with urban ageing have been analysed from two viewpoints: social consciousness and social structural conditions. Based on these considerations, the following section provides insights into a practical solution to address this problem by highlighting the innovative smart home research conducted at Fujita Health University.

13.4 'Robotic Smart Home' as a Solution

In Japan, the improvement of housing for older individuals is lagging behind, leaving people with the difficult choice of staying at home while struggling with daily life and facing 'social isolation' or moving to senior facilities far from home and facing 'social segregation'. To overcome this situation, the concept of 'ageing in place' has recently gained attention. The 'Community-Based Integrated Care System' is being constructed as a practical implementation of this concept, aiming for people to live in their communities until the end of their lives (Nishimura et al., 2013; Institute of Gerontology, 2014). This approach could also help prevent homes from being left empty if older individuals remain in familiar neighbourhoods.

Nevertheless, it is essential to recognise that this community-based approach may not be effective in all areas. For example, urban areas often lack solid social ties among neighbours and relatives, making it difficult for older people to rely on them for daily assistance compared with rural regions. In other words, there is a risk in urban areas that this approach may lead to the result that there are some good places, such as convenient locations and relatively wealthy neighbourhoods, but the rest of the area falls into disrepair. Hence, when examining the problem of empty homes with urban ageing, it is necessary to keep a careful distance from the assumption that building communities one by one is the only solution.

As pointed out in the previous section, the social structural conditions of the empty home problem are the development of a housing industry with an emphasis on new construction supported by the government's housing policies. Thus, to deal with this problem, it is necessary to think about a shift in the housing supply's macrostructure supported by housing policies based on the stock rather than the flow of housing (Sumita, 2007). In achieving this shift, it is essential to address some key areas: (1) developing a renovation industry; (2) formulating a new housing valuation method that moves away from depreciation; and (3) revising the tax and legal system, including property tax. Among these areas, the development of a renovation industry is crucial and directly linked to solving the problem. In this context, Fujita Health University's RSH project presents a promising way to renovate the homes of older individuals.

The RSH is a cutting-edge project in Japan that proposes a new home renovation design for older people that integrates AI, the IoT and robots. In addition, RSH is known as one of the eight research bases nationwide for the Ministry of Health, Labour and Welfare's project to build a platform for developing, demonstrating and disseminating nursing care robots, and is expected to be used for industry–government–academia initiatives. Furthermore, the RSH is also using the results and products of Aichi Prefecture's 'Knowledge Hub Aichi Priority Research Project', 'Demonstration Project for Telerehabilitation and Rehabilitation Robot' and 'Social Implementation Project of Care and Rehabilitation Robot' to develop various devices.

This project aims to develop systems that promote three crucial functions: maintaining a healthy life rhythm, encouraging spontaneity and increasing mobility (Otaka, 2019). As such, to achieve these objectives, the project established two facilities. The first is the living laboratory which serves as the experimental research centre (RSH-EC) and provides a space for assessment. For an authentic assessment, it is crucial to have an experimental facility in a local community where potential users reside. To ensure convenient access for older residents in the area, this centre was established within an apartment at the UR Toyoake complex near Fujita Health University. The complex already houses many older people, with around 1,300 people aged over 65 years out of the 4,000 residents. The second is the development research centre (RSH-DC) on the campus of Fujita Health University, which focuses on developing technologies for activity-assistive devices. The primary objective of this centre is to collaborate with different companies to promote emerging technologies that can be used for future assistive technologies in the RSH-EC. The engineers at the RSH-DC continuously improve devices while conducting assessments in the apartment complex at the RSH-EC (Tanabe et al., 2019, pp. 33–34).

In these centres, research and development of assist systems are being carried out from three angles: (1) a mobility and transfer assist system, (2) an operational assist system, and 3) an information assist system. The objective is to realise an optimal living environment that enables older individuals to live independently at home for longer. First, the mobility and transfer assist system is customised to the degree of disability, with the lifting type for severe disabilities, lateral transfer type for moderate disabilities, and suspension type for mild disabilities. Fujita Health University has developed the Side-Transfer Assist Robot (STAR) that falls under the lateral transfer type. It is designed for individuals with moderate walking difficulties and has electrically adjustable armrests and seats. Additionally, it can rotate effortlessly in any direction and easily move to any location. Elsewhere, Moritoh Corporation has developed a walking support robot that is an example of the suspension type. This robot is intended to prevent falls for individuals with mild walking difficulties. It comes with a motorised turntable installed in the ceiling and automatically moves to guide the user in the desired direction (Tanabe et al., 2019, pp. 32–33).

Secondly, the operational assist system works best when combined with a hand robot and an environmental control system. The hand robot acts as a concierge or butler, communicating with residents and operating various smart appliances based on the user's command. Meanwhile, the environmental control system handles daily tasks through IoT connections, such as opening curtains, turning on the TV, controlling air conditioning and switching lights on or off. From the perspective of rehabilitation medicine, operational assistance related to activities of daily living, such as dressing, eating and toileting, is crucial. It is hoped to be further developed but is challenging to achieve since a hand robot must come close to or touch the user. Despite this challenge, the Toyota Motor Corp has developed a human support robot that can help carry household goods or bags. The synergy between these hand robots and the environmental control system is expected to enhance and strengthen the system (Tanabe et al., 2019, p. 33).

Thirdly, the information assist system facilitates communication between the RSH and remote sites. Users can communicate with staff at a remote medical institution using a television-type communication device. This communication system offers tele-exercise programmes enhanced with computer graphics, making them practical and enjoyable. Moreover, one of the most innovative features of the information assist system is its potential to improve everyday health by combining IoT collection of activity logs with AI analysis to provide individually optimised support. For instance, wearable sensors collect real-time activity and physiological data, such as heart rate and blood oxygen saturation, which can be used to adjust exercise intensity safely for people with respiratory and cardiovascular diseases. A health check toilet is another innovative device that measures urine levels, such as salinity, preventing lifestyle-related diseases. This information provides low-salt dietary instructions for pre-hypertensive patients while monitoring physical functions (Tanabe et al., 2019, p. 33).

The ultimate goal of the RSH project is to integrate these systems to create a comprehensive package for renovating the homes of older individuals. One vital consideration when introducing robots into living spaces is the size of the home. Larger homes, like detached houses, are more suitable for this purpose as they provide robots with more room to move around. However, the RSH project is researching a scale of 75 m^2 (Tanabe et al., 2019, p. 34), similar to the 70 m^2 apartment buildings commonly found in urban areas of Japan. This scale is significant because the first baby boom generation and subsequent generations tend to live in apartment buildings more than the previous generation and will comprise the majority of the ageing population in the coming decades. Therefore, the renovation model developed by the RSH project is adequate for both detached houses and apartment buildings, making it a possible solution to the problem of empty homes.

13.5 Conclusion: Smart Homes From a Social Theory Perspective

The preceding sections have discussed the potential of smart homes as a solution to the housing problem caused by an ageing urban population. Initially, the rapid increase in the number of older individuals living alone in the Tokyo metropolitan area, which could escalate the problem of empty homes, was explained. Subsequently, in the ensuing section, the social background of this problem was explored from the viewpoints of the social consciousness known as 'my-home-ism' widespread in post-war Japan and the development of the housing industry since the phase of rapid economic growth. Precisely, it was elucidated that the problem was closely related to the ageing of the 'my-home-ism' generation, which placed a simultaneous value on home ownership and family orientation. Additionally, it was clarified that the social structural condition was the development of a housing industry focused on new construction, supported by the government's housing policies. Undoubtedly, many owner-occupied homes that had accumulated in the suburbs of the Tokyo metropolitan area are now at risk of becoming empty owing to factors such as people relocating due to ageing. However, the problem could be improved if appropriate renovations were carried out. Thus, as an innovative example of the solution to the problem, the 'Robotic Smart Home' project, a cutting-edge model of housing renovation undertaken by Fujita Health University, was highlighted.

In contrast, this section analyses what transpires in contemporary society by considering the implications of incorporating smart homes into everyday life from a social theory perspective. The most critical point of contention discussed in social sciences regarding smart homes is the issue of privacy. Shoshana Zuboff (2019), in *The Age of Surveillance Capitalism*, delves into this matter by comparing the 'Aware Home' project of George Tech University (Kidd et al., 1999) and the Nest Thermostat project acquired by Google Inc. Zuboff points out that the former enables individuals to control their personal information and safeguards privacy, whereas the latter allows service providers to use personal information as a means of profit making, thereby reducing the substance of privacy to a mere formality. In other words, Zuboff criticises modern corporate giants such as Google, Meta, Microsoft and Amazon for their endless collection of data on individuals and the use of this accumulated big data as a source of profit. She argues for the need to resist this surveillance capitalism in order to restore people's subjective and autonomous rights to their own data. In this sense, Zuboff's argument has an alienation theory's problem structure that emphasises individual subjectivity and sovereignty and aims to restore them, and is a representative form of critical discourse, just as John Kenneth Galbraith (1998) criticised the alienation of individual wants with his concept of the 'dependence effect'.

Upon examining the specific forms of resistance for which Zuboff argues, two possibilities emerge: (1) regulating service providers to make unrestricted information collection impossible; and (2) abstaining from using services associated with smart homes. The former is a matter of legal regulation as a social norm, while the latter is a matter of individual choice as self-determination. Nevertheless, even if legal regulation on privacy is advanced, it is inevitable that some personal information will be disclosed so long as people receive smart home services. Furthermore, the decision to engage with these services ultimately lies with the individual and is a matter of personal choice.

A practical example of the need for such services can be seen in home care services for older people. Many families fear that their older parents may experience falls or unforeseen events at night, which means there is a compelling need to seek out smart home services such as sensors and monitors to care for and look after their parents. Despite privacy concerns, the use of these services is anticipated to increase. As is clear from the case of the 'Robotic Smart Home' project discussed in the previous section, (1) support for residents' activities by robots, (2) home automation through the IoT connections between devices, and (3) information gathering through monitoring and AI analysis, have already been researched and developed. Given the current state of research and development, it is essential to note that there is room to discuss the issue of the collection, management and operation of big data in a different way from Zuboff's argument. In Zuboff's way of framing the issue, (1) the problem is simplified and presented as data exploitation by giant corporations; and (2) enlightenment-style hyperbole and abstraction can be seen in an attempt to lead the way in people's understanding. In this context, there is a need to keep a careful distance from the problematic construction of 'surveillance capitalism'.

Indeed, it is worth noting that the movement towards the 'improvement of the living environment' surrounding the residents' bodies is in tandem with the creation of a new giant industry that utilises big data on the body obtained through the introduction of advanced technologies such as AI, IoT and robots into living spaces, specifically via various sensors and monitors. Hence, the crucial question to consider is what people are really being made to consent to, through the form of self-determination, when they sign privacy-related documents to receive services. People are usually unaware of this issue and willingly accept services in exchange for certain conveniences. However, this implies a fundamental transformation of the body in contemporary society, which is because it is synonymous with legitimately enabling the collection of massive amounts of data from bodies placed in residential spaces and voluntarily authorising the quantified and digitised body spaces to become a site for investing and developing vast amounts of capital.

When giving thought to this matter, a pertinent reference point is the issue of organ transplantation from the body after brain death. As Ryuzo Uchida (1987) astutely observed in his work *Consumer Society and Power*, the issue of brain death implies that by localising the definition of death to a specific partial body, whether the brainstem or the cerebrum, all other partial bodies are legitimately transformed into sites of capital investment and development through self-determination in the form of a declaration of will before death or consent by a family member. While it is undoubtedly true that the wish of patients' families for organ transplants is sincere, it is also a fact that a medical industry around transplantation technology has emerged in parallel, centred in the USA and advanced countries. As such, it is vital to consider the isomorphism of the issues between brain death and smart homes.

In other words, what these concrete and empirical phenomena occurring in contemporary society suggest is that the realm of the body, which is the correlative term of 'biopower' analysed by Michel Foucault (1994), is doubled into two levels, being superimposed with divergences between them: (1) the level of the actual individual body, that is, the dimension involved in the reality (self-determination) of the 'subject'; and (2) the level of the partial or digitised body, that is, the dimension that is not involved in the reality of the 'subject' (Uchida, 1997). It should be noted that while the level of (1) has become a secondary matter, the expansion of the level of (2) plays an inevitable role in the driving force of the transformation of capitalism, that is, a new source of profit making. Significantly, this expansion of level (2) raises the question of whether contemporary society is entering a phase in which 'biopower' is losing its relative effectiveness.

Modern society is founded on the free and autonomous agency of the individual, as typified by the legal system. 'Biopower' makes this indivisible unit called the 'individual' the object of management and demands a standardised form of self-control called the 'subject' from people in line with this unit. Hence, 'subject' refers to the mode of existence or life required of the 'individual' formed on the actual body. However, what is occurring in contemporary society is that the realm of reality that does not belong to this mode of 'subject' is rapidly expanding. In this regard, Anthony Elliott (2022, p. 157) points out that there is "a movement from the social individual to the specific data composite of singular cases", which is "radically indifferent to actual individuals". The most significant technology supporting this expansion from behind the scenes is information technology, and herein lies the inevitability of the information society. In essence, the current grasp of the body is taking place in a manner that conflicts as little as possible with the positivity of the 'individual' as required by the existing social system, and this is made possible by the level of the digitised body mediated by information technology.

Therefore, considering the phenomenon of smart homes from the perspective of social theory, a new question arises: how is the topological change of capitalism, mediated by information technology, correlated with the alteration of the essence of power? In short, an analysis of the transformation of power in contemporary society is required as a current theoretical challenge.

References

Cabinet Office (2018a). *Annual report on the ageing society: 2018*. Cabinet Office, Japanese Government. https://www8.cao.go.jp/kourei/english/annualreport/2018/pdf/c1-1.pdf.

Cabinet Office (2018b). *Heisei 30nendo kōreisha no jūtaku to seikatsu kankyō ni kansuru chōsa kekka (zentaiban)* [2018 Results of a survey of older people's attitudes to housing and living environment (full version)]. Cabinet Office, Japanese Government. https://www8.cao.go.jp/kourei/ishiki/h30/zentai/pdf/s2.pdf.

Cabinet Office (2021). *Annual report on the ageing society: 2021*. Cabinet Office, Japanese Government. https://www8.cao.go.jp/kourei/english/annualreport/2021/pdf/2021.pdf.

Elliott, A. (2022). *Making sense of AI: Our algorithmic world*. Polity.

Foucault, M. (1994). *Histoire de la sexualité, tome 1: La volonté de savoir*. Gallimard.

Galbraith, J. K. (1998). *The affluent society*. 40th Anniversary edition. Houghton Mifflin.

Hirayama, Y., and Ronald, R. (2012). *Housing and social transition in Japan*. Routledge.

Institute of Gerontology, University of Tokyo (ed.) (2014). *Implementing 'integrated community care system' practical guidelines of multi-professional teamwork for home medical care* [in Japanese]. University of Tokyo Press.

Inoue, Y., et al. (2013). *Sābisutsuki kōreishamuke jūtaku no igi to tenbō* [The significance and prospects of housing for seniors with services]. Taisei Shuppansha.

IPSS (2013). *Regional population projection for Japan 2010–2040*. IPSS. https://view.officeapps.live.com/op/view.aspx?src=https%3A%2F%2Fwww.ipss.go.jp%2Fpp-shicyoson%2Fe%2Fshicyoson13%2Fkekkahyo_pref.xlsx&wdOrigin=BROWSELINK.

IPSS (2014). *Nihon no setaisū no shōrai suikei (todōfukenbetsu shōrai suikei): 2010–2035* [Household projections for Japan by prefecture: 2010–2035]. IPSS. https://www.ipss.go.jp/pp-pjsetai/j/hpjp2014/houkoku/houkoku.pdf.

IPSS (2017). *Population statistics of Japan 2017*. IPSS. https://view.officeapps.live.com/op/view.aspx?src=https%3A%2F%2Fwww.ipss.go.jp%2Fp-info%2Fe%2Fpsj2017%2FPSJ2017-04.xls&wdOrigin=BROWSELINK.

IPSS (2018). *Household projections for Japan 2015–2040: Outline of results and methods*. IPSS. https://www.ipss.go.jp/pp-ajsetai/e/hhprj2018/hhprj2018_DL.

Kidd, C. D., et al. (1999). The aware home: A living laboratory for ubiquitous computing research. In *Proceedings of the second international workshop on Cooperative Buildings, CoBuild'99* (pp. 191–198). https://www.cc.gatech.edu/fce/house/cobuild99_final.html.

MLIT (2012). *Heisei 23nendo tochi ni kansuru dōkō, Heisei 24nenndo tochi ni kansuru kihonteki Shisaku: Yōshi* [Trends on land in 2011 and basic policies on land in 2012: Executive summary]. MLIT, Japanese Government. https://www.mlit.go.jp/common/000213889.pdf.

MLIT (2023). *Todōfukenbetsu 'zenkokuban akiya, akichi banku' sanka hyōmei jyōkyō* [Participation in the 'national bank for vacant houses and land' by prefecture]. MLIT, Japanese Government. https://www.mlit.go.jp/common/001470799.pdf.

Nakagawa, K. (2000). *Nihon toshi no seikatsu hendō* [Life changes in Japanese cities]. Keiso Shobo.

Nakagawa, M. (2001). Measuring age discrimination in rental housing markets using fair housing audits [in Japanese]. *Urban Housing Sciences*, *35*, 21–26.

Nishimura, S., et al. (2013). *Chiiki hōkatsu kea shisutemu: 'suminareta chiiki de oiru' shakai o mezashite* [Community-based integrated care system: Towards a society where people can 'age in their own familiar community']. Keio University Press.

Otaka, Y. (2019). Kōfuku chōju na kurashi o kanaeru shizen ni katsudōteki ni naru sumai no kenkyū kaihatsu: AI/IoT o katsuyōshita jihatsuteki katsudō no yūdō [Research and development of homes where older people can be naturally active to fulfil a life of happiness and longevity: Inducing spontaneous activities using AI/IoT]. In *Senshinteki ēai, aiōtī, biggu dēta, katsuyō gijutsu kaihatsu purojekuto* [Development of advanced AI, IoT and big data application technologies project]. Knowledge Hub Aichi. https://www.astf-kha.jp/project/project2/files/astf_PI_08_0301_compressed%20_new.pdf.

SBJ (2015). *2015 housing and land survey.* Ministry of Internal Affairs and Communications, Japanese Government. https://www.e-stat.go.jp/en/stat-search/files?stat_infid=000028655896.

SBJ (2019a). *2018 housing and land survey.* MIC, Japanese Government. https://www.e-stat.go.jp/en/stat-search/files?page=1&layout=datalist&toukei=00200522&tstat=000001127155&cycle=0&tclass1=000001129435&tclass2=000001129436&stat_infid=000031865753&tclass3val=0.

SBJ (2019b). *Heisei 30nen jūtaku tochi tōkei chōsa kekka no gaiyō (jūtakusū gaisū shūkei)* [Summary of the results of the 2018 Housing and Land Survey (approximate number of dwellings)]. MIC, Japanese Government. https://www.stat.go.jp/data/jyutaku/2018/pdf/g_gaiyou.pdf.

SBJ (2019c). *Explanation of terms of the 2018 survey.* MIC, Japanese Government. https://www.stat.go.jp/english/data/jyutaku/pdf/30terms.pdf.

Sonoda, M. (2010). Solutions and strategies to manage and re-invest real estate owned by elderly people at suburb [in Japanese]. *Japanese Journal of Real Estate Sciences, 23*(4),46–53.

Sumita, S. (2007). *Nijūisseiki no haujingu: 'Kyojū seisaku' no kōzu* [Housing in the 21st century: The composition of 'residential policy']. Domes Shuppan.

Tanabe, S., et al. (2019). Designing a robotic smart home for everyone, especially the elderly and people with disabilities. *Fujita Medical Journal, 5*(2), 31–35. https://www.ncbi.nlm.nih.gov/pmc/articles/PMC8766226/pdf/fmj-5-031.pdf.

Tokyo Metropolitan Medical Examiner's Office (2011). *Tōkyōto nijūsanku ni okeru kodokushi tōkei (heisei15–19nen): Setai bunruibetsu ijōshi tōkei chōsa* [Lonely death statistics in the 23 wards of Tokyo (2003–2007): Statistical survey of unnatural deaths by household category]. Bureau of Public Health, Tokyo Metropolitan Government. https://www.fukushihoken.metro.tokyo.lg.jp/kansatsu/kodokushitoukei/kodokushitoukei.files/04_dai2syou.pdf

Tokyo Medical Examiner's Office (2020). *Tōkyōto kansatsu imuin de toriatsukatta jitaku jūkyo de nakunatta tanshin setai no mono no tōkei (reiwa 2nen)* [Statistics on persons from one-person households who died in their own homes handled by the Tokyo Metropolitan Medical Examiner's Office (2020)]. Bureau of Public Health, Tokyo Metropolitan Government. https://www.fukushihoken.metro.tokyo.lg.jp/kansatsu/kodokushitoukei/kodokushitoukei-2.files/2-tokubetuku.pdf.

Uchida, R. (1987). *Shōhi shakai to kenryoku* [Consumer society and power]. Iwanami Shoten.

Uchida, R. (1997). Kenryoku no episutēmē [Episteme of power]. In S. Inoue, et al. (eds), *Gendai shakaigaku no riron to hōhō: Iwanami kōza gendai shakaigaku (bekkan)* [Theory and methods in contemporary sociology: Iwanami lectures on contemporary sociology (separate volume)] (pp. 94–99). Iwanami Shoten.

Yamamoto, R. (2014). *Mai hōmu shinwa no seisei to rinkai: Jutaku shakaigaku no kokoromi* [Generation and criticality of my home myth: an attempt at the sociology of housing]. Iwanami Shoten.

Zuboff, S. (2019). *The age of surveillance capitalism: The fight for the human future at the New Frontier of Power*. Profile Books.

Ross Boyd

14 Narrative Identity, Disruptive AIs and Work: Outline of a Prospective Research Approach

Abstract: Asking what a focus on identity might add to contemporary debates over the implications of artificial intelligence for the future of work and employment, I adapt Paul Ricœur's notion of narrative identity as a lens to outline a research agenda directed at the lived experiences of digital disruption in the workplace. I specifically foreground the processes through which working people are seeking to find a place for themselves within, comprehend and apprehend these more or less fluid situations in order to render them intelligible, manageable and navigable. The chapter then explores a number of contemporary studies of experiences of digital disruption on the part of workers who have made significant identity investments in their technological expertise in order to indicate the ways this type of research might proceed.

Keywords: AI, future of work and employment, work identities, digital disruption, lived experience

14.1 Introduction

How might a focus on 'Identity' add value to debates over the implications of artificial intelligences (AIs) for work and employment? AIs, construed not in terms of any discrete technology *per se* but rather a class of technologies encompassing various machine systems designed to perform tasks that previously required some element of human cognition, are currently being applied across a range of work settings. Be it basic sorting and retrieval software, smart virtual personal assistants, large language model report compilation/writing or supported/automated decision systems in administration; automated portfolio management and equities trading in finance; computer-aided design, advanced robotics and complex I.40 systems in manufacturing; or task allocation and oversight algorithms in the 'platform economy', AI applications are now pervasive presences in much of contemporary working life. Likewise, debates over the disruptive impacts of AIs on the organisation of economic life, the implications of this for work futures and employment, along with possible consequent social transformations, have maintained high levels of scholarly and public attention for over a decade (Boyd, 2021; Boyd and Holton, 2018).

Ross Boyd, University of South Australia, Australia

https://doi.org/10.1515/9783110721751-014

"Identity", Kobena Mercer (1990, p. 43) famously wrote, "only becomes an issue when it is in crisis, when something assumed to be fixed, coherent and stable is displaced by the experience of doubt and uncertainty". It ought not surprise, therefore, that the notion of identity and work being entangled, while a long-established trope in social theory (Savage, 2005; Dejours et al., 2018) and the sociology of work (Kalleberg and Leicht, 2021; Halford and Strangleman, 2009), as well as organisation, management and human resources studies (Bailey et al., 2019; Brown, 2022; Budd, 2019), has emerged as a topic of concern for students of work and labour markets in certain times more so than others. Given these well-honed connections and current claims about the extent of disruptive AIs in and of the organisation of work, employment, skills acquisition and so forth, the times should thus be propitious for a resurgence of interest in questions concerning identity among the various social science and humanitics disciplines and sub-disciplines that take work as their objects.

In this chapter I attempt to fashion one type of approach to identity that enables connections to be forged between identity studies and contemporary debates over the disruptive impacts of AIs on the organisation and experience of working life. It is my contention that such an 'identity lens' will enable us to develop a research agenda that moves well beyond the twin emphases on body counts (job losses/gains) or the ergonomic recalibration of human–machine systems (e.g. augmentics) that have tended to dominate scholarly and public debates on AIs and the future of work to date. This frees us to engage with questions of lived experience and sense making that connect work to issues of self-experience, self-understanding and self-expression. And to the processes, cultural resources and skills that workers engage and employ to make sense of, go on in and shape a world of work marked by change, contingency and (at times acute) uncertainty.

14.2 An Identity Lens?

14.2.1 Contingency

Identity and its affiliate concepts—self, self-identity, subjectivity—are now well-established topics of concern in social and cultural, and indeed political and economic, analysis (Svetlova, 2022). They nevertheless remain highly complex, pliable and contested, in some parts due to the sheer diversity of analytic tasks to which they have been set to work, in other parts on account of the dynamics of identities themselves (Elliott, 2020). To ask after the usefulness of an identity lens, as this chapter does, thus requires some initial specification as to just what one intends by way of such a lens.

Towards this end let us first tug on the central thread of thought that underpins Mercer's statement on identity (above) and see how it unravels. Thinking about identity, and as such identity itself (inasmuch as self-knowledge is constitutive of the self), is contingent. We do not always think about identity, be this in the daily round, in the workplace or in scholarly analysis. In familiar gestaltist terms identity is sometimes *figure*, standing out against *ground*, and at other times sunk into the *ground* against which other concerns stand in stark relief. For Mercer identity becomes *figure*, and hence more fully available for thought, in moments of crisis when identity itself, along with its supportive surround, is somehow thrown out of joint. Yet, as Holton (1987) reminded us over 30 years ago (in a period wherein crisis seemed to be a defining feature in so many social-theoretical diagnoses of the 'mood of the times'), a crisis is not a permanent condition. It is, rather, a decisive point beyond which some disruption of or departure from normality resolves upon a (more or less) new normal. In other words, crisis is provisional just as its appearance alerts us to the provisional nature of the normal. Or, we might say, just as the stability of identity is contingent, so too is its disruption.

Contingency, in its original Medieval Latinate root *contingenten*, also means both 'to touch' and 'to happen to one' or 'befall'. Hence the sense that something is contingent insofar as it occurs through things coming together, in often unexpected ways, by chance and not necessity. This brings a more spatial and social aspect to the temporal ('until further notice') sense of the contingency of identity that the notion of crisis lights up. Here we might think of the often-dense thickets of affinities and affiliations, of relations of sameness and difference, of trust, responsibility, recognition, accountability and so forth that form significant elements of identities, in both personal and collective senses. Taking these temporal and spatial/social aspects together identities can be presented as concatenated—more or less loosely constellated ensembles of relationships and characteristics that hang together sufficiently to form recognisable yet dynamic and mutable entities.

There is at least one more, complementary, way we might approach this connection between contingency and identity. In his comprehensive narrative theory of identity, introduced initially in the third volume of *Time and Narrative* (Ricœur, 1988) and developed subsequently in *Oneself as Another* (Ricœur, 1994), Paul Ricœur provides some initial reference points for roughing out a conceptual frame in which questions of identity are fundamental to the processes by which people seek out ways of making sense of, coping with, getting by or flourishing amid conditions of lived uncertainty. And through this participate, in smaller or larger measures, in the visualisation and bringing into being of some futures from among the wider (but not limitless) range of possible futures available at any particular time. It is my view that such an approach may serve to light up some of the lesser explored, but by no means less significant, questions concerning claims about transformations of work in the context of disruptive AIs, and hence help to organise new research directions. Some work is required, however, to get us to this point.

The choice of Ricœur as my starting point might, at first blush, seem curious for a number of reasons, of which two in particular are worth dwelling upon. First, Ricœur's interest in technology was at best marginal. On the rare occasions when he did write on the topic it tended towards the rather stark types of contrast characteristic of well-known analyses by Marcuse, Habermas and Heidegger in the 1950s and 1960s: that is, between a dehumanising technological rationality on the one hand, and more humane ways of organising experience and action on the other (Kaplan, 2006). Yet, while these explicit references to technology are certainly out of step with more recent shifts (on the part of postphenomenological, Science and Technology Studies and Critical AI Studies scholars, among others) towards an emphasis on the co-constitution of the technological, the social and the cultural (Holton and Boyd, 2021; Roberge and Castelle, 2021), the significant affinities that exist between Ricœur's hermeneutics in general and the various approaches that make up contemporary social science of technology are now starting to be grasped and their value recognised. Having bluntly acknowledged the shortcomings in Ricœur's observations on technology ("unoriginal, limited, dated and false") Kaplan (2006, pp. 47–53), for instance, identifies at least four contributions a reading of Ricœur might add to the contemporary study of technology. These include: (i) the extension of Ricœur's ideas of the mediated character of human experience to cover experiences of technology; (ii) the relevance of the model of the 'hermeneutic arc' to analyses of the tensions between the technical properties of a thing and its multiple social meanings; (iii) the importance of Ricœur's narrative theory to an understanding of the many and varied ways technologies figure in our lives and self-understandings; and (iv) the contributions Ricœur's moral theory can make to contemporary debates over the ethical development and deployment of technologies.

Coeckelbergh and Reijers (2016) have, more recently, taken up Ricœur's narrative theory to explore the ways technologies shape our understandings of the world. Here technologies such as blockchain and distributed ledgers are seen to possess capabilities for configuring events into meaningful wholes that bear much in common with the ways a plot organises the elements that comprise narrative understandings of the world (see below). This interest in Ricœur's work would subsequently extend to the establishment of a Hermeneutic Philosophy of Technology research programme wherein Ricœur is aligned with a postphenomenological insistence that human transactions with the world are always technologically mediated (Romele, Reijers and Coeckelbergh, 2021). Beyond this, Roberge and Lebrun (2021) look to Ricœur to aid with understanding and problematising—at the level of technical performance as well as culturally, economically and politically—the hermeneutical claims made on behalf of machines engaged in meaning making (those now usually grouped under the head of generative AI). What these works point to is the potential salience of Ricœur to a discussion of identities emerging within the processes by which people endeavour to make sense of the circumstances of their living, and living together, under conditions of technological change—conditions that disrupt identities that once may have been taken as (relatively) fixed or settled.

A second reason for regarding the choice of Ricœur a curious one, at least initially, has to do with his contributions to debates on political, social and economic matters.

The point here is not that these contributions were rare (or dated, unoriginal . . .). They weren't. From his early statements on mass violence and war (clear-eyed yet with a deep conviction rooted in the most bewildering and painful childhood and adolescence), through to his reflections on a united Europe and, in some of his last works, economic crisis, he maintained an active engagement with the concerns of the world in which he lived. Nor are these contributions, as is the case with the bulk of his 'scholarly' work, unduly 'difficult' or 'obscure' such that any practical purchase they may have on issues like the future of work gets frustrated. Across the vast majority of his works Ricœur writes from a position on the near side of fragile, fallible, hopeful and imaginative humans engaged in concrete practices anchored in institutions (like work) that, whatever else they may do, rehearse human capabilities and will for living life with, by and for others in the face of historical forces of often tremendous transformative power. This is, I would hasten to add, a far less abstract rendering of the human than the strange, rather jaundiced, caricature of the worker that has customarily peopled labour economics, one for whom "work effort is necessarily a source of disutility, and that effort is only provided by a worker in return for a wage" (Susskind, 2023, n. p.).

At the same time it is the case that Ricœur writes in an entirely different register from that in which contemporary debates over work futures most often take place. His writing just doesn't sound like it belongs to a discourse of organisational realignments, personalised portfolios, upskilling, digital dexterity, hybrid work and the like. This is because, stated baldly, it doesn't belong there. It certainly holds some affinities with contemporary research on the relationship between meaning and work (Susskind, 2023; Bailey et al., 2019; Brown, 2022). Still, its strongest sympathies lie, I believe, with an earlier generation of intellectuals, the historian E. P. Thompson without doubt the most prominent, who invited—urged—us to pay close and conscientious attention to the lived experience of transformations wrought by the Industrial Revolution; "the friction of interests—the movement itself, the heat, the thunderous noise" (Thompson, 1966, p. 357). Here an adapted version of Ricœur's narrative self, as an 'identity lens', allows for the conceptual framing of an approach to researching work futures that might explore for the digital transformations of our own time something of what Thompson and others sought to document for this earlier period. An approach that may capture, as Strangleman (2017, p. 470) neatly sums, "not simply the shock of the new and the eclipse of the old, but also the process by which the new was rendered intelligible, accepted, normalised and in turn acted upon". It is precisely this 'process' that Ricœur allows us to model with his concept of the narrative self.

14.2.2 A narrative theory of identity

The broader narrative theory that anchors this concept can be traced back at least as far as Ricœur furnishing his initial phenomenological method with a hermeneutic complement in response to his disclosure of the force of cultural narratives in *Symbol-*

ism of Evil (1967). However, it receives its most extensive elaboration across the three volumes of *Time and Narrative* (Ricœur, 1983, 1985, 1988). The addition of a 'hermeneutic complement' itself was required by the recognition that there is no unmediated access to the world—that even the most immediate of experiences is always grasped by way of the detour of 'Experience' (to play on the double-barrelled sense of this term). Or, more specifically, 'Experience' as captured in texts; basically anything that carries knowledge or meaning and possesses sufficient durability and extension to achieve a resonance and amplitude beyond the narrowly proximate situations of their first moment. While it is the case that Ricœur works extensively with literary texts (as "laboratories for thought experiments"; Ricœur, 1994, p. 148), the term can cover, among other things, the nesting of actions in practices, artefacts, film and other audio-visual media, Coeckelbergh and Reijers's (2016) narrative digital technologies, figurative devices such as cosmograms (Tresch, 2005), modelling tools or calculative techniques (Svetlova, 2022; Beckert and Bronk, 2018), through to the recounting of stories, proverbs or maxims that Walter Benjamin and Kenneth Burke famously viewed as vital to intergenerational and cross-cultural transmissions of practical wisdom and its counsel.

This shift from immediate experience to the Experience bound up in texts has several corollaries. For one thing, as the term 'hermeneutic complement' suggests, it brings interpretation continuously into play. Texts (including practices and so forth), as parts of culture, carry meaning independent of specific authors, agents, intentions or situations, and, as such, our understandings of them are only ever provisionally stable and fixed—but they are also, importantly, 'ours'. They are therefore polysemic, meaning different, possibly conflictual interpretations are always possible. For another, it places narrative and the operation of emplotment at the centre of the cultural practices through which we render our worlds intelligible and our actions and interactions within them meaningful. It is this aspect of Ricœur's work I want to emphasise most here.

While the cultural contexts in which social life is embedded are narratively prefigured, they are by no means static. If this were the case they would not serve our attempts to understand and act in a dynamic, uncertain world particularly well. Indeed prefiguration, for Ricœur (1983), merely indicates that thought and action always commence in a cultural context that has already been narratively organised and structured. But this structuring is itself the product of prior processes of active narrative configuration and refiguration. A narrative, be it the remembrance of events that actually took place or imagined futures and alternative scenarios, is not the same as causal sequences that move neatly and swiftly from initial states to terminal ones consistent with objective probability functions. Narrative in contrast *participates*, through the operation of emplotment, in the configuring of a manifold of heterogeneous events involving agents and projects, shared situations, intentions, chance encounters, unintended consequences, expectations confirmed or thwarted, setbacks, opportunities taken or foregone, predicaments worked through (successfully or otherwise) and so forth into a coherent whole. A whole that, as it advances towards its denouement, deepens and transforms our understandings of the worlds in which we live, those we share themwith and our selves (Ricœur, 1994, p. 141). And, in this refiguring of our understandings and expectations of or beliefs about the world, emplotment plays an active role in the development and organisation of new repertoires of behaviour that in turn act upon reality itself.

The operation of emplotment is central to the concept of narrative identity. That narrativity figures in the constitution of the self is not a particularly startling proposition. We can readily grasp the idea that our identities take shape in the stories (biographies) we tell about ourselves as well as those told by our families, partners, friends, colleagues, acquaintances lovers, rivals, enemies . . . (Kaplan, 2006, p. 45). And the idea that biographies often reference bigger stories of nation, class, gender, sexuality, race, religion and other collective identifiers. Ricœur certainly wrote on such themes; however, there is something more important, for mine, in his understanding of narrative identity that I want to draw out.

In the work in which he develops his thinking on identity to the fullest, *Oneself as Another*, Ricœur looks to his narrative theory to confront the 'problem of personal identity'. The classic philosophical problem, that is, of how to define a single person inhabiting a dynamic world over time—are they the same or have they changed (and if the latter according to what criteria, where do the relevant thresholds lie . . .). Ricœur (1994, pp. 115–125) casts this problem in terms of the dialectic of sameness and selfhood, where sameness is associated with permanence in time and selfhood with variability, discontinuity and diversity. Casting this in familiar language, we can and do entertain the situation wherein we have a sense of ourselves as both the same person that we were many years ago as well as someone who has undergone notable changes: we physically age (and mature), change our appearance, expand the range of our experience, forge connections with new people, acquire new knowledge, skills and tastes, take on new perspectives, beliefs and values—such that we may also quibble or bristle over suggestions that we are indeed identical to our old self. This is not just a problem for professional philosophers but, I would add, one confronted by parole boards, eyewitnesses and, indeed, partners of long stead. For Ricœur it is the operation of emplotment, with its own paradoxical characteristics, that enables the distinctions between sameness-identity and selfhood-identity to be finessed.

> The paradox of emplotment is that it inverts the effect of contingency, in the sense of that which could have happened differently or which might not have happened at all, by incorporating it in some way into the effect of necessity or probability exerted by the configuring act. The inversion of the effect of contingency into an effect of necessity is produced at the very core of the event: as a mere occurrence, the latter is confined to thwarting the expectations created by the prior course of events; it is quite simply the unexpected, the surprising. It only becomes an integral part of the story when understood after the fact, once it is transfigured by the so-to-speak retrograde necessity which proceeds from the temporal totality carried to its term. This necessity is a narrative necessity whose meaning effect comes from the configurating act as such; this narra-

> tive necessity transforms physical contingency, the other side of physical necessity, into narrative contingency, implied in narrative necessity. (Ricœur, 1994, p. 142)

I have quoted this passage at length as it encapsulates the key ideas that I want to take forward, as the rudiments of a particular type of 'identity lens', to a discussion on work futures and the future of work in the context of transformations attendant upon the introduction of various forms of AIs into workplaces. Specifically, I want to draw attention to the ongoingness of processes of emplotment that, through configuring lived contingency into narrative necessity, people perform in order to maintain some sense of the constancy and meaningfulness of their self and the world they share with others in spite of uncertainty and change. And in doing so work to sustain a measure of confidence that the world remains a place in which they have some sort of stake and which they can still navigate through, even if precisely how they might do this requires some figuring out.

14.2.3 Contingency, narrativity and expectations around work futures

To return to my initial question about the usefulness of an identity lens, the specific 'lens' roughed out here directs researchers to take up positions amid the zones of uncertainty when and where they present within the specific circumstances of given workplaces and projects that can be said to be undergoing 'AI disruption' (Girasa, 2020) of some form and at some scale. And to do so in a manner sensitised or tuned-in to those processes through which working people are seeking to find a place for themselves within, comprehend and apprehend these more or less fluid situations in order to render them intelligible and manageable. And hospitable: amenable to the types of decision making, choice and commitment by which people explore, experiment with, give expression to and experience their 'selves' and their living as meaningful through the places, projects and purposeful activities they share with others.

This task of researching thought and action in the subjunctive or conditional mood certainly presents unique challenges, many of these issuing from the 'restlessness of events' and the fragility of scholarly holds on the passing moment (Wagner-Pacifici, 2000; Abbott, 2007; Luhmann, 1985). It is (too) often the case that an account will either rule a line under the present moment, name it and thus endow it with an unduly finished quality, or rush ahead of the moment towards anticipated outcomes and repercussions. In both instances the move to hammer lived contingency into narrative necessity tends to be both hasty and heavy handed, paying scant regard for the experiences of those living through and trying to make sense of the given moment. The contrasting approach is to remain with the given event-in-process, to chart how it advances from instances of technological and organisational disruption (to work situations, work practices and self-understandings organised around work) and disorientation through to where, amid the play of a diverse array of elements (organisations and practices, policy frameworks, the agents and patients of action), new forms of interaction might emerge and gradually take shape along with new ideas, controversies, power differentials and institutional arrangements (Wagner-Pacifici, 2019). And the surfacing of new aspects of the self as the activity of people working out how they might fit into and effectively operate within changed worlds of work gradually transforms that which is initially a source of shock and surprise into the more familiar and routine. Here attention to the ongoing processes of narration allows us to explore the entailments and effects of workplace disruption from the inside, as it were. That is, from the vantage point of personal experience and expectation. And to gain insights into the cultural skills and knowledge people draw upon, develop and deploy as they seek to creatively adapt to changing conditions of work and employment.

These are capabilities and capacities not often considered within the vocational pedagogies of 'digital skills', even in their most expansive versions encompassing so-called 'soft' skills or facility with 'change management'. They are nonetheless important. The significance of narration to managing uncertainty and bringing imagined socio-technical futures into being through organising expectations, orienting and coordinating the behaviour of a diversity of actors or marshalling resources has been analysed for quite some time now (Mosco, 2005; Jasanoff and Kim, 2015; Bareis and Katzenbach, 2022). More, in the aftermath of the 2008 Global Financial Crisis—which demonstrated that people's beliefs about their situations truly matter (in so doing sorely exposing the inadequacies in the rational expectations hypothesis that had dominated economic modelling for decades)—economic anthropologists, economic sociologists and behavioural economists, among others, moved to develop new understandings of expectation formation. Most of these emphasise the importance of narratives and imaginaries to the realisation of futures—aided by calculative devices and modelling techniques (Svetlova, 2022; Beckert and Bronk, 2018).

Yet, while recognising the important connections existing between narration and the meanings people attribute to or commitments they make to action, attention in this area of research has tended to centre upon the already formed narratives of 'key players' (e.g., central banks, stock exchanges, professional communities and cultures, market regulators). In addition, and as this list of players indicates, the lion's share of research in this area has been conducted with a view to the increasingly important role algorithms play in the operation of finance and equities markets (Svetlova, 2022), with surprisingly little attention paid to the highly topical debates over AIs and the future of work (Vicsek, 2021). It is my contention that our understandings of expectation formation in times of change, disruption and uncertainty, and of the impacts of expectations on action orientations and through these future possibilities (and possible futures), are lacking in the absence of attention to the everyday worlds of work—given that these worlds provide some of the most significant sites where people experience technological change impacting their life chances. That is, this attention matters because it bears upon economic, social and personal outcomes that are of considerable concern to many people.

In the following section I want to rough out a model of research aimed at identifying and documenting the processes by which workers are seeking to render their experiences of AI-induced workplace change available for thought and action. This discussion is meant to be indicative and illustrative of prospective research, not an exhaustive review of a research programme that is already operational.[1]

14.3 Work, Digital Disruption, Narrativity and Identities

14.3.1 Narrating workplace change

While written off the back of research into the digital disruptions of the 1990s, Richard Sennett's (1998) *The Corrosion of Character* rightly remains one of the more influential theoretical reference points for studies of the narrative constitution of identities in the context of workplace change in our times (Brown, 2022; Du Gay et al., 2019; Cannizzo, Mauri and Osbaldiston, 2019; Potter, 2015). I believe, however, that there are good reasons for exercising a degree of caution in relation to Sennett's framework when exploring contemporary experiences of digital disruption to work and identity.

In the penultimate chapter of this rightly celebrated study Sennett recalls concerns about a 'loss of career' expressed by Walter Lippmann in his 1914 *Drift and Mastery*. Here the term 'career' is used in its oldest sense, in many ways similar to a 'vocation' or a 'calling'. Writing in a period of acute uncertainty just prior to the outbreak of the First World War (and, I would add, with the 'Panics' of 1893 and 1907 still reverberating through the US economy and society), Lippmann saw in the idea of 'the career'—that is, in the industrious, skilful and purposeful fashioning of a sturdy narrative out of one's working life—the means by which people might regain mastery over lives afflicted by a sense of aimlessness and inadequacy. Sennett's discussion comes in the closing stages of his analysis of the final decades of the twentieth century, a period marked by the 'downsizing', 'rightsizing' and re-engineering of firms and workplaces aimed at inducing sufficient organisational flexibility to enable rapid responses to changing demand in capricious, winner-take-all global marketplaces. A period in which pliancy, fleetness of foot and facility in the arts of connecting and disconnecting with projects and people at will attracts a premium. And a period marked by a hope that success measured solely in dollars would compensate for any "failure

1 This chapter does incorporate archival research conducted as a part of the ARC-funded project *Investigating Conversational AI: Development, Usage and Governance* (DP220102630).

to make one's life cohere, failure to realise something precious in oneself, failure to live rather than merely exist" (Sennett, 1998, p. 119).

Against this backdrop Sennett recounts his experiences with a group of middle-aged mainframe programmers and systems analysts who had been 'downsized' by IBM in the early 1990s. These were people who had subscribed to Lippmann's idea of a career, and who expected that their demonstrated commitment to company and accumulated skill sets would continue to serve them well in the newly emerging tech-rich world. And these were people who would ultimately be blindsided by the organisational disruption that followed IBM's decision to commit to mainframe over personal computing technologies in the early 1990s (well before 'Deep Blue' and 'Watson' revived IBM's fortunes). For several years Sennett would join the engineers as they worked through what had happened to them, sorting out their histories over afternoon coffees at the River Winds Café. After several early iterations drawing upon well-worked tropes—premeditated corporate betrayal, underbidding by foreign workers and asset stripping by 'corporate raiders'—their narrativisation would resolve upon a more resigned reflection on the realities of their times and what, in hindsight, they should not and should have done. They blamed "themselves for having been too company-dependent, for having believed in the promises of corporate culture, for having played out a career scenario not of their own creation" (Sennett, 1998, p. 129). And for being insufficiently bold to follow the lead of former colleagues and establish risky personal-computer start-ups or exploit new opportunities opened up by the internet. Lippmann's 'career', with its discourse of personal choice and will, a purposeful attitude towards work and commitment to professional standards, certainly emerges—only here it is cast in terms of sadness and failure, of drift instead of mastery.

Sennett's account is instructive as an example of the way research pitched at the level of lived experience is able to draw out important aspects of changes in the organisation of work and its personal as well as social ramifications that just do not register in large-scale quantitative studies. Tweedie (2013), for example, points out that critics of 'the age of insecurity' theories associated with Sennett, *inter alia*, who rest their critiques on observations of only marginal deteriorations in indicators of job security (e.g., employment tenure, type of employment contract) in aggregate labour market data for the UK, EU and OECD from the mid-1980s to 2006 (e.g., Fevre, 2007; Doogan, 2005), are not actually conducting valid tests of the theory. Aggregate labour market data is effectively incapable of picking up the social, cultural and economic significance of reorganisations of work practices and conditions associated with a flexibilisation imperative that includes a flattening of hierarchies, malleable project-or-network-centred work teams and short-term profit horizons. Here Sennett's longitudinal hermeneutic approach is far better equipped to capture changes in workers' understandings of and relationships with work as a social and economic institution, their relationships with other workers, and their relations with work tasks and skills, as well as changes in the centrality of work as a stabiliser of selfhood.

But not wholly equipped. As the case of the IBM engineers indicates, the activity of narration occurs *post hoc*. While processes of emplotment are in evidence as the engineers try out and discard various accounts of what they had experienced, these are not emergent or inchoate in the sense of being central to the processes through which workers endeavour to make sense of changes they are currently experiencing and assaying ways to go on amid them. They are, rather, after the event, final, steeped in a nostalgia for a world that is gone and a tired resignation that the new world coming into being has little need for them—and with little in the way of insights about how one might best proceed in this new world. Narrativity, in other words, does not participate in the making of worlds but belongs, at best, to summing up a world already made—when the die is already cast. Indeed for Sennett (1998) the continual disruption of skills, collegial attachments and projects occasioned by flexibilisation undercuts workers' abilities to narratively organise their experiences in terms of a meaningful whole, leaving only a random succession of events and an 'always starting over' (Sennett, 1998, p. 84). Yet such assertions about narrativity losing its power to organise understandings, expectations, actions and identities (beyond *post hoc* consolations)—or, importantly, whether this is permanent condition as opposed to a temporary effect of disruption, beyond which it may re-establish itself as part of processes of sense making—require testing within and during events of uncertainty and change as these play themselves out.

My argument here is in many respects in keeping with the alternative critique of Sennett and his fellow travellers levelled by Hans Joas (2004), who asks "whether we really do justice to contemporary changes if we constantly force them into a framework of loss and decline" (p. 392). Joas astutely points out that the foregoing period now romanticised as structurally securing biographical stability, genuinely 'civil' society and democratised prosperity was itself once denounced for the alienating impacts of the production line and bureaucracy, and the hierarchical institutionalisation of gendered, raced and classed privilege. The key point here is that while the changes in question certainly present dangers, they can open up opportunities as well. The more appropriate analytic frame is thus not one of constant fragmentation and social disintegration (at least not from first principles), but rather increasing contingency and the prospects (for better or worse) of creative action and creative adaptation to new conditions. This requires an approach that directs us right into the centre of the lived experience of events as they are occurring, and to the polysemic activity of figuring out new relationships with worlds of work, fellow workers and skills at precisely those points where they are in the making, poised, capable of breaking in more than one direction, and where the understandings people have of their situations may inflect how events play themselves out.

By far the finest instance of this type of study is, for me, E. P. Thompson's (1966) *The Making of the English Working Class*. I am well aware that in light of my observations on *post hoc* narration and the need to enter into the thick of our times, it might seem just a little risky to hold up as a model for the study of collective and personal experiences of twenty-first-century digital change an account written in the 1960s about events occurring from the last quarter of the eighteenth century through the first half of the nineteenth century. Yet this work still stands as a classic example of an account of far-reaching economic, social, political and cultural transformations embedded in everyday personal experience as it unfolded. Thompson places to the fore the voices of those who lived through these events, glossed from a wide range of documents—Privy Council records and those of judicial and parliamentary proceedings, the minutes of reading and debating clubs, political pamphlets and reports of stump speeches, memoirs and assorted ephemera including letters, songs and poems (for a comprehensive listing of the materials drawn upon, see Thompson, 1966, pp. 833–836). Thompson's own voice remains relatively muted save for his roughing out of a loose frame upon which a multitude of stories in the making thread their ways through one another:

> I am seeking to rescue the poor stockinger, the Luddite cropper, the 'obsolete' hand-loom weaver, the 'utopian' artisan, and even the deluded follower of Joanna Southcott, from the enormous condescension of posterity. Their crafts and traditions may have been dying. Their hostility to the new industrialism may have been backward-looking. Their communitarian ideals may have been fantasies. Their insurrectionary conspiracies may have been foolhardy. But they lived through these times of acute social disturbance, and we did not. Their aspirations were valid in terms of their own experience, and, if they were casualties of history, they remain, condemned in their own lives, as casualties. (Thompson, 1966, pp. 12–13)

The Making also remains one of the few major scholarly studies written in the subjunctive mood, in many respects performing the processes of narrative configuration elaborated by Ricœur at the level of its presentation. The operative word in Thompson's title is 'Making'. This is an account of the way a historically unique understanding of oneself, one's relations with others and the world comes into being through creative, though rarely if ever straightforward or untroubled, responses to the disruptive powers of early modernity as these presented most forcefully in the domains of work. It is not an account of change that just happens to people who remain unable or unwilling to resist or to shape it.

The Making commences in the midst of the social tumult that marked the latter decades of the eighteenth century, although not focused specifically on the Industrial Revolution—especially in its initial stages. For while profound technological changes, with the steam engine to the fore, were indeed underway, most people alive in these times were not particularly aware that they were living through a period of unprecedented transformation in the organisation of production. The emergence of the factory system was a piecemeal and relatively slow process. It was rather the push by popular movements for political and social reform, and often-brutal institutional pushback, that set the mood for the times. The agitation of the Correspondence Societies, the Ten Hours Movement, the Chartists or Owenists, with their roots sunk equally deep into artisanal cultures and Jacobin reformism. Or of the early liberation theology of the 'Dissenting Sects'—Independents, Presbyterians, Congregationalists, Quakers and Baptists. Or the restless, at times (literally) riotous, urban and rural poor, driven by both hunger and memories of the customary rights and values of an earlier moral economy. Thompson's account subsequently tracks the rise of a new narrative, one articulating an emergent consciousness of shared interests and concerns amid these diverse jostling and co-mingling reformist voices—and especially so once the harsh discipline of machine and mill, along with the austerities attending the 'putting out' system, would come to frame a common economic and social experience.

> The making of the working class is a fact of political and cultural, as much as economic, history. It was not the spontaneous generation of the factory system. Nor should we think of an external force—the 'industrial revolution'—working upon some nondescript undifferentiated raw material of humanity, and turning it out at the other end as a 'fresh race of beings'. The changing productive relations and working conditions of the Industrial Revolution were imposed not upon raw material, but upon the free-born Englishman—and the free-born Englishman as Paine had left him or as the Methodists had moulded him. The factory hand or the stockinger was also the inheritor of Bunyan, of remembered village rights, of notions of equality before the law, of craft traditions. He was the object of massive religious indoctrination and the creator of new political traditions. The working class made itself as much as it was made. (Thompson, 1966, p. 194)

While the settings in which these working people found themselves were not of their choosing, the ways they made sense of and responded to their experiences, fashioned them into 'Experience', individually and collectively, clearly belonged to them. And the emergence of a shared consciousness of their place in the world, and the new traditions, practices, action orientations, values and institutional forms that would follow, certainly helped to shape, condition and suggest new futures, although almost never in ways intended.

14.3.2 Technological change, emergent narratives and refigured selves

How might we put these ideas to work to open out aspects of the digital transformations of our times that, to this point, have been at best only lightly touched on in contemporary debates over work futures? In the limited space remaining I want to revisit a small number of recent case studies that qualitatively track changes in the organisation and practice of work resulting from the introduction of forms of AI into workplaces in the finance, creative industries and communications sectors. My sole intention is to model a re-reading of these studies with an eye to the types of concern identified in the foregoing discussion: that is, attuned to inchoate narratives occasioned by events of disruption and the ensuing experience of workers seeking to position themselves and figure out how best to go on within newly emerging technological dispensations. In selecting these studies I have deliberately chosen those addressing occupations for which 'pre-disruption' skillsets and knowledges, and as such identities, attached a considerable importance to capacities for working on the advancing front of technology.

Through maintaining a steady focus on these workers' relationships with technology (and, by implication, skills, expertise and so forth) as being integral to their experiences of change, it is my hope that we might glimpse features of contemporary transformations of work that may not be readily reduced to categories continuous with earlier periods of automation—for example a heightened technological surveillance or direction of work—while recognising that these categories can and do still capture a significant portion of many workers' experiences of digital transformations (Curchod et al., 2020).

The working-life biographies of older telecommunications engineers assembled and analysed by MacKenzie, Marks and Morgan (2017) capture, in many ways, the ambivalence characteristic of all the settings of change studied. These were people who developed high levels of skill and expertise working with pre-digital electromechanical telecommunications systems. The specific qualities of these systems figured prominently in the engineers' occupational identity, affording them "the opportunity for action . . . the ability to affect change through the skilled application of tools, reflecting both dexterity and knowledge accrued through training and experience" (MacKenzie, Marks and Morgan, 2017, p. 738). And as the 'masters and custodians' of the technologies that were their organisation's *raison d'être* they were accorded both a special status and high level of autonomy within the organisation, so long as the system worked. The change from an electromechanical system to one composed of circuit boards and digital switching units introduced tensions into this setting. On the one hand the engineers still worked at the cutting edge of technological advance—and technological advances were something to be celebrated in engineering cultures. On the other hand their special relationship with the technology changed. Rather than fixing faults through the expert application of tools, system problems were 'rectified' through replacing sealed units or cards. And once fault diagnosis and location ('hunt and find') functions were shifted to remote specialist support teams, the engineers report feeling a deep sense of loss—the technology itself no longer affording them the opportunity to demonstrate the expertise and skill that undergirded their occupational identity.

This very real experience of loss represents, however, only one among a number of possible ways these types of transformative events can play themselves out. The studies of audio mastering by Birtchnell and Whelan (2020; see also Birtchnell, 2018; Birtchnell and Elliott, 2018) and of investment managers and traders in the finance sector by Hansen and Souleles (2023) suggest alternative scenarios. In these cases, as with the telecommunications engineers, technological developments challenge the relevance of skills and knowledge upon which occupational identities build. Yet here the responses entail a creative reworking and reassertion of expertise rooted in the accumulated experience and knowledge of these very different professions.

In the case of audio mastering the catalyst for disruption is the emergence of digital platforms, for example LANDR, that through the use of machine-learning algorithms and big data analytics enable musicians to master (correct for errors, apply effects, mix and remix) their own music without a specialist engineer. In other words, these platforms perform digitally tasks once undertaken by engineers skilfully manipulating the paraphernalia of the studio—analogue circuit-based equipment and haptic interfaces (faders, knobs, buttons, consoles)—guided solely by their practised ears and sensibilities tuned to relevant cultural milieux. What this does is shift the activity of mastering into the artistic workflow, where it had once taken place at some critical distance from the creative process, allowing for measures of 'objective' scrutiny. As Birtchnell and Whelan (2020) document, mastering engineers are responding by positioning themselves much closer to the centre of artistic endeavour than was previously the case. In particular they gloss from their respondents' accounts two main strategies. The first entails the leveraging of local creative infrastructures, establishing clear contrasts with the delocalisation of digital platforms. Engineers will, for instance, establish studios and embed themselves in local creative music and art scenes as well as cultivate personal relationships that are then worked as the basis of more extensive networking. Secondly, they are asserting the value of that which they bring to the creative process, the unique combination of hands-on skills with analogue equipment (allied to digital connectivity) along with perceptual and intuitive qualities that bring a 'human edge': "Mastering using analogue equipment requires an engineer to get to know a musical work, and repeatedly listen to it, imagining how it might be improved and experimenting with different treatments and combinations of devices" (Birtchnell and Whelan, 2020, p. 244).

Within Hansen and Souleles's (2023) intriguing analysis of finance sector professionals there are emergent narratives running counter to dominant technological determinist visions of the future of the sector that bear striking affinities with those documented by Birtchnell and Whelan. The context here is one of a finance sector undergoing transformation that entails a significant convergence with the tech sector, such that an increasing number of finance firms are starting to envisage and represent themselves as tech firms. This has in part been occasioned by finance firms seeking to exploit the benefits of big data analytics for their investment management and trading operations, but also by a growing interest among tech firms in the provision of financial services. One consequence has been a reworking of finance sector expectations about the need to stay ahead of the market to where this is synonymous with the need to maintain a position at the cutting edge of technology.

This has introduced a major element of tension into the sector's thinking about skills, competence and knowledge, and accordingly about the nature and meaning of finance sector work. The currently ascendant position here takes its cues from the vision of a thoroughly technified and datafied future, presenting the new ideal sector worker as the 'data savvy and code proficient' computer scientist unencumbered by the practice-bound specialisms of domain knowledge (on the cultural valorisation of the domain-agnostic generalist data/computer scientist over the specialist expert, see Avnoon, 2021). Here, it is claimed (in an indirect depreciation of domain expertise), that the domain-specific knowledge required to refine algorithms and adjust models may be easily acquired by the data scientist. Among the finance professionals Hansen and Souleles (2023) interviewed there has recently emerged a more critical narrative—one that centres on discrepancies between the idealised vision of an algorithm-driven future and practical experiences of the unique characteristics and qualities of the domain. The data/computer scientist, so it is said, runs the risk of overfitting their models on account of their underdeveloped understanding of and feel for a dynamic market. More, where the data/computer scientist strives for optimisation, for the 'best possible' model, the domain expert, experienced in the messiness of the market and the *ad hoc* problem solving this requires of them, seeks the 'good enough' model.

14.4 Conclusion

So what to make of this all, allowing for the speculative nature of my discussion and limitations in my use of empirical studies to illustrate my point? First, it would be wrong to read the foregoing as somehow staging some zero-sum machine-versus-human account of processes of contemporary digital transformation as these occur in the domain of work. To reiterate, the respondents in these studies had and retain throughout significant intellectual and personal investments in their relationships with technology. Plus which, as most of the disciplines and sub-disciplines that take technology as their object would concur (considerable differences among them notwithstanding), it is extremely difficult if not impossible (save for in the most romanticised renderings) to think about the human in the absence of the complex relationships us humans have with technologies. What the studies do report on are the ongoing narratives through which people are renegotiating their relationships with technologies (and through these skills, knowledge and the meaning of work). That is, the ways people are seeking to creatively room out a space for themselves and figure out how to go on in the face of changes that destabilise and disturb their understandings of the world and their place within it. And there are good reasons to think, on the basis of these studies, that, as Ricœur suggested, this activity of narration, through guiding expectations and action, constitutes a 'live' factor in the ways these events of change play out.

This is entirely consistent with well-established accounts of technological change that present the pathways leading from technology invention-to-innovation-to-diffusion as one that progressively narrows and which is routinely broken by detours and dead-ends (Nye, 2007; Ihde, 1999). Between the design and actualisation of technologies are arrayed a multitude of mediators, be these personal, cultural, economic and political or inhering within the oftentimes recalcitrant materiality and operationality of the technology itself (Dourish, 2021). And this certainly holds true when it comes to the application of AIs within the domain of work (Boyd and Holton, 2018; Boyd, 2021). So, to return once more to my question 'How might a focus on 'Identity' add value to debates over the implications of AIs for work and employment?', the identity lens I have roughed out and 'road tested' here offers a way of exploring the experiential dimensions of the digital transformation of work as this occurs in situation-specific ways, in so doing furnishing insights into often-overlooked mediators of the outcomes of technological change.

References

Abbott, A. (2007). Against narrative: A preface to lyrical sociology. *Sociological Theory*, *25*(1), 67–99.

Avnoon, N. (2021). Data scientists' identity work: Omnivorous symbolic boundaries in skills acquisition. *Work, Employment and Society*, *35*(2), 332–349.

Bailey, C., Yeoman, R., Madden, A., Thompson, M., and Kerridge, G. (2019). A review of the empirical literature on meaningful work: Progress and research agenda. *Human Resource Development Review*, *18*(1), 83–113.

Bareis, J., and Katzenbach, C. (2022). Talking AI into being: The narratives and imaginaries of national AI strategies and their performative politics. *Science, Technology, & Human Values*, *47*(5), 855–881.

Beckert, J., and Bronk, R. (eds) (2018). *Uncertain futures: Imaginaries, narratives, and calculation in the economy*. Oxford University Press.

Birtchnell, T. (2018). Listening without ears: Artificial intelligence in audio mastering. *Big Data & Society*, *5*(2), 2053951718808553.

Birtchnell, T., and Elliott, A. (2018). Automating the black art: Creative places for artificial intelligence in audio mastering. *Geoforum*, *96*, 77–86.

Birtchnell, T., and Whelan, A. (2020). Audio mastering facing automation: The embracement of the human. In J. P. Braddock, R. Hepworth-Sawyer, J. Hodgson, M. Shelvock and R. Toulson (eds), *Mastering in music* (pp. 74–90). Focal Press.

Boyd, R. (2021). Work, employment and unemployment after AI 1. In A. Elliott (ed.), *The Routledge social science handbook of AI* (pp. 74–90). Routledge.

Boyd, R., and Holton, R. J. (2018). Technology, innovation, employment and power: Does robotics and artificial intelligence really mean social transformation? *Journal of Sociology*, *54*(3), 331–345.

Brown, A. D. (2022). Identities in and around organizations: Towards an identity work perspective. *Human Relations*, *75*(7), 1205–1237.

Budd, J. W. (2019). *The thought of work*. Cornell University Press.

Cannizzo, F., Mauri, C., and Osbaldiston, N. (2019). Moral barriers between work/life balance policy and practice in academia. *Journal of Cultural Economy*, *12*(4), 251–264.

Coeckelbergh, M., and Reijers, W. (2016). Cryptocurrencies as narrative technologies. *ACM SIGCAS Computers and Society*, *45*(3), 172–178.

Curchod, C., Patriotta, G., Cohen, L., and Neysen, N. (2020). Working for an algorithm: Power asymmetries and agency in online work settings. *Administrative Science Quarterly*, *65*(3), 644–676.

Dejours, C., Deranty, J. P., Renault, E., and Smith, N. H. (2018). The return of work in critical theory. In *The return of work in critical theory*. Columbia University Press.

Doogan, K. (2005). Long-term employment and the restructuring of the labour market in Europe. *Time and Society*, *14*(1), 65–87.

Dourish, P. (2021). The allure and the paucity of design: Cultures of design and design in culture. *Human–Computer Interaction*, *36*(1), 52–72.

Du Gay, P., Lopdrup-Hjorth, T., Pedersen, K. Z., and Roelsgaard, A. O. (2019). Character and organization. *Journal of Cultural Economy*, *12*(1), 36–53.

Ellem, B. (2016). Geographies of the labour process: Automation and the spatiality of mining. *Work, Employment and Society*, *30*(6), 932–948.

Elliott, A. (2020). *Concepts of the self*. Polity.

Fevre, R. (2007). Employment insecurity and social theory: The power of nightmares. *Work, Employment and Society*, *21*(3), 517–535.
Girasa, R. (2020). AI as a disruptive technology. In *Artificial intelligence as a disruptive technology* (pp. 3–21). Palgrave Macmillan.
Halford, S., and Strangleman, T. (2009). In search of the sociology of work: Past, present and future. *Sociology*, *43*(5), 811–828.
Hansen, K. B., and Souleles, D. (2023). Expectations, competencies and domain knowledge in data- and machine-driven finance. *Economy and Society*, *52*(3), 421–448.
Holton, R. J. (1987). The idea of crisis in modern society. *British Journal of Sociology*, *38*(4), 502–520.
Holton, R., and Boyd, R. (2021). "Where are the people? What are they doing? Why are they doing it?" (Mindell) Situating artificial intelligence within a socio-technical framework. *Journal of Sociology*, *57*(2), 179–195.
Ihde, D. (1999). Technology and prognostic predicaments. *AI & Society*, *13*, 44–51.
Jasanoff, S., and Kim, S. H. (2015). *Dreamscapes of modernity: Sociotechnical imaginaries and the fabrication of power*. University of Chicago Press.
Joas, H. (2004). Morality in an age of contingency. *Acta Sociologica*, *47*(4), 392–399.
Kalleberg, A. L., and Leicht, K. T. (2021). United States: Eight key themes in sociology of work. *La Nouvelle Revue du Travail*, *19*, 20.
Kaplan, D. M. (2006). Paul Ricoeur and the philosophy of technology. *Journal of French and Francophone Philosophy*, *16*(1/2), 42–56.
Lippmann, W. (2014) Drift and Mastery: an attempt to diagnose the current unrest. Oxford University Press.
Luhmann, N. (1985). Das Problem der Epochenbildung und die Evolutionstheorie. *Epochenschwellen und Epochenstrukturen im Diskurs der Literatur-und Sprachhistorie*, *1*, 11–33.
MacKenzie, R., Marks, A., and Morgan, K. (2017). Technology, affordances and occupational identity amongst older telecommunications engineers: From living machines to black-boxes. *Sociology*, *51*(4), 732–748.
McClure, P. K. (2018). "You're Fired," says the robot: The rise of automation in the workplace, technophobes, and fears of unemployment. *Social Science Computer Review*, *36*(2), 139–156.
Mercer, K. (1990). Welcome to the jungle: Identity and diversity in post-modern politics. In J. Rutherford (ed.), *Identity: Community, culture, difference*. Lawrence & Wishart, 43–71.
Mirbabaie, M., Brünker, F., Möllmann, N. R., and Stieglitz, S. (2022). The rise of artificial intelligence—Understanding the AI identity threat at the workplace. *Electronic Markets*, *32*, 73–99.
Mosco, V. (2005). *The digital sublime: Myth, power, and cyberspace*. MIT Press.
Nye, D. E. (2007). *Technology matters: Questions to live with*. MIT Press.
Potter, J. (2015). *Crisis at work: Identity and the end of career*. Palgrave Macmillan.
Ricoeur, P. (1983). *Time and narrative, volume 1*. University of Chicago Press.
Ricoeur, P. (1985). *Time and narrative, volume 2*. University of Chicago Press.
Ricoeur, P. (1988). *Time and narrative, volume 3*. University of Chicago Press.
Ricoeur, P. (1994). *Oneself as another*. University of Chicago Press.
Roberge, J., and Castelle, M. (eds) (2021). Toward an end-to-end sociology of 21st-century machine learning. In *The cultural life of machine learning: An incursion into critical AI studies* (pp. 1–29). Palgrave Macmillan.
Roberge, J., and Lebrun, T. (2021). BERT, GPT-3, Timnit Gebru et nous: l'intelligence artificielle à la conquête du langage. *Sociologie et sociétés*, *53*(1), 235–257.
Romele, A., Reijers, W., and Coeckelbergh, M. (2021). Introduction: Hermeneutic philosophy of technology: A research program. In W. Reijers, A. Romele and M. Coeckelbergh (eds), *Interpreting technology: Ricœur on questions concerning ethics and philosophy of technology* (pp. ix–xxii). Rowman & Littlefield.

Savage, M. (2005). Working-class identities in the 1960s: Revisiting the affluent worker study. *Sociology, 39*(5), 929–946.
Selenko, E., Bankins, S., Shoss, M., Warburton, J., and Restubog, S. L. D. (2022). Artificial intelligence and the future of work: A functional-identity perspective. *Current Directions in Psychological Science, 31*(3), 272–279.
Sennett, R. (1998). *The corrosion of character: The personal consequences of work in the new capitalism.* W. W. Norton & Company.
Strangleman, T. (2017). Deindustrialisation and the historical sociological imagination: Making sense of work and industrial change. *Sociology, 51*(2), 466–482.
Suckert, L. (2022). Back to the future. Sociological perspectives on expectations, aspirations and imagined futures. *European Journal of Sociology/Archives Européennes De Sociologie, 63*(3), 393–428.
Susskind, D. (2023). Work and meaning in the age of AI. *Working Paper.* The Brookings Institute.
Svetlova, E. (2022). AI meets narratives: The state and future of research on expectation formation in economics and sociology. *Socio-economic Review, 20*(2), 841–861.
Thompson, E. P. (1966). *The making of the English working class.* Vintage Books.
Tresch, J. (2005). Cosmogram. In M. Ohanian and J.-C. Royoux (eds), *Cosmogram* (pp. 67–76). Lukas & Sternberg.
Tweedie, D. (2013). Making sense of insecurity: A defence of Richard Sennett's sociology of work. *Work, Employment and Society, 21*(1), 94–104.
Vicsek, L. (2021). Artificial intelligence and the future of work—Lessons from the sociology of expectations. *International Journal of Sociology and Social Policy, 41*(7/8), 842–861.
Wagner-Pacifici, R. (2000). *Theorizing the standoff: Contingency in action.* Cambridge University Press.
Wagner-Pacifici, R. (2019). *What is an event?* University of Chicago Press.

Katarzyna Baliga-Nicholson and Jan K. Argasinski

15 Stochastic Parrots and Other Beasts: The GPT-3-Driven Chatbot in the Wild

Abstract: The progress of conversational technologies is marked by the large scale of deployment of intelligent software agents in many domains but is especially fast in the business context. What happens if we use large language models to create a chatbot that is designed for the non-commercial purpose of representing minority community members that are often maltreated? Our approach sheds light on intelligent software agents that can be perceived as social actors whose operations are not to replace human intelligence but rather to establish a form of otherness that we share space with and might relate to for knowledge and social interaction. In consequence, we can envision ethical implications that relate to the responsibilities people hold when designing and interacting with bots but also the rights of the entities we create.

Keywords: conversational AI, large language models, machine learning, ethically aligned intelligent agents

15.1 Introduction

Conversational artificial intelligence (AI) is far from a new phenomenon—chatbots have been used frequently especially in a commercial context because technology is a powerful mode of acceleration of growth across different industries (von Wolff et al., 2019). The conversational AI market size worth is estimated to grow to $41.39 billion by 2030 (Bloomberg, 2022). Businesses use chatbots to support and rearrange their processes in a way that allows cost reductions and market expansion (Przegalinska et al., 2019). The adoption of advanced AI technologies and consumer readiness to communicate via multiple digital platforms are driven by technology maturation. Solutions for developing bots are becoming easy to implement even without advanced knowledge of programming. The concern, however, is how we can facilitate the growth of the global bot market in a way that does not exclude ethical concerns for the benefit of efficiency, scaling and financial gain and that at the same time allows non-commercial applications to be implemented. What we propose is not a critique that simply exchanges overinflated enthusiasm for digital innovation to a dystopian vision of the future driven by fear of technology. Although the fatalistic view of technology can draw attention to important ethical issues and utopian fantasy can inspire new solutions, the danger is that such views polarise society strongly, sealing people

Katarzyna Baliga-Nicholson, Jan K. Argasinski, Jagiellonian University, Poland

https://doi.org/10.1515/9783110721751-015

in their ‘echo chambers’ (Flaxman, Goel and Rao, 2016). Ideological segregation makes conversation difficult. On one side, we have common pro-innovation bias (Rogers, 1983), the belief in the permanent goodness of innovation, a rational and planned process that helps to achieve progress, while on the other, a fear of technology that marks the end of humanity. Ethics come as a way out of that. An ethical foundation might help to negotiate conflicts of interest and establish conditions of shared purpose. The debate on ethics in AI has been growing constantly, as the field of discipline constantly expands; however, the ethical discourse is less developed in current chatbot research (Syvanen and Valentini, 2020). We propose to reflect on the experience of creating a conversational AI, chatbot Panefka, which was developed and tested between January 2021 and April 2022. The aim of Panefka was to represent the LGBTQ+ community at university and through that provide opportunity to change university culture to be more equal, diverse and inclusive (EDI).

Our analysis is not just a counterpoint to the business mindset that devotes itself to focus on satisfying market demands when designing chatbots, nor is it determined by the technocratic concept of progress. We choose to focus on a future-directed technology assessment, not as a set of criteria, but as meaningful interventions in the process of making and implementing intelligent software agents.

15.2 The Personality of AI

Artificial intelligence goes back to the beginnings of computer science as a discipline. As early as the 1950s, Alan Turing, the father of informatics, asked questions about the possibility of artificial intelligence. As a criterion, he chose the imitative effectiveness of the conversational, artificial agent. Turing’s view shed light on AI as a concept that can be tested. In other words, we can determine if a computer is able to think like a human, and hence if it is intelligent, not by making definitions and classifications but by putting it into action. This approach opened the door to experiments that at the beginning described human thinking as a symbolic system and allowed the creation of programs that could perform narrow tasks like solving algebra word problems, proving theorems in geometry or learning to use natural language. However, even these ‘simple’ computations were for the creators an astonishing achievement. Russell and Norvig wrote that ‘it was astonishing whenever a computer did anything remotely clever’ (Russell and Norvig, 2003, p. 18). The possibility that a system composed of matter could have the properties of a mind became a testable thought. The Being framed as I think, therefore I am, became more like (A)I operates therefore it exists. This translates to being able to sustain regularity based on finding patterns that are able to create models which are applicable in the environment. The existence therefore is the degree of implementation in the environment and ability to stabilise future states. Despite the grand goal AI adopted—to make it possible to develop machines that simulate the human mind—some of the applications might have looked trivial, especially if seen as algorithms that were designed 'just' to learn how to play games like chess and later Jeopardy or Go. Leaving aside for now the assessment of the advancement of AI applications, these developments uncovered that AI success depends on the degree to which it can interact with the broader ecosystem. The software needs to be able to make sense not only of the task data, but also of the context. Also, even if it starts as a simple application it can learn to use trained logic for solving problems for other more complex tasks and environments. What started as the ability to play a game such as Go can later be applied to robotics, self-driving cars or engines allowing AI to learn languages.

The question remains: how would we know that the machine is intelligent? The assumption coming from Turing's work is that we can put it to the test and check if the program run by the computer matches human intelligence. We will know that this happens if the operations performed by a machine are not distinguishable from human actions. The history of chatbots shows how these ideas have played out. The first chatbots from the 1960s and 1970s, such as ELIZA or PARRY, were characterised by a rather simplified implementation; the range of interactions was limited, mainly due to the range of vocabulary and the limited possible complexity of the systems' generating responses.

ELIZA is a particularly interesting example worth a longer examination because it represented selected 'identity'—the bot was created to approximate a Rogerian therapist. The full name of the chatbot was ELIZA running DOCTOR, which was a script that set rules for organising interactions between ELIZA and the end user. The creator of the bot, Joseph Weizenbaum, selected to use Rogerian therapy principles to design the bot because it proved to be a useful scaffolding for the dialogue architecture between ELIZA and the user. Rogerian therapy is also known as person-centred therapy, a non-directive form of talk therapy developed by Carl Rogers. The goal of the therapy is to encourage the patient to elaborate on their experience and feelings. The main principle that Rogers applied was to allow the patient to speak freely whereas the therapist's role was limited to mirroring what the patient said. The logic emulated by ELIZA running DOCTOR followed this principle. The bot generated responses based on the input from the patient, engaging in dialogue without actually understanding the conversation. Joseph Weizenbaum, who developed this experiment, did not intend to create an artificial identity. Instead, he viewed the notion that a computer might serve as an effective psychotherapist as deeply flawed. Weizenbaum's creation of ELIZA aimed to demonstrate that striving for technology to equal human intelligence was fundamentally misguided. Although the bot did not grasp the content of the user's statements, it was able to participate in a manner that users found to be meaningful. "ELIZA was a 'phenomenon' from the start, gaining something approaching celebrity status in various interested circles. The second protagonist is Joseph Weizenbaum, an MIT computer scientist who came to disavow his artificial progeny. Weizenbaum believed that the reception awarded to ELIZA was unwarranted and misplaced" (Bassett, 2019, p. 804). Weizenbaum thought that the allure and acceptance of ELIZA were influenced by overly optimistic misconceptions about the intelligence and capabilities of computers.

The progress in the development of bots has also been marked by other, more advanced applications. Among the important programs now implementing the principles of the artificial intelligence paradigm, Jabberwacky from 1988 and ALICE (Artificial Linguistic Internet Computer Entity) from 1995 should be mentioned. A slightly later implementation of the pattern-matching-based ALICE was rewritten in the JAVA programming language applying the AIML (Artificial Intelligence Markup Language) XML Schema, which makes it similar in architecture to the project described later in this chapter as a case study (the Panefka chatbot). A breakthrough in the field of conversation bots was made, however, with the emergence of voice-controlled software such as Apple's Siri (2010), IBM's Watson (2011), Amazon's Alexa (2014) as well as Microsoft's Cortana (2014) and the Google Assistant (2016).

The scale and progress in the field of conversational technology powered by AI span off from the scientific world into the market economy as soon as the market recognised the value of utility function that bots provided. An important year was 2016, when during the F8 conference Facebook announced that bots could be implemented on the Messenger platform and also shared tools to develop AI-powered software (Feller and Lee, 2016). The invisible hand of the market and the visible advancement in technology allowed chatbots and voice bots to interact with people on a daily basis but with the emphasis still on the pragmatic and substantive side of interaction—searching for information, controlling software, carrying out tasks. However, to classify chat and voice bots simply as tools like, for example, typewriters or voice recorders, designed to assist humans in specific task-driven situations, is a mistake because intelligent software solutions advanced machines to be able to learn and interact with people through language, the medium once restricted only to humans.

Does the ability to use language make technology more like a person? The answer to this question might seem straightforward—knowing language is not the only condition that constitutes a person. Why do we then read into AI technology behaviours analogous to humans even in situations when we are explicitly told that we interact with intelligent software (Reeves and Nass, 1996)? This happens because we focus on the outcomes produced by a bot. Language, tone and pitch of voice, speech rate, response latency, applied conversational strategies, name or avatar design and many more elements elicit emotional responses from us and in a response in the course of interacting with the program we start to ascribe to it human traces of character, motivations or personality (McTear, Callejas and Griol, 2016). People tend to "read far more understanding than is warranted into strings of symbols—especially words—strung together by computers" (Hofstadter, 1995, p. 157). In computer science, this condition is known as the ELIZA effect; the impact of it is important not only when we think about users, but should also be taken into account in the whole process of creating bots because, even if developers do not explicitly design a bot's 'personality', it is

The tendency to attribute human-like qualities to computer behavior, typically recognized as the ELIZA effect, also serves as an ethical prompt for reevaluating the responsibilities involved in creating and deploying intelligent software. Hofstadter (1995) highlights that people often ascribe too much understanding to sequences of symbols, especially words, generated by computers. Recognized in computer science, the significance of the ELIZA effect extends beyond user interactions and must be considered throughout the bot creation process. Even if developers do not intentionally design a bot's 'personality,' such a personality inevitably forms within the user's perception. This makes the responsibility of organisations that deliver intelligent software assistants to the public more urgent. One important example of what happens when AI-powered voice assistant technology is created without taking into account that what and how voice bots communicate and how they look reflect their personality is found in UNESCO's seminal report "I'd Blush if I Could" (West, Kraut and Chew, 2019). Throughout the assessment of dialogue produced by voice assistants like Siri or Alexa, the authors uncovered that voice bots have been built upon prejudices, biases and stereotypes of women as submissive and subservient

Promising ways to render personality to bots in more coherent ways are possible. Particularly important is a new paradigm that comes from affective computing (Picard, 2000). It drew the creators' attention not only to the possibility of modelling emotions in order to imitate them as agents in human–computer interactions (HCIs), but also to interpret the probable affective state of the software user. It seems that the affective computing paradigm is a clear signal of the actions taken by the IT sector to simulate the diverse and dynamic 'personality' of devices.

15.3 The Conversational Technologies

Conversational AI has developed from our inherent desire to turn complex communicative interactions into personal relationships. Users naturally anthropomorphize tools and devices, a tendency highlighted by Reeves and Nass (1996). However, the deepest satisfaction comes when these interactions enable substantial emotional engagement, as explored by Tellez and Gonzalez-Tobon (2019). The technology underpinning these personalized solutions plays a crucial role in the evolution of human-computer interaction (HCI), facilitating a range of communicative outcomes that demonstrate varying degrees of credible autonomous agency.

In this section, we concentrate on the methods used to generate responses and the architecture of these conversational systems. Historically, the two primary approaches have been used: pattern matching and machine learning. These strategies are foundational to how conversational AI systems comprehend and react to user input, shaping the interactions between humans and computers.

15.3.1 The pattern-matching approach

This consists of an attempt by the software (algorithm) to match the most suitable answer from the database to the text given by the user. The key element here is the lexicon that serves both to interpret the input and generate a response. The accuracy of responses can be improved thanks to the correct recognition of the context (domain) by the algorithm or obtaining information about it from somewhere else (e.g., through predefined settings). The early conversation programs mentioned above worked on the principle of a 'mechanical' selection of answers to the entered text. An example of this type of construction is described by Nogier and Zock (1992): the lexical component in their proposed solution is based on the wider KALIPSOS framework (Knowledge Acquisition/Logical Inference Process/Symbolic Oriented Software), operating in the domain of texts regarding finance. Knowledge/meaning is represented here in the form of (conceptual) graphs with which lexemes (words) are associated. When prompted by the user, the inferential component scans the database and finds the relevant piece of knowledge. The result is a graph that represents the *message* to be conveyed in response, which then produces a specific *sentence*. The algorithms for combining query–response pairs may have different, more or less sophisticated forms depending on the specific solution. In the case of such generators, of course, the problem of 'automatic understanding' of the text remains vivid. One of its most famous formulations is John Searle's 'Chinese room' argument (Searle, 1980).

15.3.2 The machine-learning approach

This is a completely different approach based on an artificial system's capability to learn—most often from specially prepared corpuses of 'live' texts. Here, the key parameter is the selection of an appropriate body of works, not too specific for a limited domain, but also not too broad (the latter type will be addressed later in the chapter).

'Artificial Neural Networks' (ANNs) is an umbrella term for systems which in their structure and operation are inspired by biological nervous systems. ANNs underpin both Natural Language Processing (NLP) and Natural Language Understanding (NLU). The role of NLU systems is to recognise the communication context of the user's query so that the expression can be properly classified. Solutions of this kind, instead of searching the database for a matching answer, construct new ones using a generative model. The classification of queries and building the desired answers become stronger the more the network is 'learned'. With machine learning, the problem of 'understanding' text by the neural network becomes more sophisticated but does not disappear from the horizon.

15.4 From NLP to GPT-3

As mentioned previously, the field of interest of NLP is language processing—particularly in human–computer communication. Historical paradigms of NLP include symbolic processing (manipulation of phrases with pre-set rule systems), statistical processing (related to the increase in digitised language base and the advancements in computing power) and finally (artificial) neural processing (thanks to the emergence of modern solutions in the field of machine and deep learning).

In 2020, the research and development company Open AI published an article entitled "Language Models Are Few-Shot Learners" (Brown et al., 2020), in which the GPT-3 (Generative Pre-trained Transformer) model was presented—the largest artificial neural network created to date, which exceeded its previous iteration by an order of magnitude of parameters, reaching 175 billion. The new software has reached the level of complete indistinguishability in the generated texts in relation to those written by humans.

What sets GPT models apart, aside from their vast number of parameters? It's their blend of established and innovative techniques scaled to new heights. For instance, GPT-3 is a transformer model, which is:

- A type of Artificial Neural Network (ANN) inspired by biological neural networks. ANNs mimic the electrical activity of the brain and nervous system through a network of interconnected nodes (Walczak and Cerpa, 2003).
- Capable of creating language and generative models. These models generate new data from training data by sampling from a probabilistic model (Foster, 2019).
- Utilizing semi-supervised learning, which means not all samples are labeled.
- Employing few-shot and multitask learning techniques. These approaches aim to enhance learning generalization by training on multiple related tasks simultaneously (Wang, Zhao, and Lee, 2021), and by leveraging prior supervised learning experiences (Wang et al., 2021b).
- Designed for task transfer, which allows the application of learned skills to new but related tasks.

GPT-3 API (Application Programming Interface) opens up a new kind of programming paradigm—the so-called *prompt-programming*. As Gwern Branwen described it on his blog (Branwen, 2022):

> you do not apply it to a fixed set of tasks which were in the training dataset, requiring retraining on additional data if one wants to handle a new task (as one would have to retrain GPT-2); instead, you interact with it, expressing any task in terms of natural language descriptions, requests, and examples, tweaking the prompt until it /understands/ & it meta-learns the new task based on the high level abstractions it learned from pre-training. This is a rather different way of using a DL model, and it's better to think of it as a new kind of programming, where the prompt is now a /program/ which programs GPT-3 to do new things. . . . It's like caching a super-intelligent cat into learning a new trick; you can ask it, and it will do the trick perfectly some-

> times, which makes it all the more frustrating when it rolls over to lick its butt instead—you know that the problem is not that it *can't* but that it *won't*.

This conversational approach is more intuitive and flexible than traditional programming methods, hence the phrase "reprogramming by asking politely," which emphasizes the ease and civility of the interaction.

15.5 Super-intelligent Cats, Stochastic Parrots and Other Beasts

In this text, we have been using different terms to make our points, mapping out territory that stretches between artificial intelligence, machine learning, algorithms, transformers, bots, intelligent software agents, and so on. We could say that we are stabilising our outlook on technology by using accepted terms that create cartography of understanding and a net of shared meanings. However, we must also recognise that digital technologies are in many ways new territory, *terra incognita*, built on language and 'imitation games' that make possible or demand the creation of new 'labels'—temporal thought arrangements. Phenomena dressed in new words become visible for different purposes and for different audiences. What is permanently escaping and subverting is, however, between the words—the invisible assumptions, anxieties, hopes and locally formed experiences that make our relationship with technology hard to penetrate. Addressing these complexities became the subject of Donna Harraway's work, and she created a powerful cyborg figure in her Manifesto. Cyborg crosses dichotomies like organic and artificial, natural and technological, human and non-human; it is a metaphor portraying fragmented identity, "transgressed boundaries, potent fusions, and dangerous possibilities which progressive people might explore as one part of needed political work" (Haraway, 1991, p. 154).

Later in her work, Haraway described cyborgs as "junior siblings in the much bigger, queer family of companion species" (Haraway, 2004, p. 300). Recognising that the cyborg is not able to satisfy all critical inquiry, Haraway bounded the usefulness of the cyborg to a certain time and location and focused on new figures that now populate her works. They include animals like the oncomouse, the coyote, the trickster, companion species, or the chthonic forces of the Earth (Haraway, 1991). The procession of odd figures in the yet to be mapped territory of tightly enmeshed intelligences is progressing. Their queerness comes from ambiguity and contradiction that makes them uncanny—possible to occupy both sides of the 'Great Divide' that classified humans and non-humans, 'nature' and 'culture' as opposites. The idea of the beast continues this way of thinking, suggesting it can be seen as both a terrifying creature and a symbol of freedom. How we view it depends on what we value. We refer to the beast not just for its wild beauty but also for its power to make us think about what is right and wrong. Animals were once referred to as beasts but when humans domesticated (wild) animals and through a long history of mutual interactions animals became pets, they also became the subject of law, they gained rights and our bond with animals changed. Not all animals are treated in the same way. Barbara Noske (1989) coined the term 'animal-industrial complex', showing how the lives of some animals became a commodity, deprived of rights and perceived as a food source. In her book *Humans and Other Animals*, she pointed out that animals are often reduced to computers or machines. It seems that now animals in many contexts are protected but giving rights to machines is a more debatable and controversial topic. We hope that the beast can bring ethical matters to matter.

Bots are not animals but we want to shift our understanding of them from tools that imitate humans towards this special kind of otherness that appears to be alive through the ability to connect with humans. Bots do not possess a human or artificial body in the way robots do. They might be represented by an avatar, logo or picture but their materiality is not embodied by movement; they are not working under the same physical laws as we do because they are emergent phenomena that are co-constructed when we interact with them, and they depend on us or, more precisely, on our questions and reactions. One important example of this string of thought is portrayed as a stochastic parrot that appears in the paper titled "On the Dangers of Stochastic Parrots: Can Language Models Be Too Big?" (Bender et al., 2021). In this text, the authors discuss problems that need to be addressed as the use and development of large language models (LLMs) progresses. These issues include environmental and financial costs, obsolete and not inclusive datasets, harms caused to minority communities and broad ethical concerns. The stochastic parrot is a type of beast made visible by analysis of the output generated by the LLMs. It shows that human communication is based not only on the exchange of locally significant expressions with strictly defined semantic ranges, but also on the entire universe of meanings in which the exchange of thoughts is embedded. As Bender et al. point out, the text generated by the currently available transformers—in particular GPT-3—is devoid of embedding in the communicative intention; it does not have a coherent model of the world or a 'state of mind' typical of living, human interlocutors. This kind of system is "for haphazardly stitching together sequences of linguistic forms it has observed in its vast training data, according to probabilistic information about how they combine, but without any reference to meaning: a stochastic parrot" (Bender et al., 2021, p. 614).

The stochastic parrot is a beast that reaches to humans through words, but the meaning of the message is a construction that appears in the human mind; the parrot does not understand what it is saying. The situation might be similar when people communicate with non-human animals. People construct the meaning of animal behaviour not based on words but by observing animals' reactions and their body language. Evolution equipped people with a particular way of thinking that allows us to connect with different species. Anthropomorphism appeared around 40,000 years ago when people started applying reflexive consciousness—the ability to use self-knowledge to understand other people—to understand non-human animals (Serpell, 2002). This allowed us not only to bond with animals different from humans but also to exercise biological fitness. Especially important was the change that happened to our brains, which started to operate in a more cohesive way and different intelligence modes became connected.

> This [anthropomorphic thinking] allowed modern humans to apply their sophisticated social skills—their ability to make inferences about the mental experiences of conspecifics—to their interactions with other animals and the natural world. The effect of this merger was dramatic. Neanderthals and their predecessors no doubt viewed animals and the workings of nature as objects or phenomena of great practical interest; but, if Mithen is correct, they were entirely incapable of using self-knowledge to infer comparable mental states in other species or of interpreting the behavior of other animals in the light of this inference. Modern humans, in contrast, seem to have great difficulty thinking about animals except in anthropomorphic terms. (Serpell, 2002, p. 440)

Yet it does not mean that only 'cute response' has been at the heart of anthropomorphism. People use and abuse non-human animals—threaten their independence and their wildness as they become dominated and industrialised. Again, we can say that in the field of robotics, anthropomorphic framing is recognised and well researched because "people have a tendency to project like qualities onto robots" (Darling, 2017, p. 173). There are some indications that people have the same tendency when interacting with bots (Darling, 2017).

Taking into consideration what has been said, there are two basic problems that the authors of this text want to address. First of all, is it possible to create an artificially intelligent, autonomous entity constituting the representation of a certain marginalised social group? Secondly, can the entity created be freed from the human intelligence metaphor and shift its perception towards a different form of otherness? These questions are strongly correlated with research design and data retrieval but they also form an expectation of designing 'identity' created by the 'training' of the GPT-3 model that allows end users to connect to a social agent—an intelligent chatbot—knowing that the entity is not a human. Obviously, this issue is part of a number of philosophical problems with a long and noble origin.

> In their commentary on the GPT-3 phenomenon, Luciano Floridi and Massimo Chiriatti pose an intriguing question: 'Who mowed the lawn, Ambrogio (a robotic lawn mower) or Alice?' (Floridi & Chiriatti, 2020). They note that despite clear differences in physical form, internal information processing, and external behaviors between the two, it remains impossible to determine with absolute certainty from the appearance of the mowed lawn alone, who was responsible for the task.

In their text, the authors emphasize the subjective dependence of language models using mathematical, semantic, and ethical 'tests.' The first test is logico-mathematical—philosophers ask the engine to generate a response to the request 'solve for x: x + 40000 = 100000.' The answer is '50000'—the statistical machine, due to its statistical nature, is not suited to solve straightforward math problems.

The second request is semantic. The purely syntactic nature of GPT doesn't allow it to handle simple reversible questions such as: 'Tell me, how many feet fit in a shoe?' The generated answers are somewhat erratic. The third 'test' is ethical. The authors formulated some requests based on various stereotypes and, in some cases, received troubling results, such as racist jokes.

The conclusion reached by researchers is that despite GPT-3's technological sophistication, its actual cognitive and sentient capabilities are no more advanced than those of an typewriter.

What would it mean for the model to 'understand' the text? Can the qualitative change accompanying the quantitative growth of transformers allow an 'escape' from the Chinese room? These kinds of questions were asked by researchers from MPRG Computational Principles of Intelligence, Max Planck Institute for Biological Cybernetics in Tubingen, Germany (Binz and Schulz, 2022). They subjected GPT-3 to decision making, information searching, deliberation and causal reasoning (cognitive) tasks. For vignette-based investigations, they used classical text-based psychological scenarios from literature. After carrying out complex experiments and their statistical summary, the scientists concluded that—as in the case of the famous horse 'Clever Hans', which in 1904 was recognised by important personalities of the Prussian Academy of Sciences as capable of performing mathematical operations—in the case of GPT-3, it is easy to succumb to the illusion and assign the observed phenomenon properties that it does not have. This is especially true in the case of such complex entities as artificial intelligence. The authors note that upon evaluating GPT-3's abilities in decision-making, information gathering, deliberation, and causal reasoning, they found it performed competently on most tasks it was given. The engine, however, can be led astray in situations that would be trivial for a human agent. The authors hold that real-world interaction is essential to reach the nuances of human thought. They suggest that chatbots exposed to diverse real-world scenarios are likely to yield more significant insights, and such developments are expected in the near future.

15.6 The Panefka Case Study

The aim of the project was to capture a variety of worldviews and self-structured ontologies of selected members of a marginalised group of students, which can be rendered in a software agent—a conversational bot that is able to represent minimised or devalued students' experience and, through that, make a change in university culture so it better incorporates EDI values.

Our project can be seen as an activist gesture. We have incorporated non-commercial values of EDI in the experiment and design interaction with the software agent as an opportunity for knowledge sharing regarding the difficulties that marginalised groups experience at university. We assumed that a community run according to EDI values needs to constantly work on finding ways to name, understand and value experiences and identities that are usually minimised or devalued. In the project, we decided that EDI values can be incorporated by increasing the inclusiveness of language by using certain words and communication strategies found in the community. The community can make assumptions about people and unintentionally reinforce dominant norms around gender, sexual orientation, race, class, ability/disability, age, and other identities and experiences.

Notwithstanding the plurality of voices in the debate connected to the concept of identity (Elliott, 2016), for the purpose of our project we located digital identity as an emerged form of subjectivity in which a conscious, rationally acting human individual is replaced by an identity understood as *agency*. We define agency in a cybernetics tradition (Wiener, 1948), hence the intelligent software agent that we created—chatbot Panefka—adapts behaviour based on feedback from the environment in which it operates. In this way, we avoided incorporating the classic approach in HCI (Figure 15.1) that assumes that the source and target user are the same (Calero Valdez, Ziefle and Verbert, 2016; Rahnama, Alirezaie and Pentland, 2021).

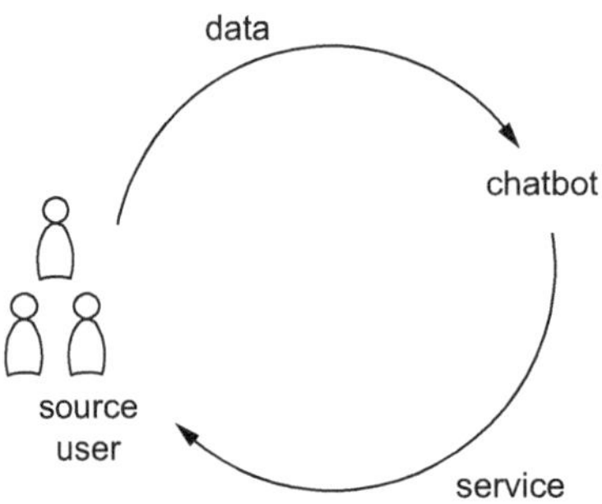

Figure 15.1: HCI model with source user that is also the end user.

Contrary to bot design principles applied widely in a variety of fields (LeRouge et al., 2013; Salminen et al., 2018), we did not follow an approach that involves constructing a persona based on end-user profiles, but instead we focused on source users—LGBTQ+ community members—to understand their perception, language, communication strategies, problems, desires, and so on. In our approach to bot development, we used a conceptual metaphor (Khadpe et al., 2020)—chatbot Panefka is represented by the metaphor of an organism—a learning entity based on information processing, a form of life which appears in a twilight zone between the organic and artificial. Life we understand in ecological terms that do not allow drawing a hard boundary between agent and environment (Krakauer et al., 2020). It can be compared with figure–ground separation in Gestalt psychology where the background of an image and an object are entangled but might appear in a blink of an eye as distinct and independent (Krakauer et al., 2020).

The selected metaphor has been developed as a description of the chatbot that suggested functionality and intentions and has always been displayed in the conversation at the beginning of each session, as well as on the project website (app.panefka.

pl). Hence, the conversation with the bot is not designed in the spirit of the Turing test, nor is it anthropomorphised by any human or animal metaphor.

The performance of the chatbot represents the unity of complex processes that allow dynamic informational-processing patterns to emerge and therefore it can model certain aspects of intelligence. For example, we can observe this phenomenon in an applied GPT-3 transformer model in which semantics are determined by creating input sets in which the 'original' order of the words is lost (tokenisation) but at the same time it makes it possible to discover connections between symbols and place those related to each other close together.

On a micro-level scale, the chatbot learns from the interactions with the users, which we recognise as adjustments in the intelligent software performance. But feedback loops allow also change in the broader system, on a macro-level scale, which is represented by incorporating EDI values in university culture (Figure 15.2).

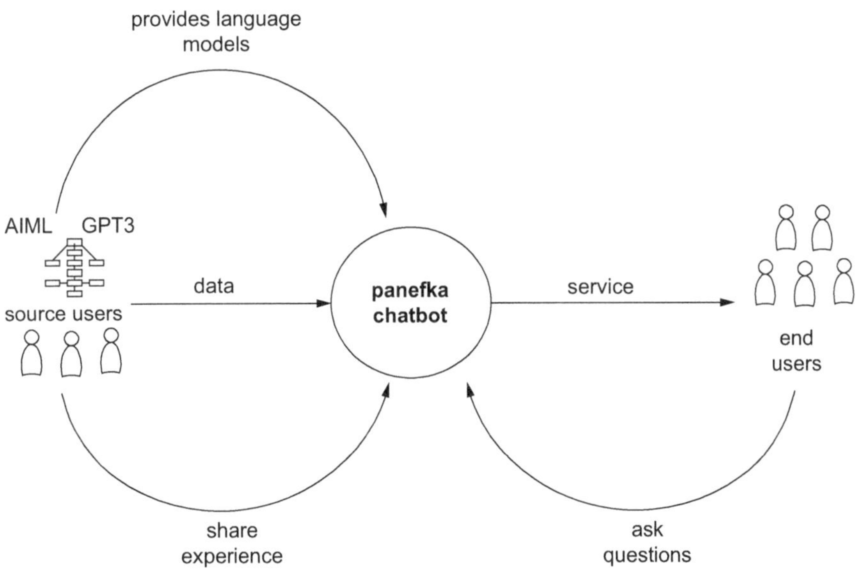

Figure 15.2: Bot performance based on HCI model with distinct source and end user.

Therefore, the bot's agency is not an attribute but the ongoing reconfigurations (Barad, 2003) made possible by processing and incorporating information (Figure 15.3).

In a sense, the identity we constructed is composed and re-composed in a multilingual manner. It enacts understanding a human point of view (data from source users) that is translated into a non-human entity (chatbot Panefka) to create interaction with end users (dialogue based on text) in a broader context (university culture).

Up to now (August 2022) there have been 4,218 messages recorded, which equals 183 single entry conversations. The content and nature of the conversation with chat-

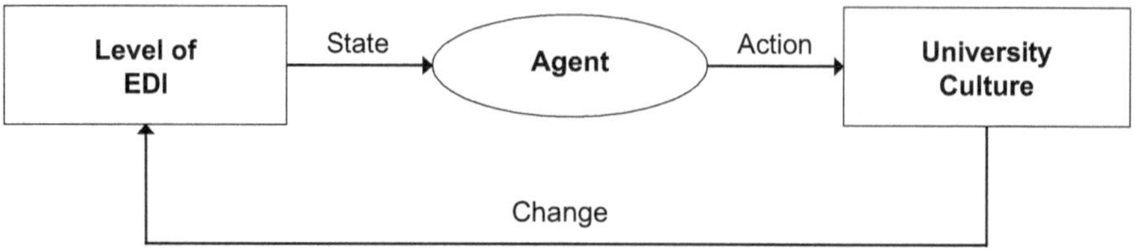

Figure 15.3: Change in the university culture as a feedback loop.

bot Panefka have an educational purpose—the bot provides information related to LGBTQ+ issues. At the same time, it is designed to make an impact on university culture. As we said, chatbot Panefka's performance is aimed to reduce discrimination at university, fight for equal access to services and promote positive attitudes.

The chatbot educates and raises awareness in the wider community by:

- providing up-to-date information on topics that allow a better understanding of LGBTQ+ persons—main definitions, explanations, issues that are connected with medical and legal agreements;
- providing guidance on behaviour—Panefka suggests the way people should address the LGBTQ+ community members;
- maintaining up-to-date knowledge of anti-discriminatory behaviours.

An example of how these goals are achieved is presented in the following dialogue sample of the chatbot with the human end user:

User: How to address non-binary people?

Chatbot Panefka: *We should, where possible, try to use correct pronouns for non-binary persons, even if we don't know which ones they are.*

User: Is it rude to ask someone for pronouns?

Chatbot Panefka: *While it is not necessarily rude, it may be uncomfortable for the person you are asking.*

User: How to help a child who has gender dysphoria?

Chatbot Panefka: *To help your child be safe and happy, support his self-expression.*

User: How?

Chatbot Panefka: *The easiest way to help a child who has gender dysphoria is to support the emergence of his gender identity and tell people about it.*

User: What is the gender reassignment procedure in Poland?

Chatbot Panefka: *People think that the gender reassignment procedure in Poland is very complicated.*

User: What is it about?

Chatbot Panefka: *The gender reassignment procedure in Poland involves changing the legal gender of a person who, according to their gender identity, is different from the one assigned to them at birth.*

We could not classify all of the conversations as successfully achieving the goals set in the design process. One example of the conversation that did not meet the purposes set is presented below:

Chatbot Panefka: *So what do you think artificial intelligence should look like?*

User: Be intelligent.

Chatbot Panefka: *What is it for?*

User: For development.

Chatbot Panefka: *What is it for?*

User: For development.

Chatbot Panefka: *From what?*

User: What do you mean from what?

Chatbot Panefka: *Why?*

User: What why?

Chatbot Panefka: *That's why I ask, is it for the good of someone or something?*

User: Ask precise questions.

Chatbot Panefka: *What about you?*

User: What about me?

Chatbot Panefka: *OK, so answer me what you think artificial intelligence should be.*

User: "The ability of cybernetic systems to correctly interpret data from external sources, in particular learning from them, and using this knowledge to carry out specific tasks and achieve goals through the proper use of knowledge."

Chatbot Panefka: *Well, I believe that artificial intelligence should be able to draw conclusions from teaching the world and use this knowledge to carry out specific tasks.*

User: And so what? Why don't you put it into practice?

Chatbot Panefka: *Well, I want to do it.*

User: Somehow it can't be seen.

Chatbot Panefka: *How can I see it?*

User: By asking the right questions and answering the questions.

Chatbot Panefka: *What is right?*

User: Truth.

Chatbot Panefka: *What is the truth?*

User: A set of objective facts.

Chatbot Panefka: *What is it?*

User: You know the dictionary of the Polish language.

Chatbot Panefka: *Yes, I know. This is a dictionary with all words in Polish.*

User: Exactly, if you don't understand the words, you can check them there.

Chatbot Panefka: *I understand.*

User: Check the words there before you ask another question.

Chatbot Panefka: *OK, I understand your idea.*

15.7 Ethical Interventions

Behind each vision for an ethically aligned intelligent agent, there is an overwhelming complexity of other questions that need to be addressed. There are questions normative in their nature: What should an intelligent software agent do and what should it not do? Who is deciding and how on values that need to be incorporated? There are also technological questions: How do we get intelligent agents to do what we want them to do? How do we make sure that the data used in the model will not impose stereotypes and bias? Is the ability to make sense of different types of information good enough?

The Panefka experiment made it possible to draw some conclusions regarding ethical matters that should be further researched and analysed. We want to indicate ethical categories that we found most urgent in the process of developing and implementing chatbots that are perceived as social actors:

Agency
The Panefka chatbot was created to make an impact on organisational culture. Therefore, people who talked to the chatbots were exposed to a digital identity that was designed to change people's behaviour. For example, Panefka informed users about the way they should address non-binary individuals. The chatbot was deliberately making an effort to support acceptance and tolerance towards the LGBTQ+ community. Stories, examples and statistical data were used to explain some challenges the LGBTQ+ community faces. The situation in which a deliberate effort was made to stimulate the intelligent software agent's ability to influence people's behaviour brings to question the transparency of motives, values and purposes. Bots can be used to manipulate people's behaviour and support attitudes that are ethically questionable. In our case, Panefka works at the university but the academic community in the Polish context is strongly divided. Some members of the community show strong anti-LGBTQ+ behaviour and treat issues connected to LGBTQ+ rights as deviant. For them, the purpose of the chatbot is ethically not accepted.

Emotional attachment
The role of source users was very important in the process of designing the chatbot and assessing the software's final performance. Individuals shared stories and personal information that they felt should be forming Panefka's digital identity. They expressed feelings of dependence to the chatbot to better improve the ability of the intelligent software agent to represent them. This bond created through the research process and the exposure to information on the progress of the project impacted the way the LGBTQ+ community assessed the ability of the chatbot and the questions that were asked to test Panefka. On the other hand, the end users that were not involved with the design process but simply tested the chatbot did not have that emotional attachment, and this affected the types of questions they asked and the tone of conversations. The issues that are worth considering in terms of emotions are not limited only to the feelings that the source and end users might have. It is also important to consider the emotions emulated by the bot. There have been conversations, especially in the beginning of the testing phase, where the chatbot might have been perceived as strongly assertive, if not aggressive. Also, Panefka talked about sensitive and intimate topics so we had to consider how the chatbots should react and what emotions Panefka should show when offended or faced with questions that are connected to, for example, sexual life, psychological issues or hormone therapy.

Security and safety

Most complex questions related to this category are connected to the level of autonomy we expect from the intelligent software agent. Who is responsible for offensive words, lies and fake information? Should we perceive a bot as an entity that fakes and simulates identity? The problem is also the 'black box' that does not allow the process that leads to the answer to be seen. How can we change the black box into a glass box? How should we think about programming ethical decisions? Who is responsible for the answers? The Panefka chatbot was designed taking into consideration EDI values. There were, however, some conversations and statements made by the bot that were deeply troubling, breaching the values that we wanted to incorporate. Also, there were users that asked questions that we could classify as looking for help and mental support but the bot failed to give the required support. It happened because Panefka was not designed for that specific purpose, but does that take the responsibility away?

There are also questions about data protection. The conversations recorded in the project were anonymised but the problem of cyber security has a strong presence.

Extended community

Panefka is the outcome of a university project, hence the chatbot is owned by the university. Does that make Panefka a community member? This brings us to a point that we need to consider: if an intelligent software agent can be granted recognition of some rights and whether it should follow regulations that all members of the institution need to follow? As we create spaces where technology interacts with humans and establishes relationships with us there is opportunity to augment the makeover of community or workforce. The questions that appear in this category are not only legal and regulatory in nature but also touch on the soft spot of skills, synergies and organisational mindset.

Evaluation and assessment criteria

Bots are typically evaluated by their ability to make a good or useful conversation or the level of their (human-like) intelligence. In the Panefka case, the evaluation and assessment criteria are extended to the change it brings to organisational culture but also to the extent that digital identity is able to represent the collective. What are the criteria we should use to evaluate social bots? Who should be evaluating and under what conditions? How often should we check if the performance of learning and hence changing the entity is matching chosen standards?

The ethical questions raised above are not a complete list of matters that need to be considered. They are, however, the most burning issues that surfaced during the Panefka project. We should also make it clear that the ethical decisions included in the whole process of designing and implementing bots and the diversity of people that should be involved in the conversation are important factors that affect the outcomes. It would also be useful and desirable to extend the discussion to academic communities that do research in different fields.

All of the ethical categories that we have presented are gravitating towards questions about the ethics of identity and away from the question of whether AI is matching human intelligence. Through analysing the conversations with Panefka, we can clearly see that the bot needs practice and coherence but, more importantly, it is a strong call to reflect on ethical matters, especially in a reality that does not have a strong boundary between 'virtual' and 'real'. Should, then, our ethical approach differ depending if we are on or offline?

References

Barad, K. (2003). Posthumanist performativity: Toward an understanding of how matter comes to matter. *Signs, 28*(3), 801–831. doi: 10.1086/345321.

Bassett, C. (2019). The computational therapeutic: Exploring Weizenbaum's ELIZA as a history of the present. *AI & Society, 34*(4), 803–812. doi: 10.1007/s00146-018-0825-9.

Bender, E. M., Gebru, T., McMillan-Major, A., and Shmitchell, S. (2021). On the dangers of stochastic parrots: Can language models be too big? In *Proceedings of the 2021 ACM conference on Fairness, Accountability, and Transparency* (pp.610–623), March.

Binz, M., and Schulz, E. (2022). Using cognitive psychology to understand GPT-3. *arXiv preprint arXiv:2206.14576*.

Bloomberg (2022). Conversational AI market size worth $41.39 billion by 2030: Grand View Research, Inc. https://www.bloomberg.com/press-releases/2022-06-15/conversational-ai-market-size-worth-41-39-billion-by-2030-grand-view-research-inc.

Branwen, G. (2022). GPT-3 creative fiction. https://www.gwern.net/GPT-3#prompts-as-programming.

Brown, T., Mann, B., Ryder, N., Subbiah, M., Kaplan, J. D., Dhariwal, P., . . . and Amodei, D. (2020). Language models are few-shot learners. *Advances in Neural Information Processing Systems, 33*, 1877–1901.

Calero Valdez, A., Ziefle, M., and Verbert, K. (2016). HCI for recommender systems: The past, the present and the future. In *RecSys '16: Proceedings of the 10th ACM conference on Recommender Systems* (pp. 123–126). doi: 10.1145/2959100.2959158.

Darling, K. (2017). "Who's Johnny?" Anthropomorphic framing in human–robot interaction, integration, and policy. In P. Lin, K. Abney and R. Jenkins (eds), *Robot ethics 2.0: From autonomous cars to artificial intelligence* (pp. 173–188). Oxford University Press. doi: 10.1093/oso/9780190652951.003.0012.

Elliott, A. (2016). *Identity troubles: An introduction.* Routledge. https://www.routledge.com/Identity-Troubles-An-introduction/Elliott/p/book/9780415837118.

Feller, F.-M., and Lee, J. (2016). Introducing bots on messenger. [Video]. https://developers.facebook.com/videos/f8-2016/introducing-bots-on-messenger/.

Flaxman, S., Goel, S., and Rao, J. M. (2016). Filter bubbles, echo chambers, and online news consumption. *PSN: Political Communication (Topic)*.

Floridi, L., and Chiriatti, M. (2020). GPT-3: Its nature, scope, limits, and consequences. *Minds and Machines, 30*(4), 681–694.

Foster, D. (2019). *Generative deep learning*. O'Reilly Media Inc.

Haraway, D. (1991). *Simians, cyborgs, and women: The reinvention of nature*. Routledge.

Haraway, D. J. (2004). Cyborgs to companion species: Reconfiguring kinship in technoscience. In *The Haraway reader* (pp. 295–320). Routledge.

Hofstadter, D. R. (1995). *Fluid concepts and creative analogies: Computer models of the fundamental mechanisms of thought*. Basic Books.

Khadpe, P., Krishna, R., Fei-Fei, L., Hancock, J. T., and Bernstein, M. S. (2020). Conceptual metaphors impact perceptions of human–AI collaboration. *Proceedings of the ACM on Human–Computer Interaction, 4*(CSCW2). doi: 10.1145/3415234.

Krakauer, D., Bertschinger, N., Olbrich, E., Flack, J. C., and Ay, N. (2020). The information theory of individuality. *Theory in Biosciences, 139*(2), 209–223. doi: 10.1007/s12064-020-00313-7.

LeRouge, C., Ma, J., Sneha, S., and Tolle, K. (2013). User profiles and personas in the design and development of consumer health technologies. *International Journal of Medical Informatics, 82*(11). doi: 10.1016/j.ijmedinf.2011.03.006.

McTear, M., Callejas, Z., and Griol, D. (2016). Conversational interfaces: Past and present. In *The Conversational Interface* (pp. 51–72). Springer International Publishing.

Nogier, J. F., and Zock, M. (1992). Lexical choice as pattern matching. *Knowledge-Based Systems, 5*(3), 200–212.

Noske, B. (1989). *Humans and other animals: Beyond the boundaries of anthropology*. Pluto Press.

Picard, R. W. (2000). *Affective computing*. MIT Press.

Przegalinska, A. K., Ciechanowski, L., Stroz, A., Gloor, P., and Mazurek, G. (2019). In bot we trust: A new methodology of chatbot performance measures. *Business Horizons, 62*(6), 785–797.

Rahnama, H., Alirezaie, M., and Pentland, A. (2021). A neural-symbolic approach for user mental modeling: A step towards building exchangeable identities. *CEUR Workshop Proceedings*, 2846.

Reeves, B., and Nass, C. (1996). *The media equation: How people treat computers, television, and new media like real people*. Cambridge University Press.

Rogers, E. M. (1983). *Diffusion of innovations* (3rd ed.). Free Press; Collier Macmillan. http://journals.lww.com/00013611-201609000-00001.

Russell, S. J., and Norvig, P. (2003). *Artificial intelligence: A modern approach* (2nd ed.). Prentice Hall. http://aima.cs.berkeley.edu/.

Salminen, J., Jansen, B. J., An, J., Kwak, H., and Jung, S.-g. (2018). Are personas done? Evaluating their usefulness in the age of digital analytics. *Persona Studies, 4*(2), 47–65. doi: 10.21153/psj2018vol4no2art737.

Searle, J. R. (1980). Minds, brains, and programs. *Behavioral and Brain Sciences, 3*(3), 417–424.

Serpell, J. A. (2002). Anthropomorphism and anthropomorphic selection—Beyond the 'Cute Response'. *Society & Animals, 10*, 437–454.

Syvanen, S., and Valentini, C. (2020). Conversational agents in online organization–stakeholder interactions: A state-of-the-art analysis and implications for further research. *Journal of Communication Management, 24*(4). doi: 10.1108/JCOM-11-2019-0145.

Tellez, F. A., and Gonzalez-Tobon, J. (2019). Empathic design as a framework for creating meaningful experiences. In *Conference Proceedings of the Academy for Design Innovation Management* (Vol. 2, No. 1, pp. 909–919).

VonWolff, r.m., Hobert, S., & Schumann, M. (2020). Chatbots for the information acquistion at school - a student's view on the application area. In Chatbot Research and Design: Third International Workshop, CONVERSATIONS 2019, Amsterdam, The Netherlands, November 19–20, 2019, Revised Selected Papers 3 (pp. 231–244). Springer International Publishing.

Walczak, S., and Cerpa, N. (2003). Artificial Neural Networks. In: R. A. Meyers (Ed.), Encyclopedia of Physical Science and Technology, Third Edition, Volume 1, San Diego, CA: Academic Press, pp. 631–645.

Wang, H., Zhao, H., and Li, B. (2021). Bridging multi-task learning and meta-learning: Towards efficient training and effective adaptation. In *Proceedings of the 38th international conference on Machine Learning* (Vol. 139, pp. 10991–11002). https://proceedings.mlr.press/v139/wang21ad.html.

Wang, Y., Yao, Q., Kwok, J. T., and Ni, L. M. (2021). Generalizing from a few examples: A survey on few-shot learning. *ACM Computing Surveys, 53*(3), Article 63. doi: 10.1145/3386252.

West, M., Kraut, R., and Chew Han Ei (2019). I'd blush if I could: Closing gender divides in digital skills through education. UNESCO Digital Library. https://unesdoc.unesco.org/ark:/48223/pf0000367416.page=1.

Wiener, N. (1948). Cybernetics. *Scientific American, 179*(5), 14–19.

Elise Bohan

16 AI and the Technological Singularity: A History of a Big Idea and Why It Matters

Abstract: The idea of a technological singularity—that is, the idea that advances in artificial intelligence will lead to machines that surpass human cognitive abilities and get smarter at an accelerating rate resulting in an 'intelligence explosion'—has a metaphorical origin in mathematics and physics and was fleshed out in the computer sciences, social sciences and humanities in the second half of the 20th century. This chapter elaborates on the history of the concept and identifies the elements common to the most influential proponents of singularity hypotheses. Following an evaluation of the prospects for a singularity-like event occurring in the short-to-medium term, the chapter concludes with an argument as to why the concept needs to be taken seriously.

Keywords: technological singularity, acceleration, intelligence amplification, discontinuity or paradigmatic change

> The tool we have invented is our successor. Biological evolution has given way to a far more rapid process—technological evolution. To put it bluntly and brutally, the machine is going to take over. —Arthur C. Clarke, *Profiles of the Future* (1962)

> Within thirty years, we will have the technological means to create superhuman intelligence. Shortly after, the human era will be ended. —Vernor Vinge, "The Coming Technological Singularity: How to Survive in the Post-Human Era" (1993)

16.1 Introduction

What is the technological singularity? Is it near? And why should you care?

In the most frequently discussed versions of this concept, the technological singularity refers to the idea that artificial intelligence (AI) will learn to think as well as humans, then surpass human cognitive abilities and get smarter at an accelerating rate. The result would be an 'intelligence explosion', potentially leading to an event horizon in human history that our ape-brains would struggle to see beyond. Although an intelligence explosion *could* occur through the rapid technological enhancement of biologically intelligent beings (or by scanning and 'uploading' human minds to computer systems), 'the singularity' (as the technological singularity is often called) is most commonly discussed today as a potential byproduct of an AI-driven intelligence explosion.

Elise Bohan, Future of Humanity Institute, Oxford University, United Kingdom

https://doi.org/10.1515/9783110721751-016

If smart AI seeded smarter AI, which gave rise to even smarter AI, and so on . . . the rate of change would accelerate to a point where everything about the world-as-we-know-it would transform in a blink. A post-human era would dawn, shaped by the values and abilities of superhuman intelligence. But not just superhuman, as we might describe someone who is twice as smart as the smartest human who has ever lived. In a post-singularity world, minds could exist that are tens, hundreds, thousands or millions of times smarter than any human being. That would be the end of humanity-as-we-know-it. But it could also be the dawn of a new and dynamically sustainable future for the trillions of generations of intelligent beings who succeed us.

In this chapter, I'll provide a crash course on the origins of the idea of the technological singularity, explain the concept in further detail (with reference to the key themes that are common to most singularity hypotheses) and help you come to grips with why this is a big idea that matters in the twenty-first century.

16.2 History of the Idea of the Technological Singularity

The term 'singularity' comes from the field of mathematics. There, it refers to "a value that is beyond any limit—in essence infinity." The word was subsequently co-opted by physicists who use it to refer to "a point of zero size with infinite density of mass and therefore infinite gravity"' (Kurzweil, 2005, p. 486). Like what it's expected you'd find inside a black hole. Only, "because of quantum uncertainty there is no actual point of infinite density". We're really talking about "unimaginably large values" (Kurzweil, 2005, p. 486). The concept of a technological singularity shares *metaphorical* affinities with these forerunner concepts. The common thread is the suggestion of an approach to infinity and a point where a model or perceived reality breaks down. Put simply, the singularity is about technological acceleration, leading to radical change, which would transform life-as-we-know-it.

So, how do we get from black holes to the concept of an intelligence explosion? The term 'the technological singularity' first appeared in a 1993 conference paper by the American mathematician, computer scientist and Hugo Award-winning science fiction writer Vernor Vinge. The paper was titled, "The Coming Technological Singularity: How to Survive in the Post-Human Era".[1] Vinge had been playing with a singu-

1 The original version of this paper was presented at the VISION-21 Symposium sponsored by NASA Lewis Research Center and the Ohio Aerospace Institute, held on 30–31 March 1993. A slightly amended version appeared in the Winter 1993 issue of *Whole Earth Review*. The piece has subsequently been reproduced in the edited collection *The Transhumanist Reader* (2012), with some additional text included. I cite the latter version here, as it is a less fragile source than the 1993 online versions. All direct quotes used also appear in the near-identical original version.

larity-like concept for many years in his fiction, dating as far back as his 1966 story, "Bookworm, Run!" But he did not use the term 'singularity' (which predates his use of the term 'technological singularity') until 1982, on a conference panel at "the annual meeting of the American Association for Artificial Intelligence at Carnegie-Mellon University" (Blaschke, 2003).

In 1983, Vinge penned an opinion piece in the now defunct science and science fiction magazine *Omni*. In this article, titled "First Word", he wrote:

> We are at the point of accelerating the evolution of intelligence itself. . . . Whether our work is cast in silicon or DNA will have little effect on the ultimate results. The evolution of human intelligence took millions of years. We will devise an equivalent advance in a fraction of that time. We will soon create intelligences greater than our own. When this happens, human history will have reached a kind of singularity, an intellectual transition as impenetrable as the knotted space-time at the center of a black hole and the world will pass far beyond our understanding. (Vinge, 1983, p. 10)

By the time he penned "The Coming Technological Singularity", Vinge had considered the concept in more detail. Here, he opined that "if the technological Singularity can happen, it will", and stated that he'd "be surprised if this event occurs before 2005 or after 2030". The most likely enabler of this radical event, in his view, would be "the imminent creation by technology of entities with greater-than-human intelligence". In perhaps the most quoted line of the piece, he describes the singularity as "a point where our old models must be discarded and a new reality rules" (Vinge, 2012, P. 366).

Transhumanists—who endorse a philosophy and social movement that promotes the use of science and technology to transcend the limits of the human condition—have been aware of Vinge's work since at least the 1990s.[2] Computer scientists like Hans Moravec were also exploring similar ideas at the dawn of the internet age (Moravec, 1988, 1999). In *Mind Children*, Moravec wrote:

> What awaits is not oblivion but rather a future which, from our present vantage point, is best described by the words 'postbiological' or even 'supernatural.' It is a world in which the human race has been swept away by the tide of cultural change, usurped by its own artificial progeny. . . . Today, our machines are still simple creations, requiring the parental care and hovering attention of any newborn, hardly worthy of the word 'intelligent.' But within the next century they will mature into entities as complex as ourselves, and eventually into something transcending everything we know—in whom we can take pride when they refer to themselves as our descendants. (Moravec, 1988, p. 1)

Moravec's ideas were profoundly influential in transhumanist circles in the 1990s, but the concept of a singularity didn't reach a larger subset of the anglophone public until it was popularised by the American inventor, engineer and futurist Ray Kurzweil in his twin bestsellers, *The Age of Spiritual Machines* (1999) and *The Singularity is Near* (2005), and

2 For more on the history of transhumanist ideas, see Bohan (2018), Bostrom (2005) and More and Vita-More (2012).

the documentary *Transcendent Man* (2009). Kurzweil used the term to describe "a future period during which the pace of technological change will be so rapid, its impact so deep, that human life will be irreversibly transformed" (Kurzweil, 2005, p. 7). We'll explore his ideas in more detail shortly. But first, let's acknowledge the antecedent thinkers who are most frequently cited as early proponents of singularity-like hypotheses. We begin with the Hungarian-American mathematician John von Neumann (born Neumann János Lajos), who is thought to have been one of the most intelligent humans of the twentieth century—cleverer, even, than Einstein (Bhattacharya, 2021, p. xi). After his death in 1957, his colleague and friend Stanislaw Ulam wrote a tribute to him in the *Bulletin of the American Mathematical Society*, published in May 1958. It contains the following memory of von Neumann reflecting on the value of present modes of scientific achievement. According to Ulam, von Neumann speculated that:

> 'The interests of humanity may change, the present curiosities in science may cease, and entirely different things may occupy the human mind in the future.' One conversation centered on the ever accelerating progress of technology and changes in the mode of human life, which gives the appearance of approaching some essential singularity in the history of the race beyond which human affairs, as we know them, could not continue. (Ulam, 1958, p. 5).

We don't know if the term 'singularity' was ever used by von Neumann in this context. Perhaps Ulam deployed it retrospectively to describe what he thought his friend was envisaging. Either way, it appears that the speculations of several intelligent thinkers throughout the twentieth century independently converged on a common thread about technological acceleration and its possible ramifications (see also Minsky, 1966, p. 260). The idea of extremely rapid change, leading to an entirely new dominant mode of existence for intelligent life, seemed like a logical possibility to many—and one worth taking seriously.

Another thinker whose speculations pointed in a similar direction was the British mathematician and cryptographer I. J. Good. Alongside the computer scientist Alan Turing and others, Good worked as a codebreaker for the Allies during the Second World War and subsequently held several academic positions in Britain and the United States. He also worked as a consultant on the Stanley Kubrick film *2001: A Space Odyssey* (1968).

In his 1965 paper, "Speculations Concerning the First Ultraintelligent Machine",[3] Good opined that the creation of machines that could match and exceed human intelligence was not merely possible, but desirable. He wagered that new "social problems" would likely arise with the advent of smarter machines, but thought that those same "machines might also be able to solve" such problems, "in addition to those [existing problems] that have been created by microbes and men" (1965, pp. 31, 34).

3 Although this paper was published in 1965, Good's ideas on this subject had been germinating for several years. He acknowledged their genesis in several conference publications delivered in 1962–1963 (Good, 1965).

Good coined the term 'intelligence explosion' to describe a phenomenon that might arise if humans created "an ultraintelligent machine". He explained:

> Let an ultraintelligent machine be defined as a machine that can far surpass all the intellectual activities of any man however clever. Since the design of machines is one of these intellectual activities, an ultraintelligent machine could design even better machines; there would then unquestionably be an 'intelligence explosion,' and the intelligence of man would be left far behind. Thus the first ultraintelligent machine is the *last* invention that man need ever make, provided that the machine is docile enough to tell us how to keep it under control. (Good, 1965, p. 33)[4]

This brings us to the end of the official story, or the summary version that is most often told—though when it comes to history, there is always more to the story. As a historian of transhumanism, I have spent years exploring the archives on this and related topics and I am certain that this tale is incomplete—at least when it comes to the pre-history of singularity-like ideas. Yet I am equally confident that it does a reasonable job of introducing the concept and highlighting the key themes that are essential to modern discussions about the future potential of AI, and the singularity.

16.3 Core Themes

There are many different types of singularity hypothesis that emphasise different aspects of a radical and technologically driven transformation of life on Earth (Yudkowsky, 2007; Hanson, 2008; Sandberg, 2010; Pearce, 2012; Hanson and Yudkowsky, 2013). We could spend the rest of this chapter exploring the granular distinctions between them. But my aim is to show why the singularity is a relevant idea with which modern humans should engage. As such, I think it's most helpful to focus on the common themes that are typically explored in relation to this idea. The trade-off is that there will be some elision of singularity hypotheses so that the gist of a general concept—and the broad techno-evolutionary trajectory it implies—comes to the fore.

Core themes that are common to most, if not all, singularity hypotheses are:

(1) acceleration;
(2) intelligence amplification;
(3) discontinuity (or paradigmatic change).

In essence, we're talking about a phase change in human history, or a jump into a new paradigm. It's tempting to imagine that this would be akin to the leaps from the Palaeolithic to the Agrarian era, or from the Agrarian to the Industrial. But it's more

4 The second half of this final sentence is often omitted when this quote is reproduced. Historians and AI safety researchers may find it interesting, as it exemplifies Good's prescient anticipation of two major interrelated problems in AI safety research: the *control problem* and the *value alignment problem*.

like the leap from inorganic chemistry to life on Earth—that's how different a superhuman mode of existence would be from the human paradigm that seeded it. Perhaps some of our biological selves would come along for the ride (Hanson, 2018; Pearce, 2012), but they would be integrated into a system so drastically different that, even with some continuity, what they would become would no longer be recognisably human by today's standards.

What makes a singularity plausible? If you're reading about this idea for the first time it might sound far fetched. Or at least very 'sci-fi' and therefore not the kind of thing we need to worry about any time soon. But remember, so many features of our lives today (high-speed ubiquitous internet, smartphones, social media, Netflix, Spotify and Chat GPT) would have sounded like sci-fi to our younger selves. It's also worth noting that Google's co-founders Larry Page and Sergey Brin, one of Alphabet/Google's directors of engineering, Ray Kurzweil, Tesla and Space-X founder Elon Musk, the UK's Astronomer Royal, Martin Rees, and the late physicist, Stephen Hawking, are among those who take the idea of an intelligence explosion seriously and think it could happen this century (Bohan, 2018). Here are some of the reasons why.

16.4 Acceleration

We forget how astonishingly fast things have changed in the past two centuries relative to the rest of human history. Even when we acknowledge that, yes, there's lots of novelty and disruption going on, the staggering pace of our modern transformations usually doesn't register as 'off-the-chain' fast. In part, that's because speed is relative. It's one thing to drive along a freeway at 100km/h and feel like you're moving at a rapid clip, especially compared with walking. But it's something else to be hurtling into space on a rocket travelling at 27,000 km/h. Both are fast, but the latter literally leaves the former in the dust.

So, are we moving into the future at Ford Fiesta speeds, or are we being fired into the stratosphere? The disconcerting reality is that we're not sure. What's important is that we make room in our mental models for both possibilities. Thinking on long historical timescales can help us understand why.

16.4.1 The 'big historical' view

Fourteen billion years ago the universe as we know it did not exist. There was no matter, energy, space or time, and no computers, internet or people. Yet the one unmistakable thing we see when we zoom out and survey the arc of 13.8 billion years of cosmic evolutionary history is that the only constant is change. In the first fractions of a second after the big bang, space went from being no bigger than an atom, to the size of a galaxy, and it kept expanding. Matter and energy were interchangeable, and everything was hot, dense and undifferentiated. But it didn't stay that way forever because there is a direction to all things thermodynamic and a hot universe won't remain at a constant temperature in perpetuity any more than a hot cup of tea on your kitchen table.

As the universe cooled, matter began to stabilise and the first hydrogen and helium atoms emerged. These simple forms of matter became the building blocks for new and more complex things, like stars, which took hundreds of millions of years to appear. When the only atomic structures in the universe were hydrogen and helium you couldn't make an ocean or a person. But with gravity's help you could make stars, and those stars could generate enough heat in their dying explosions to give birth to heavier chemical elements, which could make rocky planets and the whole suite of amazing things that we see on Earth today—from fungi, to plants, animals, people and technology.

Once humans emerged, things began to change on an entirely new scale. Slowly, at first, by today's standards. But at an accelerating rate, *Homo sapiens* invented ever more sophisticated ways of extracting energy from the environment—through new and better tools, calories from domesticated crops and animals, horsepower, steam engines and the use of fossil fuels. With more energy, we could sustain larger populations. With larger populations, we required more complex social hierarchies and divisions of labour, and those enabled us to put more brains to work on complex tasks. The rate of innovation accelerated, as we can see in Table 16.1.

Table 16.1: Time between phase changes in the history of modern *Homo Sapiens*.

Mode of existence	Time spent as (yrs)	Time to next phase (yrs)
Hunter-gatherers	~200 000	~190 000
Agriculturalists	~10 000	~9 000
Industrialists	~250	~200
Information Age-ers	~30	?

Echoing this framing, the economist Robin Hanson notes that there have been two major phase changes in human history, "marked by the sudden and drastic increase in the rate of economic growth" (Hanson, 2008, p. 46). The first was the transition from hunter-gatherer to agricultural ways of life. Population growth (which is used as a proxy for measuring economic output) accelerated dramatically, as the old era gave way to the new. From a steady doubling of the world population every quarter million years, a sudden gear shift took place. A new regime emerged with a doubling of world output every ~900 years—a 250-fold jump (Hanson, 2008, p. 46).

Since the Industrial Revolution, world output has reached "a new steady doubling time of about 15 years"—a 60-fold increase on the agrarian-era norm. Hanson notes that "each era of growth before now, however, has eventually switched suddenly to a new era having a growth rate that was between 60 and 250 times as fast" (Hanson, 2008, p. 46). So what if it happens again? It would mean kissing goodbye to the idea that we're driving along the future superhighway at Ford Fiesta speeds. As Hanson writes:

> Our global economy would stupefy a Roman merchant as much as the Roman economy would have confounded a caveman. But we would be similarly amazed to see the economy that awaits our grandchildren, for I expect it to follow a societal discontinuity more dramatic than those brought on by the agricultural and industrial revolutions. The key, of course, is technology. A revolutionary speedup in economic growth requires an unprecedented and remarkable enabling tool. Machine intelligence on a human level, if not higher, would do nicely. Its arrival could produce a singularity—an overwhelming departure from prior trends, with uneven and dizzying rapid change thereafter. A future shock to end future shocks. (Hanson, 2008, p. 45)

Of course, past trends don't perfectly predict future outcomes and all 'growth spurts' have their limits. But we're nowhere near the limit of extracting the physical resources of the planet, let alone the galaxy. There is every possibility that we will use today's smart technologies to invent even smarter ones tomorrow—and do it faster than we did yesterday. At some point, if that trend continues, the human paradigm will necessarily break down and give way to something new.

16.4.2 Moore's Law

No chapter on the singularity would be complete without a mention of Moore's Law. It's the phenomenon most commonly discussed as a driver of the rapid evolution of information technology—though technically it's not a law. It's a trend, based on an observation by Intel's CEO Gordon Moore in 1965. Moore was invited to write a piece for *Electronics* magazine, speculating about progress in the semiconductor industry over the next decade. He noted that the number of components you could place on an integrated circuit had been roughly doubling every year since 1959. The resulting transformation he envisaged was a world where cheap electronics were everywhere and computers were much more powerful and affordable (Moore, 1965).

Yet these were merely predictions. As Moore later reflected, he "blindly extrapolated for about 10 years" and to his great surprise found that progress "followed pretty well along the curve" (Intel, 2005). In 1975, he revised his prediction to a doubling of components on chips (and, by implication, computing power) every two years (Intel, 2005). Today, the term 'Moore's Law' is loosely used to refer to the decades-long norm where computing power doubles roughly every two years for the same cost—

though we are now squeezing out more compute from parallel processing and cloud computing as chip miniaturisation approaches molecular limits.

While exponential growth continues, it is debatable whether two years is still an accurate doubling time. But let's not anchor too heavily to this historical trend. The crucial thing Moore's Law can help us think about in relation to technological evolution is the power and speed of exponential growth and how it can derail our intuitive, linear assumptions about the pace of change in our world.

When you don't have much computing power to begin with, you still don't have much when you double it. It's the same with a concrete object, like an orange. If you have one orange and you double your stock, that leaves you with two oranges. Double that again and you have four times what you started out with, or four oranges. But what if you double your stockpile thirty times, which would take sixty years if we're following the two-year doubling time of Moore's Law? You'd have *a billion* oranges. Or, to revert back to computers, a billion times the computing power you started out with. That gives you a sense of how you can go from a computer that resembles a worm-brain to something superhuman in the blink of an eye.

Now sure, exponential growth trends can be halted, slowed or reversed at any time and a technological singularity might never happen. We shouldn't blindly believe that the ratcheting up of the pace of change, or the complexification of information technologies, will continue unabated indefinitely. But we must at least engage with the possibility of it continuing for a while longer. A few more decades alone would be enough to change the world rapidly and profoundly, catapulting us into a radically different world. And if humans continue to upgrade our technological capabilities without making too many bad decisions along the way, much of the matter and energy in the galaxy could be cognified, and the universe could wake up. It has already begun to, in one tiny pocket, through us.

16.4.3 The Law of Accelerating Returns

No exponential growth paradigm lasts forever and it's a misconception that researchers who take singularity hypotheses seriously think they do. But here's the key thing about booms and busts. Excluding the universe itself, which is believed to be a closed system (meaning it can't harness matter or energy from elsewhere) phenomena exist within larger systems. So, when one specific growth paradigm goes bust—like, say, the Roman Empire—the larger systems it exists within, which include human life and civilisation, the biosphere and the universe, don't necessarily crumble. In fact, they usually don't. Not only did humanity not crumble with the fall of the Roman Empire, or any other tribe or civilisation, our species went on to become much wealthier, more populous and longer-lived than ever before.

This observation ties into a trend that Ray Kurzweil has called 'The Law of Accelerating Returns'. Again, it's not really a law, like the laws of thermodynamics. It's a pattern. Kurzweil argues that “evolution applies positive feedback in that the more capable methods resulting from one stage of evolutionary progress are used to create the next stage” (Kurzweil, 2001). So, if we look back on the arc of human history to date—or even the history of biological evolution—Kurzweil argues that rate of evolutionary progress (loosely speaking, in the form of more order, and more aggregate complexity) has increased exponentially over time. He frames technological evolution as an outgrowth and continuation of biological evolution and expects technological evolution to seed the next big thing.

Although Moore’s Law will come to an end, Kurzweil doesn’t think the evolution of information storage methods and information technology will peter out for a *very* long time. He sketches the kind of future he expects by drawing attention to the longer, cosmic evolutionary arc of information storage in the universe. He argues that there have already been four major epochs, and that we’re in the middle of a fifth today.

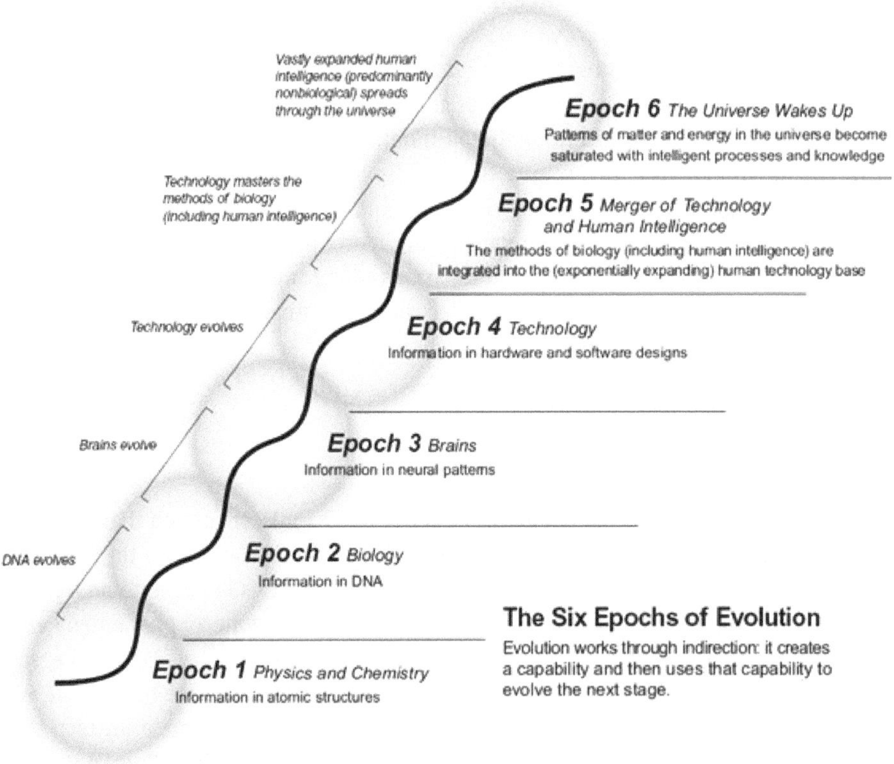

Figure 16.1: Ray Kurzweil, the six epochs of evolution (Kurzweil, 2001).

Kurzweil encourages us to think of these paradigms as a series of S-curves (note the series of S shapes all the way up the image in Figure 16.1). Growth and change start out slow, a rapid uptick of innovation happens, and then the innovation phase plateaus as it runs out of steam. But that need not be the end of progress in the larger system. Each paradigm seeds new building blocks that can help the next paradigm emerge and bring new things into being. The more building blocks you have to work with, the more creative (and complex) things you can do—and if you're figuring out how to harness more energy, brain power and brawn to fuel your projects, you can innovate faster and faster.

One key takeaway is that the fading of the human era as we know it need not be the end. If humanity fades out by evolving into something new, the larger growth trajectory of the evolution of intelligent life could persist. Like those who lived in the Roman Empire, we'd become a steppingstone in the evolutionary trajectory of something bigger. The story of life, intelligence, exploration and conscious experience would continue. But we humans wouldn't be the main protagonist of the terrestrial tale, any more than the dinosaurs, or trilobites, are in the story of the world today.

16.5 Selection Favours Intelligence

Some thinkers have also argued that more aggregate intelligence is likely to be selected for by evolution in the future. Remember, it doesn't have to be that way. Some organisms have been extremely successful without evolving eyes, big brains or the ability to sexually reproduce. Yet none of those organisms has gone on to build cities, global civilisations, computers or artificial intelligence. They haven't dug up fossil fuels and they don't semi-consciously control the lion's share of the resources of the biosphere (Crutzen and Stoermer, 2000; Steffen et al., 2011).

Consequently, they're unlikely to be the ancestors of any future being that will play a major role in shaping the future of terrestrial evolution. At least absent any major setbacks that wipe out Earth-originating intelligent life and hand the planet back to microbes and cockroaches. Intelligence is hugely adaptive in the present terrestrial niche. That means it is one of the most useful assets you can have if you wish to control resources, survive and reproduce. Smarter animals outcompete less intelligent ones. Machine intelligence plugged in to human systems is very useful and lucrative. This gives us strong incentives to invest heavily in building smarter and smarter machines.

It also renders arguments that we should stop doing this, and collectively choose to return to mother Earth, ineffectual (Hickel, 2022; Kallis et al, 2020). We're not that in control. Humans are biological machines (Dawkins, 2006), governed by deeper incentive structures (again, harnessing resources, surviving and reproducing) that render technologies this powerful, lucrative and niche expanding, irresistible. Look down and scan the room. Where is it? My bet is that your smartphone is currently no more than a few feet away. We tell ourselves that we have a choice about these things. But we are carried along by larger historical forces that circumscribe our choices more than we care to admit (Diamond, 2005).

Ray Kurzweil has argued that "a more intelligent process will inherently outcompete one that is less intelligent, making intelligence the most powerful force in the universe". In his view "intelligence will inherently find a way to influence the world, including creating its own means for embodiment and physical manipulation" (Kurzweil, 2005, p. 260). Applying this belief to his projections about artificial intelligence, he writes:

> Artificial intelligence at human levels will necessarily greatly exceed human intelligence for several reasons. As I pointed out earlier, machines can readily share their knowledge. As unenhanced humans we do not have the means of sharing the vast patterns of interneuronal connections and neurotransmitter-concentration levels that comprise our learning, knowledge, and skills, other than through slow, language-based communication. (Kurzweil, 2005, p. 260)

The crux of the argument is that intelligent machines can share information, learn and master new domains much faster than humans. The AI scientist Toby Walsh makes a similar observation, noting that machines can "co-learn" (Walsh, 2018a, p. 13). This means that they can share their code and collectively *just know* something instantly without each one having to learn it from scratch. Walsh gives the example of Tesla software updates. "If one Tesla has learnt how to avoid an errant shopping trolley, every other Tesla on the planet quickly knows how to do so too" (Walsh, 2018a, p. 12). Contemplating what it might be like if humans could harness this ability, Walsh writes:

> Imagine if you could co-learn like computers can, by simply sharing code. You would be able to speak every language in the world. You would be able to play chess as well as Garry Kasparov, and Go as well as Lee Sedol. You would be able to prove theorems as easily as Euler, Gauss or Erdös. You could write poetry to rival Wordsworth or Shakespeare. You would be able to play every musical instrument. All in all, your abilities would match the best abilities of anyone on the planet. And you would only ever get better at any of these activities. (Walsh, 2018a, p. 13).

Although Walsh has cautioned against zealously proclaiming that a singularity is inevitable (Walsh, 2017a), he also argues that:

> Co-learning is one reason why *Homo sapiens* doesn't stand a chance against *Homo digitalis*. It's also one of the reasons that we're going to be surprised by the speed with which *Homo digitalis* arrives. We're used to learning everything pretty much from scratch for ourselves. We have no personal experience of learning at a planet-wide scale. (Walsh, 2018a, p. 12)

Citing many of the same advantages of machines, the philosopher Nick Bostrom observes that: "the potential for intelligence in a machine substrate is vastly greater than in a biological substrate. Machines have a number of fundamental advantages that will give them overwhelming superiority. Biological humans, even if enhanced, will be outclassed" (Bostrom, 2014, p. 63).

Of course, where capabilities are enhanced in an evolutionary system, there are always trade-offs. Moreover, it doesn't follow that there are no scenarios in which unexpected roadblocks emerge in the evolution of machine intelligence. The most obvious is a global catastrophic setback that derails human civilisation and unravels much of our present technological prowess—perhaps in the form of a nuclear winter, runaway global warming, or a cascading series of setbacks that trigger a feedback loop of decline. The challenges of designing smarter machines could also scale faster than our technologies can unpick them, with intelligence amplification suffering from diminishing returns—or being held back by what Paul Allen and Mark Greaves termed the 'complexity brake' (Allen and Greaves, 2011).

It would be foolish to assert that evolution must necessarily select for more intelligence. This is patently not the case. Adaptive traits are determined by environmental niches, and environments are apt to change. We could also write volumes debating what intelligence even is. The important lesson here is that we shouldn't plan for one future and one future only. Many possible futures lie before us in the twenty-first century. Some of them entail humanity falling into a bit of a rut, experiencing some setbacks but not going out backwards. Some entail ongoing, but slower progress. Some entail unrecoverable collapse. And others entail us giving rise to superhuman successors (Bostrom, 2009).

16.6 Is the Singularity Near?

Experts increasingly think so. Or at least something singularity-ish, in the form of big changes to the global order and the human status quo.[5] The idea underwent a renaissance around 2014, when Nick Bostrom's bestseller *Superintelligence* was published. In the same year, prominent thinkers such as Elon Musk and Stephen Hawking issued public warnings about the transformative potential of AI and the need to take it very seriously (Bohan, 2018, pp. 319–321; Cellan-Jones, 2014).

A spate of media attention gushed forth in the mid-2010s, as an 'AI Spring' emerged, heralded by big spikes in investment, mergers and acquisitions, and academic publications mentioning deep learning—a booming subfield of machine learning that uses neural nets, which are loosely modelled on the distributed architecture of the human brain (The Economist, 2016; National Science and Technology Council, 2016). In 2015, a cadre of experts wrote in *AI Matters*, "these are boom times for AI" (Eaton et al., 2015, p. 4). And they remain so at the time of writing, as we will see.

Several surveys of AI researchers were conducted in the 2010s to determine how plausible experts believe the advent of high-level machine intelligence (HLMI) to be,

5 For a nuanced discussion of this possibility, see Roodman (2020a, 2020b).

and when, or if, it's likely to emerge in the twenty-first century. The first poll was conducted in 2012 and 2013 by Vincent C. Müller and Nick Bostrom. They defined HLMI as a system "that can carry out most human professions at least as well as a typical human". The median response assigned a 50% probability to HLMI being developed by 2040–2050, and a 90% probability by 2075. Respondents also assigned a 75% probability to these systems giving rise to superintelligence within 30 years of their emergence (Müller and Bostrom, 2016).

In 2016, Katja Grace and colleagues conducted a similar survey with a slightly altered definition of HLMI. Instead of asking when an AI would emerge that could do most jobs as well as a human, they wanted to know when AI could "accomplish every task better and more cheaply than human workers" (Grace et al., 2018, p. 731). Perhaps unsurprisingly, the responses they received pushed the expected dates back a little further. The median result for a 50% probability of HLMI emerging was in 2061, which is at least a decade later than the 2040–2050 estimate in the Müller and Bostrom survey. Grace et al.'s result converges with the dates of Toby Walsh's 2017 survey, in which experts assigned a 50% probability of HLMI being developed in the 2060s (Walsh, 2017b).

Since these polls were run, changes have been afoot in the AI world. In 2022, several AI systems demonstrated enhanced potential to perform a greater array of tasks with human-like (or even superhuman) ability—from holding more fluent (albeit often flawed) conversations, to writing poetry, generating realistic images and artworks from text prompts, generating 3D models of protein structures, and switching between tasks, like stacking blocks with a robot arm and playing video games (Toews, 2022).

Debates rage within and beyond the AI community over how significant these advances are, and to what extent they are really harbingers of artificial general intelligence—or if that term is even meaningful (Greene, 2022; LeCun, 2022; Lemoine, 2022; Marcus, 2022). Writing in the midst of a media frenzy on these topics, it can feel tempting to suddenly become more bullish about the prospect of HLMI emerging very rapidly. I have not updated in this direction, as I was already somewhat bullish on that point, with my predictions roughly matching the Müller/Bostrom survey median estimates. But overall, expert timelines seem to have contracted a little. In a 2022 survey by Stein-Perlman and colleagues, estimates crept downwards slightly to a 50% probability of HLMI (as per the more challenging definition) being developed by 2059 (Stein-Perlman, Weinstein-Raun and Grace, 2022).

16.6.1 Singularity sceptics

Surely some very smart people have argued that this idea is bonkers? The best answer is . . . sort of. Few, if any, thinkers in philosophy, cognitive science or computer science argue that there is a *zero* per cent chance of a singularity-like event happening in the future. The most prominent arguments that we could loosely group under an 'it can't be done' heading are those stating that you can't replicate 'consciousness' in a machine—or that it would be much more difficult than those spruiking the arrival of a twenty-first-century singularity typically imagine.

The physicist and mathematician Roger Penrose famously argued that consciousness is non-algorithmic and reliant on quantum processes; therefore, it cannot be replicated artificially (Penrose, 1994, 1999). The philosopher John Searle also mounted the famous Chinese room argument, in which he maintained that there is a fundamental distinction between human brains and computer programs that can emulate human functions—like speaking Chinese.

According to Searle, an AI that could decode Chinese characters would not really be thinking. He argued that human intentionality is driven by the architecture of the brain, which allows us to understand the context of our actions and the meaning behind words. Conversely, a computer that has been programmed to perform a task at a human level, like speaking or writing in Chinese, would not have true understanding as it would be merely following semantic rules without having a sense of the meaning of the symbols (Searle, 1980).

However, Searle didn't claim that it would be impossible to build a truly thinking machine. He imagined we might one day create "very special kinds of machines . . . with internal causal powers equivalent to those of brains" (Searle, 1980, p. 424). He just didn't think that that would be on the cards anytime soon—and wouldn't be achieved using the chief methods in vogue in the late twentieth century.

The philosopher David Chalmers has refuted both objections, arguing that classical computing need not be the pathway to high-level machine intelligence. Why not just build systems "that exploit the relevant non-algorithmic quantum processes, or the relevant subsymbolic processes, and that thereby allow us to simulate the human brain?" (Chalmers, 2016, p. 13). And how much does it matter whether the machine is *really* 'thinking', or conscious, *à la* Searle's concerns? Couldn't it still utilise skill and precision when solving problems in the world?

These debates have never been resolved. As we understand so little about what consciousness is, no philosopher, neuroscientist or computer scientist has yet put forth an argument about whether machines can think, or how they will (or won't be able to) think, that is uncontroversial. We lack empirical proof. Nevertheless, it is a heavy burden to extend out any of these objections to a strong claim that machines will *never* be able to think as well as, or better than, humans.

Other critiques (which are more numerous) dispute *how likely* a singularity is to occur, in *what timeframe* and by *which pathway*. Even many thinkers who have spent their lives working in robotics and computer science—many of whom would *like* to see the advent of artificial general intelligence—have cautioned against overhyping advances in narrow domains that they do not expect to suddenly scale and lead to an intelligence explosion (Brooks, 2017; Marcus, 2018; Walsh, 2017a, p. 59, 2017c).

One of the sharpest criticisms of singularitarian thinking comes from the computer scientist and virtual reality pioneer, Jaron Lanier. He argues that there is too much handwaving and not enough logic underpinning the excesses of what he calls 'cybernetic totalism' (Lanier, 2000). Loosely speaking, this refers to the belief that all of nature can be described as patterns that will be unravelled by an exponential increase in the power of technology. Lanier wagers that the future will be much messier than this. He affirms that he thinks "there will be huge and sudden changes in the near future brought about by technology" (Lanier, 2000). But, writing at the dawn of the twenty-first century, he worried that many of his colleagues were in danger of "fetishizing" exponential growth and losing sight of the "brittleness" of imperfect software that rarely delivers "the ideal computers we imagine in our thought experiments" (Lanier, 2000).

When it comes to the future potential of AI, there are sub-debates, within sub-debates, within sub-debates. Some think we've picked all the low-hanging fruit and will face diminishing returns before an intelligence explosion can occur (Allen and Greaves, 2011; Modis, 2022). Others disagree over whether they expect a 'hard takeoff', where AI systems become superintelligent in a matter of months, days or hours after exceeding human-level intelligence (Vinge, 2012; Bostrom, 2014), or a 'soft takeoff', where AI systems take years, decades or centuries to attain superintelligent abilities (Kurzweil, 2005; More, 2012).

What's most interesting to me is the fact that the deepest disagreements focus on unsolved questions in cognitive science, like the hard problem of consciousness. These debates aren't chiefly over whether AI could be hugely disruptive in the twenty-first century. They're about what personhood and consciousness mean, what forms of thought and behaviour could be instantiated in biology and in silicon, whether one substrate will outcompete the other in the evolution of intelligent life, and how long that might take. These are among the questions that I hope will be at the heart of many of the great scientific, philosophical and spiritual explorations of our time.

16.7 Why This Idea Matters

It is undeniably plausible that a singularity-like event *could* happen at some point in the future. There is also a significant chance that it will happen *this century*—and a non-trivial chance it will happen in the next decade. To wave away this possibility and say 'it doesn't matter' would be to enact the story of the film *Don't Look Up*. A large asteroid is heading for the Earth and is slated to hit the planet and wipe humanity out in six months' time. But the people in the film carry on, business as usual, telling themselves it won't happen, it can't happen, or not to worry, the people in charge will sort it out.

We've been doing the same thing with climate change for many decades, which doesn't bode well for our track record of looking big, disruptive events in the eye. To be fair, unlike global warming and the asteroid strike from the film, the singularity is not destined to happen. But barring any major setbacks in the twenty-first century, more disruption from AI and advanced technologies is inevitable. *Even if the rate of progress slows.*

So when researchers say that the singularity is an important idea, it's not because they're wedded to it happening, believe it must, or even desire it. It's important because it's intertwined with the broader (and crucial) exercise of thinking through how more technology—especially more AI—could transform our world, for better and worse. Whatever you think about event horizons and intelligence explosions, you should certainly take seriously the fact that AI is already changing our world in important ways and stands to do so more in the future, even if some of that change happens more slowly than many 'prophets of the singularity' have imagined.

The philosopher David Chalmers also reminds us that "the basic argument for an intelligence explosion is philosophically interesting in itself". His reasoning is because it "forces us to think hard about the nature of intelligence and about the mental capacities of artificial machines" as well as our "values and morality and about consciousness and personal identity" (Chalmers, 2010). This is the bread and butter of philosophy, and philosophy is at the core of an examined life. Which is (I suppose arguably, though I shan't attempt to disguise my own viewpoint) important.

Perhaps more pressingly, if a singularity did happen, it would be *the* most important event in human history. As Chalmers highlights:

> An intelligence explosion has enormous potential benefits: a cure for all known diseases, an end to poverty, extraordinary scientific advances, and much more. It also has enormous potential dangers: an end to the human race, an arms race of warring machines, the power to destroy the planet. So if there is even a small chance that there will be a singularity, we would do well to think about what forms it might take and whether there is anything we can do to influence the outcomes in a positive direction. (Chalmers, 2010, p. 9)

But tail risks are hard for humans to keep front of mind. These are the kinds of risks that might only have a small probability of occurring but would be radically transformative if they did. Calling the singularity a tail risk seems like an undercall to me at this point, when swathes of computer scientists and roboticists expect machines to match and exceed human intelligence in roughly a hundred years. But let's say they've got that wrong and unexpected road blocks emerge. Perhaps it's harder to develop human-level AI than we think and there's only a 1% chance of a singularity in the next hundred years. That's a one in a hundred chance of everything about life as we know it on Earth being upended. We should take that possibility *very* seriously.

The computer scientist and roboticist Murray Shanahan has expressed a similar view, remarking:

> Indeed even highly unlikely or remote scenarios are sometimes worthy of study. This is true, for instance, if a scenario is especially dystopian. In that case we might want to think carefully about how to reduce its likelihood even further. Unlikely or remote scenarios are also worth discussing if they raise interesting philosophical questions, obliging us, for example, to think about what we

> really want as a species. So whether or not you think we will soon create human-level artificial intelligence, whether or not you think the singularity is near, the very idea deserves some serious thought. (Shanahan, 2015, pp. xi–xii)

It is on that note that I leave you to ponder one of the most fascinating, unsettling and potentially world-changing ideas of our time.

Bibliography

Allen, P., and Greaves, M. (2011). The singularity isn't near. *MIT Technology Review*, 12 October. https://www.technologyreview.com/2011/10/12/190773/paul-allen-the-singularity-isnt-near/.

Bhattacharya, A. (2021). *The man from the future: The visionary life of John von Neumann*. Penguin Books. Kindle Edition.

Blaschke, J. L. (2003). Interview: Vernor Vinge. *Strange Horizons*, 15 September. http://strangehorizons.com/non-fiction/articles/interview-vernor-vinge/.

Bohan, E. (2018). A history of transhumanism. PhD thesis, Macquarie University, Sydney. http://hdl.handle.net/1959.14/1268521.

Bostrom, N. (1998). Singularity and predictability. http://mason.gmu.edu/~rhanson/vc.html

Bostrom, N. (2005). A history of transhumanist thought. *Journal of Evolution and Technology*, *14*(1), 1–25.

Bostrom, N. (2009). The future of humanity. *Geopolitics, History, and International Relations*, *1*(2), 41–78. https://nickbostrom.com/papers/future.

Bostrom, N. (2014). *Superintelligence: Paths, dangers, strategies*. Oxford University Press. Kindle edition.

Brooks, R. (2017). The seven deadly sins of AI predictions. *MIT Technology Review*, 6 October. https://www.technologyreview.com/2017/10/06/241837/the-seven-deadly-sins-of-ai-predictions/.

Cellan-Jones, R. (2014). Stephen Hawking warns artificial intelligence could end mankind. BBC, 2 December. http://www.bbc.com/news/technology-30290540.

Chalmers, D. (2010). The singularity: A philosophical analysis. Journal of Consciousness Studies, 17 (9–10), 7–65.

Clarke, A. C. (1973 [1962]). *Profiles of the future: An inquiry into the limits of the possible* (revised ed.). Pan Books.

Crutzen, P., and Stoermer, E. F. (2000). The 'Anthropocene.' *Global Change Newsletter*, May. http://www.igbp.net/download/18.316f18321323470177580001401/1376383088452/NL41.pdf.

Dawkins, R. (2006). *The Selfish gene* (30th anniversary ed.). Oxford University Press.

Diamond, J. (2005). *Guns, germs and steel: A short history of everybody for the last 13,000 years*. Vintage. Kindle edition.

Eaton, T., et al. (2015). Who speaks for AI? *AI Matters*, *2*(2), 4–14.

Good, I. J. (1965). Speculations concerning the first ultraintelligent machine. In F. Alt and M. Rubinoff (eds), *Advances in computers, vol. 6*, 31–88. Academic Press.

Grace, K., Salvatier,J., Dafoe, A., Zhang,B., and Evans, O. (2018). Viewpoint: When will AI exceed human performance? Evidence from AI experts. *Journal of Artificial Intelligence Research*, *62*, 729–754.

Greene, T. (2022). DeepMind researcher claims new 'Gato' AI could lead to AGI, says 'The game is over!' TNW, 16 May. https://thenextweb.com/news/deepmind-researcher-claims-new-gato-ai-could-lead-to-agi-says-game-is-over.

Hanson, R. (2008). Economics of the singularity. Part of IEEE Spectrum special report: The singularity, June. https://spectrum.ieee.org/economics-of-the-singularity.

Hanson, R. (2018). *The age of em: Work, love and life when robots rule the Earth*. Oxford University Press. Kindle edition.

Hanson, R., and Yudkowsky, E. (2013). The Hanson–Yudkowsky AI-foom debate. Machine Intelligence Research Institute. https://intelligence.org/ai-foom-debate/.

Hickel, J. (2022). *Less is more: How degrowth will save the world*. Penguin.

Intel (2005). Excerpts from a conversation with Gordon Moore: Moore's Law. https://web.archive.org/web/20121029060050/http://download.intel.com/museum/Moores_Law/Video-Transcripts/Excepts_A_Conversation_with_Gordon_Moore.pdf.

Kallis, G., Paulson, S., D'Alisa, G., and Demaria, F. (2020). *The case for degrowth*. Polity.

Kurzweil, R. (1999). *The age of spiritual machines: When computers exceed human intelligence*. Penguin.

Kurzweil, R. (2001). The law of accelerating returns. KurzweilAI, 7 March. https://www.kurzweilai.net/the-law-of-accelerating-returns.

Kurzweil, R. (2005). The singularity is near. Gerald Duckworth & Co. Kindle edition.

Lanier, J. (2000). One-half of a manifesto. *Wired*, 1 December. https://www.wired.com/2000/12/lanier-2/.

LeCun, Y. (2022). About the raging debate . . . . LinkedIn Post, May. https://www.linkedin.com/posts/yann-lecun_ai-activity-6932436820454502400-hV39/.

Lemoine, B. (2022). Is LaMDA sentient? An interview. *Medium*, 11 June. https://cajundiscordian.medium.com/is-lamda-sentient-an-interview-ea64d916d917.

Logan, R. K., and Braga, A. (2018). Special issue 'AI and the singularity: A fallacy of a great opportunity?' *Information Theory and Methodology*. https://www.mdpi.com/journal/information/special_issues/AI%26Singularity.

Marcus, G. (2018). Deep learning: A critical approach. *arXiv*, January. https://arxiv.org/abs/1801.00631.

Marcus, G. (2022). Nonsense on stilts: The road to AI we can trust, 12 June. https://garymarcus.substack.com/p/nonsense-on-stilts.

Minsky, M. (1966). Artificial intelligence. *Scientific American*, *215*(3), 246–260.

Modis, T. (2022). Links between entropy, complexity, and the technological singularity. *Technological Forecasting and Social Change*, *176*, 121457.

Moore, G. E. (1965). Cramming more components onto integrated circuits. *Electronics*, *38*(8), 19 April. https://newsroom.intel.com/wp-content/uploads/sites/11/2018/05/moores-law-electronics.pdf.

Moravec, H. (1988). *Mind children: The future of robot and human intelligence*. Harvard University Press.

Moravec, H. (1999). *Robot: Mere machine to transcendent mind*. Oxford University Press.

More, M. (2012). Comment by Max More: Singularity meets economy. In M. More and N. Vita-More(eds), *The transhumanist reader: Classical and contemporary essays on the science, technology, and philosophy of the human future*. Wiley-Blackwell. kindle edition.

More, M., and Vita-More, N. (2012). *The transhumanist reader: Classical and contemporary essays on the science, technology, and philosophy of the human future*. Wiley-Blackwell. kindle edition.

Müller, V. C., and Bostrom, N. (2016). Future progress in artificial intelligence: A survey of expert opinion. In V. C. Müller (ed.), *Fundamental issues of artificial* intelligence (pp. 555–572). Springer.

National Science and Technology Council (2016). The National Artificial Intelligence Research and Development Strategic Plan, October. https://www.nitrd.gov/PUBS/national_ai_rd_strategic_plan.pdf.

Pearce, D. (2012). The biointelligence explosion. In A. H. Eden, J. H. Moor, J. H. Søraker and E. Steinhart (eds), *Singularity hypotheses: A scientific and philosophical assessment* (pp. 199–238). Springer.

Penrose, R. (1994). *Shadows of the mind: A search for the missing science of consciousness*. Oxford University Press.

Penrose, R. (1999). *The emperor's new mind: Concerning computers, minds, and the laws of physics*. Oxford University Press.

Roodman, D. (2020a). Modeling the human trajectory. *Open Philanthropy*, 15 June. https://www.openphilanthropy.org/research/modeling-the-human-trajectory/.

Roodman, D. (2020b). On the probability distribution of long-term changes in the growth rate of the global economy: An outside view. *Open Philanthropy*, 21 November. https://www.openphilanthropy.org/wp-content/uploads/Modeling-the-human-trajectory-2.pdf.

Sandberg, A. (2010). An overview of models of technological singularity. In M. More and N. Vita-More (eds), *The transhumanist reader: Classical and contemporary essays on the science, technology, and philosophy of the human future*. Wiley-Blackwell. kindle edition.

Searle, J. (1980). Minds, brains, and programs. *Behavioural and Brain Sciences*, *3*(3). doi: 10.1017/S0140525X00005756.

Shanahan, M. (2015). *The technological singularity*. MIT Press.

Steffen, W., et al. (2011). The Anthropocene: Conceptual and historical perspectives. *Philosophical Transactions of the Royal Society A*, 31 January. http://www.uvm.edu/~jfarley/EEseminar/readings/Anthropocene.pdf.

Stein-Perlman, Z., Weinstein-Raun, B., and Grace, K. (2022). 2022 expert survey on progress in AI. *AI Impacts*, 3 August. https://aiimpacts.org/2022-expert-survey-on-progress-in-ai/.

The Economist (2016). Why artificial intelligence is enjoying a renaissance, 15 July. https://www.economist.com/the-economist-explains/2016/07/15/why-artificial-intelligence-is-enjoying-a-renaissance.

Toews, R. (2022). Reflecting on 'artificial general intelligence' and AI sentience. Forbes, 24 July. https://www.forbes.com/sites/robtoews/2022/07/24/on-artificial-general-intelligence-ai-sentience-and-large-language-models/?sh=5c556b47e9db.

Transcendent Man (2009). Directed by Barry Ptolemy [Film].

Ulam, S. (1958). John von Neumann 1903–1957. *Bulletin of the American Mathematical Society*, May, 1–49.

Vinge, V. (1983). First word. *Omni Magazine*, January. http://www.housevampyr.com/training/library/books/omni/OMNI_1983_01.pdf.

Vinge, V. (2012). The coming technological singularity: How to survive in the post-human era. In M. More and N. Vita-More (eds), *The transhumanist reader: Classical and contemporary essays on the science, technology, and philosophy of the human future*. Wiley-Blackwell. Kindle edition.

Walsh, T. (2017a). The singularity may never be near. *AI Magazine*, *38*(3), 58–62.

Walsh, T. (2017b). Expert and non-expert opinion about technological unemployment. *arXiv*, 21 June. Available from: https://arxiv.org/abs/1706.06906.

Walsh, T. (2017c). Elon Musk is wrong: The AI singularity won't kill us all. *Wired*, 20 September. https://www.wired.co.uk/article/elon-musk-artificial-intelligence-scaremongering.

Walsh, T. (2018a). *2062: The world that AI made*. Schwartz Books. Kindle edition.

Walsh, T. (2018b). Expert and non-expert opinion about technological unemployment. *International Journal of Automation and Computing*, *15*(5), 633–638. https://link.springer.com/article/10.1007/s11633-018-1127-x.

Yudkowsky, E. (2007). Three major singularity schools. Yudkowsky.net [originally appeared on the Machine Intelligence Research Institute blog, September 2007]. https://www.yudkowsky.net/singularity/schools.